Recent Advances in Single-Particle Tracking: Experiment and Analysis

Recent Advances in Single-Particle Tracking: Experiment and Analysis

Editors

Janusz Szwabiński
Aleksander Weron

MDPI • Basel • Beijing • Wuhan • Barcelona • Belgrade • Manchester • Tokyo • Cluj • Tianjin

Editors
Janusz Szwabiński
Wrocław University of Science and Technology
Poland

Aleksander Weron
Wrocław University of Science and Technology
Poland

Editorial Office
MDPI
St. Alban-Anlage 66
4052 Basel, Switzerland

This is a reprint of articles from the Special Issue published online in the open access journal *Entropy* (ISSN 1099-4300) (available at: https://www.mdpi.com/journal/entropy/special_issues/single-particle_tracking).

For citation purposes, cite each article independently as indicated on the article page online and as indicated below:

LastName, A.A.; LastName, B.B.; LastName, C.C. Article Title. *Journal Name* **Year**, *Volume Number*, Page Range.

ISBN 978-3-0365-3485-5 (Hbk)
ISBN 978-3-0365-3486-2 (PDF)

Contents

About the Editors

Janusz Szwabiński obtained his B.Sc. in Physics from the University of Wrocław, Poland, and his Ph.D. in Science from Saarland University in Saarbruecken, Germany. He held postdoctoral positions at the University of Wrocław, Poland, and the University of Geneva, Switzerland. He is currently a university professor in the Department of Applied Mathematics at the Wrocław University of Science and Technology and the chair of the Polish chapter of EU-MATH-IN, a European Service Network of Mathematics for Industry and Innovation. He is mainly interested in complex systems and has a track record in multidisciplinary research, with applications in statistical physics, biology, social science end economy.

Aleksander Weron obtained his B.Sc. in Mathematics from the University of Wrocław, Poland, and his Ph.D. in Mathematics from Wrocław University of Science and Technology, Poland. He held postdoctoral positions at Tbilisi State University, Georgia 1971–1972, Visiting Research Professorship at Center for Stochastic Processes, Univ. of North Carolina, Chapel Hill, USA 1983–1984, and Senior Research Fellowship under the Fulbright Program at University of California: UCSB Santa Barbara and UCLA Los Angeles, USA 1995–1996. He is currently a full professor in the Department of Applied Mathematics at the Wrocław University of Science and Technology and has been recently awarded the status of professor magnus for his scientific excellence. He is also the head of the Hugo Steinhaus Center for Stochastic Methods. Professor Weron is an author or editor of 13 books and has published over 145 research papers on probability theory, stochastic processes, and their applications to physics, biology, and economy. His recent research interests include anomalous diffusion, ergodicity testing, fractional dynamics, and molecule imaging.

Preface to "Recent Advances in Single-Particle Tracking: Experiment and Analysis"

Studying diffusion is not the newest topic in the field of statistical physics. The idea is commonly linked to Robert Brown, who investigated the motion of pollen grains in water in 1827. However, the transportation of particles and their dynamics had already been analyzed by Jan Ingenhousz. His description of the movement of coal dust suspended in alcohol dates back to 1785. Since then, interest in the molecular phenomenon of diffusion has remained practically unbroken, because it appears in many domains, including physics, chemistry, biology, sociology, economics, and finance.

Starting with the pioneering experiments by Perrin and Nordlung in the 1910s, a quantitative analysis of microscopy images of diffusing particles has become an important technique for various disciplines. Over time, this method has evolved into what is now known as single-particle tracking (SPT). The concept has deeply penetrated molecular biology and statistical and chemical physics because it helps to unveil the local physical properties of molecules and their environment. It has also become a popular field in applied mathematics.

The growth of single-particle tracking as a research topic is enormous. A recent query for SPT in one of the popular scientific databases resulted in almost 2 million direct hits. Advances in recent years have led to a vast array of new scientific lines of inquiry. Given the deluge of information available, the idea behind this Special Issue was to summarize the recent findings in single-particle tracking and bring them to a broader audience. The 13 contributions focus on different aspects of SPT, both experimental and theoretical. A short summary of the topics covered by the papers can be found in the following.

The first three contributions cover some experimental aspects of SPT. Scheda et al. introduce an original pipeline for the segmentation and analysis of phase-contrast images of the wound-healing scratch assay acquired in a time-lapse. They use an ensemble of pseudo-particles to represent the wound edge. By tracking their stochastic motion, they are able to overcome some limitations of standard approaches due to the change in the shape and density of cells during migration. Speckner and Weiss focus on transport phenomena in intermediate systems, i.e., biochemically active cell extracts. They have performed extensive SPT experiments on beads in native and chemically treated Xenopus laevis extracts to show that the beads feature an anti-persistent subdiffusion that is consistent with fractional Brownian motion. Finally, Zhang and Welsher present a novel 3D single-particle tracking system with a 20% increase in precision compared to traditional approaches. They use smart off-center sampling patterns for the optimal utilization of photons coming from illumination. Their method may be of particular importance for studying biological samples, where photons are often limited to small amounts.

The most commonly used method for the analysis of SPT trajectories is based on the mean-squared displacement (MSD) of particles. Although quite simple in principle, the method is known to have some drawbacks, mainly related to the short lengths of the experimental trajectories, their heterogeneity (i.e., several types of motion within a single path), and the presence of noise. Consequently, there is still a need for more robust analytical methods that go beyond MSD and allow for a proper interpretation of the experimental results. The next five papers in our collection fit into this research direction. Balcerek and Burnecki present a rigorous statistical test to detect a multifractional Brownian motion (i.e., a fractional Brownian motion with a time-dependent Hurst exponent) within the trajectories. Their approach is based on the covariance function and should be helpful in the analysis of anomalous diffusion. Hidalgo-Soria et al analyze the two-state "jumping diffusivity" model. They show, with the help of the perturbation theory, that a non-analytical behavior (a cusp) may be found in the distribution of displacements within the model in the short time limit. Korabel et al. study the heterogeneous intracellular transport of endosomes. To improve the prediction power of the local analysis, they split the ensemble of trajectories into fast and slow subsets prior to the actual analysis. This step allows for a separate treatment of different motion regimes. Lanoiselée et al. propose a new structural approach to detect transient trapping of particles. Their method is based on the recognition of block structures along the diagonal of the recurrence matrix. Stanislavsky and Weron use the conjugate Bernstein function theory to find a connection between the tempered subdiffusion and the diffusion-limited aggregation. Since the model allows for the detection of confined random walks within the trajectories, it may be applied to SPT data.

In recent years, machine learning (ML) has been employed for the analysis of single-particle tracking data. In contrast to standard algorithms, where the user is required to explicitly define the rules of data processing, ML algorithms can directly learn those rules from a series of data. Three papers in this Special Issue cover different aspects of this approach to SPT. Gajowczyk and Szwabiński use the deep recurrent neural network for the classification of trajectories (i.e., the detection of diffusion modes). Loch-Olszewska and Szwabiński tackle the same problem with a more traditional ML approach. Compared to deep learning, the feature-based methods they use do not work with raw trajectories; instead, they require a set of human-engineered features for each trajectory in order to feed a classifier. Although deep learning performs a little better, the traditional method was better in terms of interpretability. Szarek et al. combine the two methods. They use a neural network, together with features calculated from trajectories (autocovariance function), in order to estimate the anomalous exponent from the trajectories. Their approach outperforms the analytical one.

Last but not least, the two remaining contributions cover some topics related to diffusion. Dinis and Parrondo show how to optimally extract work from a Brownian particle that is reversibly contained in an optical tweezer potential. Their model of a molecular motor should work, at least in principle, for systems with much shorter response time of measurements than the systems' relaxation times. Lachowicz and Debowski investigate models that may lead to diauxic growth at the mesoscopic scale. They may help us to understand some complex motion patterns of bacterial cells.

We express our thanks to the authors of the contributions, and to the journal Entropy and MDPI for their support during the preparation of this Special Issue. We hope that you will enjoy reading it, whether you are a newcomer to the field and looking for a place to start, or already working in the field and looking for stimulation. We also hope that you will recommend this issue to your colleagues.

Janusz Szwabiński, Aleksander Weron
Editors

 entropy

MDPI

Article

Study of Wound Healing Dynamics by Single Pseudo-Particle Tracking in Phase Contrast Images Acquired in Time-Lapse

Riccardo Scheda [1,†], Silvia Vitali [2,*], Enrico Giampieri [3], Gianni Pagnini [2,4] and Isabella Zironi [1]

[1] DIFA-Physics and Astronomy Department, University of Bologna, Viale C. Berti Pichat 6/2, 40127 Bologna, Italy; riccardo.scheda@studio.unibo.it (R.S.); isabella.zironi@unibo.it (I.Z.)
[2] BCAM-Basque Center for Applied Mathematics, Alameda de Mazarredo 14, 48009 Bilbao, Spain; gpagnini@bcamath.org
[3] eDIMESlab, Department of Experimental, Diagnostic and Specialty Medicine, University of Bologna, Via Irnerio 49, 40126 Bologna, Italy; enrico.giampieri@unibo.it
[4] Ikerbasque-Basque Foundation for Science, Plaza Euskadi 5, 48009 Bilbao, Spain
* Correspondence: svitali@bcamath.org
† Current affiliation: Department of Electrical, Electronic and Information Engineering "Guglielmo Marconi", Viale del Risorgimento 2, 40136 Bologna, Italy.

Abstract: Cellular contacts modify the way cells migrate in a cohesive group with respect to a free single cell. The resulting motion is persistent and correlated, with cells' velocities self-aligning in time. The presence of a dense agglomerate of cells makes the application of single particle tracking techniques to define cells dynamics difficult, especially in the case of phase contrast images. Here, we propose an original pipeline for the analysis of phase contrast images of the wound healing scratch assay acquired in time-lapse, with the aim of extracting single particle trajectories describing the dynamics of the wound closure. In such an approach, the membrane of the cells at the border of the wound is taken as a unicum, i.e., the wound edge, and the dynamics is described by the stochastic motion of an ensemble of points on such a membrane, i.e., pseudo-particles. For each single frame, the pipeline of analysis includes: first, a texture classification for separating the background from the cells and for identifying the wound edge; second, the computation of the coordinates of the ensemble of pseudo-particles, chosen to be uniformly distributed along the length of the wound edge. We show the results of this method applied to a glioma cell line (T98G) performing a wound healing scratch assay without external stimuli. We discuss the efficiency of the method to assess cell motility and possible applications to other experimental layouts, such as single cell motion. The pipeline is developed in the Python language and is available upon request.

Keywords: wound healing dynamics; single pseudo-particle tracking; phase contrast image segmentation

Citation: Scheda, R.; Vitali, S.; Giampieri, E.; Pagnini, G.; Zironi, I. Study of Wound Healing Dynamics by Single Pseudo-Particle Tracking in Phase Contrast Images Acquired in Time-Lapse. *Entropy* **2021**, *23*, 284. https://doi.org/10.3390/e23030284

Academic Editors: Janusz Szwabiński and David Holcman

Received: 24 November 2020
Accepted: 23 February 2021
Published: 26 February 2021

Publisher's Note: MDPI stays neutral with regard to jurisdictional claims in published maps and institutional affiliations.

1. Introduction

In this paper, we propose a pipeline for the segmentation and analysis of phase contrast images acquired in time-lapse in the wound healing scratch assay, to overcome some limitations of standard approaches due to the change in shape and density of the cells during migration.

Cellular migration is a fundamental process for animal's physiology during both the period of development and that of maturity. Cells migrate to shape organs and tissues and, in the case of damage, regenerate them. Furthermore, motility is a primary skill in cancer metastatic processes and in the immune responses [1,2]. The capability to migrate is a highly regulated process in which cells respond to external and internal mechanical, electrical, and chemical stimuli by complex physiological processes that promote, enhance, or suppress cell motility [3,4]. Cells can be induced to move in a particular direction by positive and negative guidance signals, while in the absence of external guidance, cells move randomly [5,6].

In cutaneous wound healing, which is a complex cellular and biochemical process necessary to restore structurally damaged tissue, skin cells migrate from the wound edges towards the empty space to restore skin integrity. In this case, the cohesive group of cells organized in a layer modifies the classical characteristics of single cell migration, and the presence of the wound induces peculiar migration behaviors. In fact, while a certain freedom of movement is maintained inside the tissue, the cells along the edge of the wound (front) move preferentially toward the gap. Such a process involves dynamical interactions between both the contacting cells (which are absent in single cell migration) and the extracellular matrix. These interactions regulate motility enhancement or suppression [7,8].

The wound healing scratch assay is a widespread experimental tool applied to study the collective migration of cells cultured in vitro. Standard protocols provide that a highly confluent monolayer of cells is scratched by a fine pipette tip to create a gap, which is then allowed to heal. As a protocol of analysis, the area of the scratch is measured as a function of time to determine the speed of the closure. This method is meant to simulate a natural wound, and the procedure is simple and easy to set up, but it is difficult to analyze and produce precise and reproducible results [9].

The mathematical continuum models that focus on the collective properties of cells can explain the requirements for the onset of movement and some typical characteristics of cell motility, but are usually limited to small space-time scales. Therefore, they provide little information on how the integration of the lamellipodium protrusion, the retraction of the posterior part, and the transduction of force on the extracellular matrix lead to the long-term prolonged movement of the entire cell. This process is characterized by alternating phases of direct migration and changes of direction and polarization. The coordinated interaction of these phases suggests the existence of intermittency, strong space-time correlations, and a close relationship between units (cell-cell interaction). It is therefore an important question whether the long-term movement of the entire cell can still be understood as a simple diffusive behavior such as Brownian movement or a random walk or whether more advanced dynamic modeling concepts should be applied [10–12].

The change in shape and density of the cells during migration make it difficult to apply standard automatic single particle tracking (SPT) pipelines to extract the cell migration trajectory in phase contrast images acquired in time-lapse. These difficulties are even greater when collective motion is considered and a dense agglomerate of cells is present. To overcome such limitations, here, we propose a pipeline for segmentation and SPT extraction in phase contrast images of the wound healing scratch assay. The pipeline is original and follows the principle of Occam's razor, based on a simple measure as linear binary patterns (LBPs), which results in being sufficient to classify the texture as cells or background by using a principal component analysis (PCA) and Gaussian mixture classification, the code is available at the git-hub repository https://github.com/riccardoscheda/AnomalousDiffusion (accessed on 1 November 2020). We chose the manual segmentation performed over one experiment as the ground truth. We further compared the performance of our pipeline with segmentation by Otsu thresholding [13] without manual adjustment of the parameters for different frames of the same image. For all the cases, the wound edge is approximated to a unique membrane and its dynamics approximated by the stochastic motion of a point on the membrane, i.e., a pseudo-particle. This choice is motivated by the fact that in the experiment under study, faster cells do not separate from the borders during wound closure. Therefore, the dynamics and the heterogeneity of the process are characterized through the collection of such SPT trajectories.

The paper is organized as follows: in the the Methods Section, we present step by step our pipeline for phase contrast image processing and SPT; in the Results Section, we show the trends of the pseudo-particle trajectories' statistics for a wound healing scratch assay, performed with glioblastoma T98G cells; in the Conclusions Section, we discuss the performance and the SPT statistics of our pipeline, in comparison with the corresponding measurements obtained through the professional tool ImageJ [14].

2. Methods

2.1. Data

Glioma cells (T98G), derived from brain human tumor glioblastoma multiforme (GBM), were plated at a density of 1×10^5 cells/cm^2 on 35 (⌀) mm sterile Petri dishes with a 10 (⌀) mm glass microwell (MatTek Corporation, Ashland, MA, USA) suitable for optical microscopy. The cell culture, with a population doubling time (PDT) approximately of 28 h, as reported by the ATTC Company, which provided the cell line, was maintained in Gibco^{TM} Minimum Essential Medium (MEM) with Earle's salts (Fisher Scientific, Milano, Italy) supplemented with 10% fetal bovine serum, 1% L-glutamine, 1% sodium pyruvate, and antibiotics (1% penicillin and 1% streptomycin) inside the incubator at 5% of CO_2 and 37 °C. All chemicals were purchased from Merck KGaA (Darmstadt, Germany). After 48 h from seeding, the population covered the entire surface as a monolayer of confluent and tightly contacting cells. Using a sterile pipette tip for Gilson (10–200 µL), a scratch ranging 200–400 µm along the middle axis was done. Right after, the specimen was placed into the pre-heated microscope stage incubator in the motorized table of the inverted optical microscope Eclipse Ti (Nikon, Bologna, Italy). The phase-contrast micrographs of multiple visual fields, pre-selected along the narrow scrape by the NIS Elements AR 4.0 (Nikon, Bologna, Italy) software, were acquired at 100× magnification for 20 h at the rate of 4 frames/hour. The setup allowed the acquisition of time-lapse images of living cultured cells maintained in standard conditions for the entire duration of the experiment.

2.2. Image Processing

The aim of the pipeline was to identify the wound edges, which correspond to the free edge of the two cell layers, separated by the wound. The procedure was done in the following steps, which should be applied to all the frames of an experiment: (i) equalization, to make all the frames comparable; (ii) binarization, to separate the background regions from the cell layers; (iii) wound edge identification; and (iv) storage of the coordinates. We describe here two alternative procedures of binarization, the first based on texture classification and the second on hand drawing the wound edges over the images by using the professional tool ImageJ [14].

2.2.1. Equalization of the Frames

To improve the difference of the wound borders from the background regions, we applied to all the frames contrast limited adaptive histogram equalization (CLAHE) [15].

The image was divided into small blocks (tiles), with a tile size of 50 × 50, to enhance the difference between the cell border and the background (tile size is 8 × 8 by default in [16]). Then, each of these blocks was histogram equalized. Therefore, in a small area, the histogram would be confined to a small region (unless there was noise). If noise was there, it would be amplified. To avoid this, contrast limiting was applied. If any histogram bin was above the specified contrast limit (by default, 40 in [16]), those pixels were clipped and distributed uniformly to other bins before applying histogram equalization. This procedure increased the image contrast and enhanced the texture patterns (e.g., Figure 1b) by equalizing pixels' intensity distribution of all the frames to a fixed range wider than the original ones.

2.2.2. Image Binarization by Texture Analysis

Image binarization was performed by dividing each frame into 10,000 subimages (12 × 16 pixel subimages in 1200 × 1600 pixel image) and by classifying each of them as the background or cell layer on the bases of a score. The score was built to characterize the texture of each subimage and corresponded to the distribution of the local binary pattern (LBP) values for all the pixels of the subimage (scikit-image Python library [17], skimage.feature.local_binary_pattern). We considered grayscale images; thus, the LBP of each pixel corresponded to a scalar value. We calculate the LBP for a pixel by comparing the pixel with its 8 first neighbors. To each couple was assigned a score: if the central pixel value was greater than or equal to the neighbor pixel value, we assigned 1, otherwise, if the central pixel value was less than the neighbor pixel value, the score was 0. The LBP value of the pixel corresponded to the sum of these scores, ranging from 0 to 8, and contained information about the 3×3 square of pixels. The frequency of such LBP scores for each subimage was an array of 9 values representing a texture feature of the subimage. Therefore, each frame (image) was characterized by a matrix of 10,000 (sub-images) $\times$ 9 (LBP score) values. Principal component analysis (PCA) [18] was performed over the 9 dimensions of the texture score to separate the 10,000 subimages into two clusters: one corresponding to the background regions and one containing the cell layer regions (Figure 1c). Taking the first 5 principal components, the points belonging to the two clusters were classified and labeled (0 or 1) using the Gaussian mixture model clustering algorithm (scikit-learn Python library [19], sklearn.mixture.GaussianMixture). Each point in Figure 1c corresponds to a subimage; hence, the obtained binary color labels (yellow or blue) were used as binarized intensities for the corresponding subimages to build the binarized image (Figure 1d).

The performance of the algorithm with respect to the size of the subimages was studied in terms of the Pearson correlation of the segmented fronts with the ground truth for squared and rectangular shapes of different sizes. For complete tessellation of the image, it supported the choice of the subimages' size of 12×16 pixels (see the Supplementary Material).

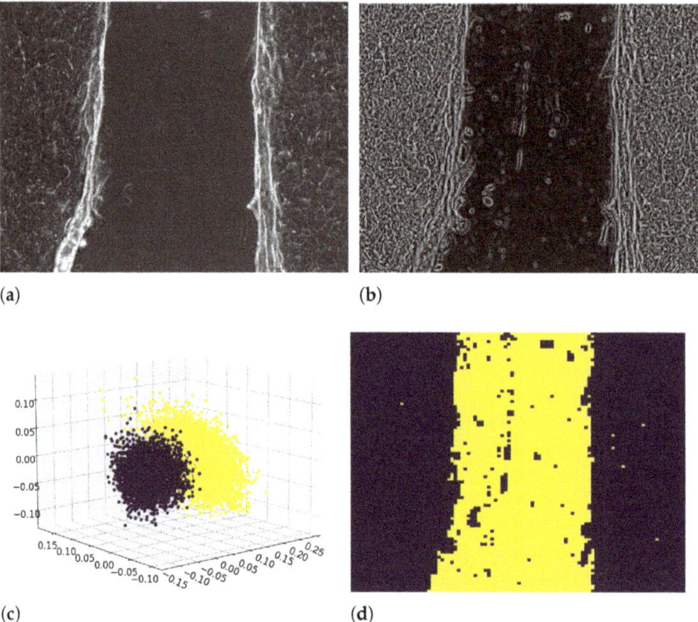

Figure 1. Original image (**a**); transformed image by using adaptive histogram equalization (**b**); 3D scatter plot of the first 3 principal components of the linear binary pattern (LBP) score PCA (**c**); the data points in the scatter plot are clustered by the Gaussian mixture model clustering algorithm; color labels refer to cells (blue) or background (yellow); binarized frame image by using texture analysis (**d**).

2.2.3. Wound Edge Recognition from Binarized Images

Contour lines can be easily recognized in a binarized image as the contour of 0 or 1 regions. We applied a function for contour identification, returning a list of all the contours in an image (OpenCV-Python library, findContours). In the frames, the longest contour line refers to the central part of the image until the two cellular fronts remain separate, identifying at the same time the background regions an the two borders of the cell layers (Figure 2a).

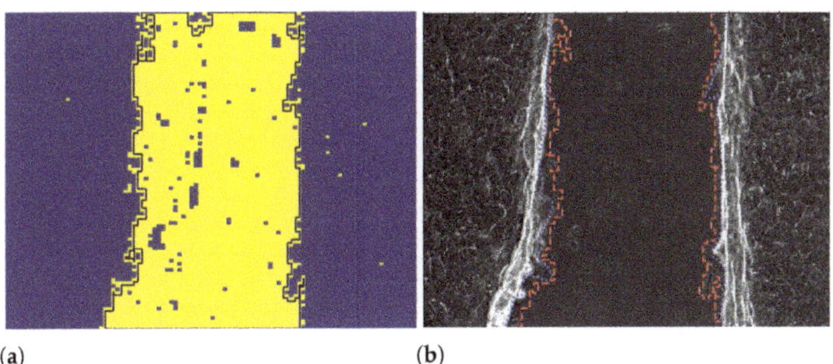

(**a**) (**b**)

Figure 2. Image binarized by texture analysis with borders recognized by OpenCV-Python (**a**); example comparison of the borders obtained through the professional tool ImageJ (blue line) and the texture analysis (red dashed line) method superposed on the original image frame (**b**).

2.2.4. Wound Edge Recognition with the ImageJ Professional Tool

The extraction of the fronts with the professional tool ImageJ for image analysis was performed by hand drawing the line of the front over each image (Figure 2b) and then by saving the corresponding coordinates for each frame and for left (L) and right (R) front in a .txt file [14].

2.2.5. Wound Edge Recognition with Otsu Thresholding

For the simple thresholding of the images, we performed an adaptive histogram equalization (CLAHE) to improve the difference of the wound borders from the background, then we blurred the image with OpenCV Gaussian Blur, in order to have better results for the Otsu thresholding. Then, we applied Otsu thresholding on the image [17]. Then, we applied morphological transformations in order to have a smoother border of the wound. After morphological transformations, we collected the coordinates of the borders (OpenCV-Python library, findContours).

2.3. *Pseudo-Particles' Trajectories*

The wound edge (L and R) of the cell layers was considered as a single homogeneous elastic membrane. The movement of such a membrane was tracked by means of N points uniformly distributed along its length as pearls on an elastic necklace. To derive the coordinates of the N pseudo-particles, we interpolated the wound edges' 2D coordinates as a function of the front length (scipy Python library, scipy.interpolate.interp1d), and then, we computed the coordinates of the N pseudo-particles uniformly distributed along its length [20].

The collection of the N points constituted the collection of pseudo-particles, and for each of them, an SPT was built by considering its coordinates in the time sequence of the experiment frames. The SPTs in 2D were allowed to cross because of invaginations and protrusions of the front, despite the cells being attached to each other and the membrane of the wound edge being considered as a unicum. However, the average displacement along the membrane was approximately zero because it was constrained by the geometry of the system and the microscope field.

The pseudo-particle n at time t_k was defined by the coordinates of the n-th pseudo-particle in the k-th frame of the image. The collection of the coordinates of the pseudo-particle n for all the frames represented the trajectory of the pseudo-particle n. Thus, the dynamics of the membrane could then be tracked by working on a matrix $N \times M$, where N is the number of tracked pseudo-particle and M is the number of frames. The latter represent the time steps of the sampling.

2.4. SPT Statistics

To study the SPT statistics, we applied the discrete version of mean squared displacement (MSD) and autocorrelation functions (ACFs). In fact, discrete SPT statistics can be performed directly on an SPT N (pseudo-particles)$\times M$ (frames) matrix dataset, one for the x coordinate and one for the y coordinate, where the frame index $k = 0, 1, 2 \ldots, M-1$ corresponds to the sampling time, i.e., the time step of the process, and $n = 1, 2 \ldots, N$ corresponds to the index of the pseudo-particle. Statistics on the single trajectory could be performed as matrix operations along the M columns, while the ensemble average could be performed by averaging over the N rows of the transformed matrix. In the present work, we considered only the movement toward the free edge of the layers for the sake of simplicity, i.e., the x component of the quantities of interest. For statistical analysis, we applied a shift to the pseudo-particle position such that $\langle X(t=0) \rangle = 0$. Moreover, due to the lack of long stationary trajectories, we considered here only ensemble averages:

$$\mathbb{E}(Y(k)) = \frac{1}{N} \sum_{n=1}^{N} y_n(k) \,, \tag{1}$$

where $y_n(k)$ is the value of the variable Y for the n-th pseudo-particle at time $t = k$. The velocity of the pseudo-particle is defined as the increment of the pseudo-particle position X per unit sampling time (0.25 h):

$$V(\tau) = X(\tau+1) - X(\tau) \,, \quad \tau = 0, 1, 2 \ldots, M-2 \,. \tag{2}$$

The increments of the velocity per unit sampling time are defined as the following:

$$A(\tau) = V(\tau+1) - V(\tau) \,, \quad \tau = 0, 1, 2 \ldots, M-3 \,. \tag{3}$$

The autocorrelation function ACF_Y for the generic variable Y reads:

$$ACF_Y(\tau) = \frac{\mathbb{E}[(Y(t_0) - \mu_{t_0})(Y(t_0+\tau) - \mu_{t_0+\tau})]}{\sigma_{t_0}\sigma_{t_0+\tau}} \,, \quad \tau = 0, 1, 2 \ldots, M-1 \,, \tag{4}$$

where the initial time is $t_0 = 0$ and σ_k and μ_k represent respectively the standard deviation and the mean of the variable Y at time $t = k$.

The mean squared value (MSY) for the generic variable Y reads:

$$MSY(\tau) = \mathbb{E}[(Y(t_0+\tau) - Y(t_0))^2] \,, \quad \tau = 0, 1, 2, \ldots, M-1 \,, \tag{5}$$

where the initial time is again $t_0 = 0$.

2.5. Fit Procedure

All the fits were performed through a ordinary least squares (OLS) regression (scipy Python library, scipy.optimize.curve_fit), which returned the optimized parameters of the model and their matrix of covariance [20]. Poissonian uncertainty for counts in histograms was considered. We further compared (results not shown) the parameters estimated by OLS with the ones obtained by the maximum likelihood estimate (MLE) (stats Python library, scipy.stats.rv_continuous.fit).

3. Results

We considered a single field in an experiment of the wound healing scratch assay (without external stimuli applied to the cell substrate) as the test image.

The trends of the N SPTs obtained through the texture analysis for the experiment under study are compared with the ones obtained by using the professional tool ImageJ in Figure 3. SPTs from the right front are also mirrored.

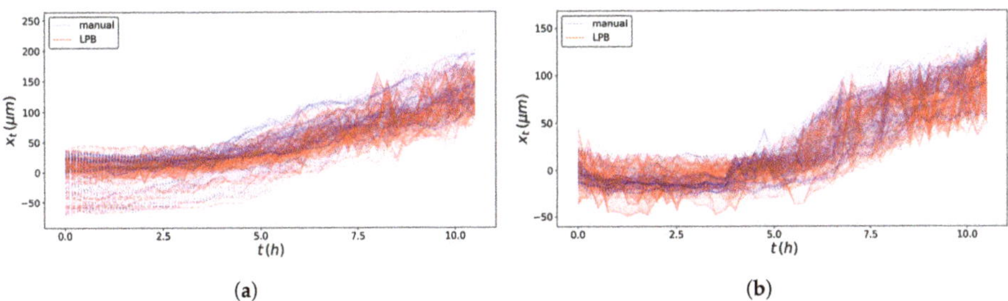

Figure 3. Comparison of the single particle trajectories obtained with the professional tool ImageJ (blue line) and the texture analysis (red or dashed line) method for the left wound edge (**a**) and the right wound edge (**b**), for $N = 10^3$.

In Figure 4, we display the temporal trends of the area between the wound edges during wound closure, estimated by using the texture analysis in comparison with the wound edges recognized manually.

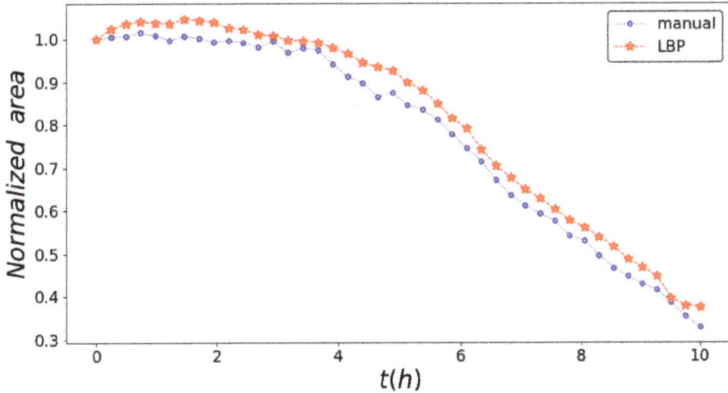

Figure 4. Normalized area between the wound edges during wound closure as a function of time for the texture analysis (blue line) and the professional tool ImageJ (red line).

The pseudo-particle average position and average velocity for the two methods of analysis are displayed in Figures 5 and 6. The ACFs of the pseudo-particle position and velocity along the x coordinate are compared for the two methods of analysis in Figures 7 and 8, respectively. A regime with stationary increments of the velocity was identified for the time range between 5 h and 8 h (Table 1), corresponding to the duration of the regime with constant drift velocity in the ensemble averaged position (Figure 5). The medium could be roughly approximated as viscous, and a constant velocity implies constant force, on average, applied against friction by the cells. This stationary regime with constant drift velocity was supported by zero correlation in the VACF(Figure 8) and by the symmetric distribution of velocity increments with a zero average. For such a time range, the distribution of the instant acceleration (velocity increments) along the x coordinate is shown in Figure 9 for the two methods of image segmentation. The tails of these distribution are compatible with both the exponential and the Gaussian scaling, with comparable characteristic scales (Table 2). However, the linear decay of the tails in Figure 10 suggests that a truncated-exponential decay is more plausible. To estimate the consistency between the two methods, we computed the Pearson's correlation (Table 3) for their estimates of the pseudo-particles' coordinates, i.e., the entire collection of estimated position for the x and y coordinates, the average position $\langle X \rangle$, the average velocity $\langle V \rangle$, and the average velocity increments $\langle A \rangle$ for $N = 10^3$ and $M = 40$.

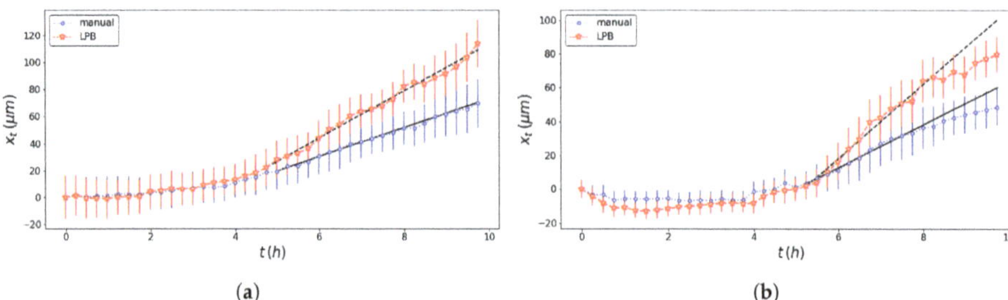

Figure 5. Comparison of the average position of the pseudo-particle obtained with the professional tool ImageJ (blue dotted line) and the texture analysis (red dashed line) and their linear OLS best fit (black dashed line; see Table 1 for details) for the left wound edge (**a**) and the right wound edge (**b**).

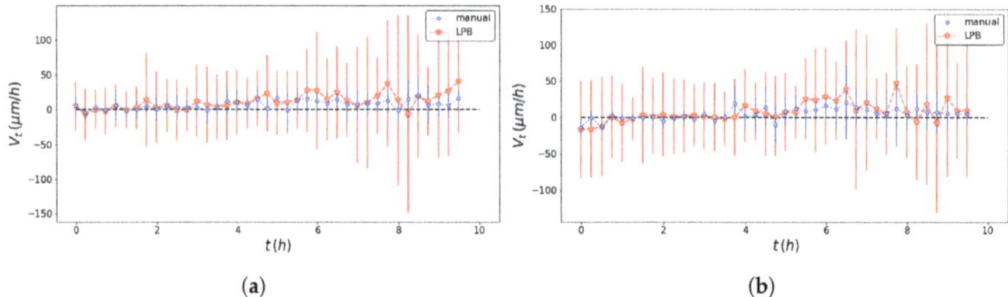

Figure 6. Comparison of the ensemble averaged pseudo-particle velocity obtained with the professional tool ImageJ (blue dotted line) and the texture analysis (red dashed line) for the left wound edge (**a**) and the right wound edge (**b**).

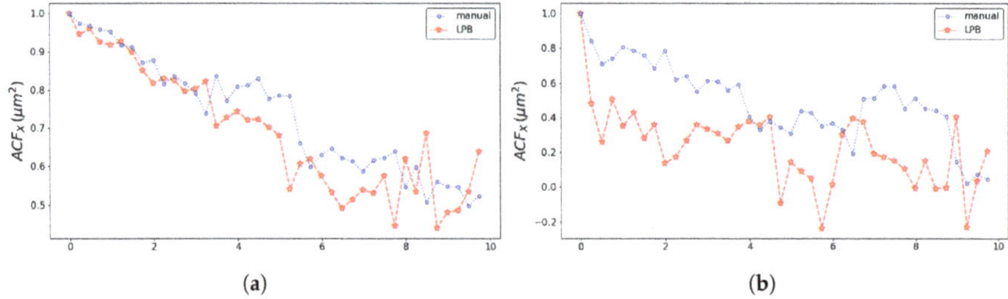

Figure 7. Comparison of the coordinate X autocorrelation function obtained with the professional tool ImageJ (blue dotted line) and the texture analysis (red dashed line) for the left wound edge (**a**) and the right wound edge (**b**).

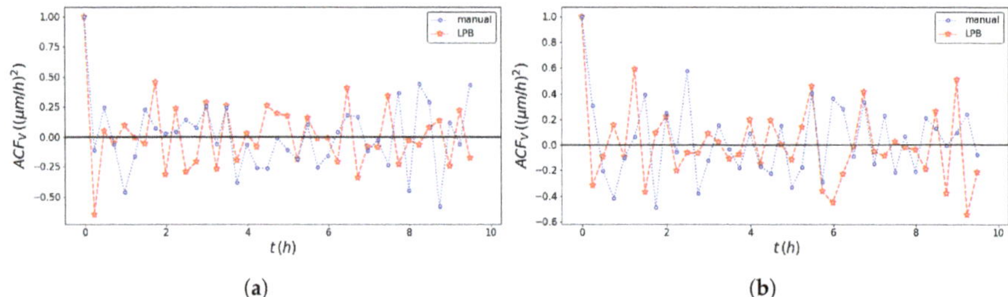

Figure 8. Comparison of the velocity autocorrelation function obtained with the professional tool ImageJ (blue dotted line) and the texture analysis (red dashed line) method for the left wound edge (**a**) and the right wound edge (**b**).

Table 1. Estimated average drift velocity of the cell front v_d and the lag time of quiescence τ_1 with their corresponding standard error obtained through OLS regression of the model and goodness of fit (Adj. R-squared). The second time scale $\tau_2 = 8$ h is estimated from Figure 5b.

Model	Front	Method	$v_d \pm$ SD (µm/h)	$\tau_1 \pm$ SD (h)	Adj. R-Squared
$x(t)\|_{\tau_1<t<\tau_2} = v_d \cdot (t - \tau_1)$	L	ImageJ	19.5 ± 0.3	4.49 ± 0.03	0.990
	L	texture analysis	17.4 ± 0.3	4.75 ± 0.03	0.984
	R	ImageJ	23.9 ± 0.4	5.25 ± 0.03	0.987
	R	texture analysis	21.8 ± 0.5	5.25 ± 0.03	0.975

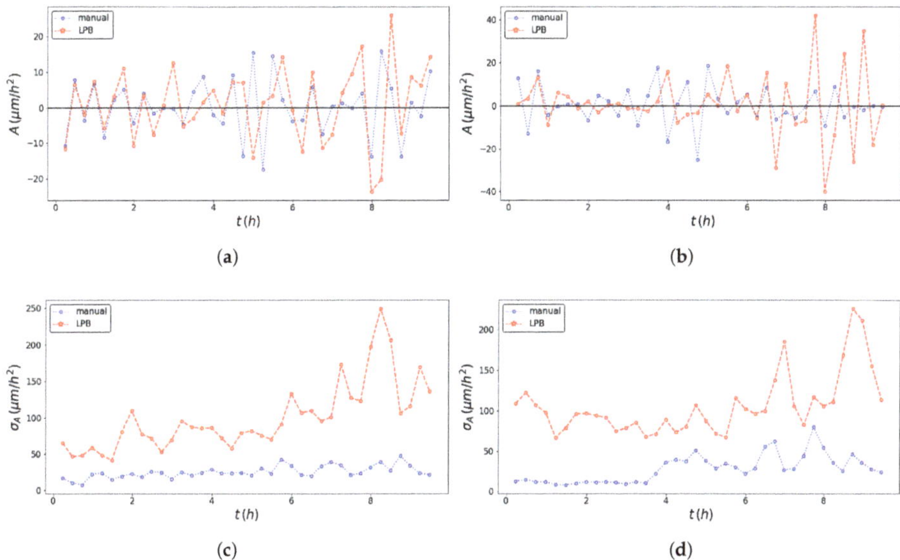

Figure 9. Comparison of the ensemble averaged pseudo-particle acceleration trajectory obtained with the professional tool ImageJ (blue dotted line) and the texture analysis (red dashed line) method for the left wound edge (**a**) and the right wound edge (**b**); comparison of the standard deviation of pseudo-particle acceleration of the ensemble of pseudo-particles obtained with the professional tool ImageJ (blue dotted line) and the texture analysis (red dashed line) method for the left wound edge (**c**) and the right wound edge (**d**).

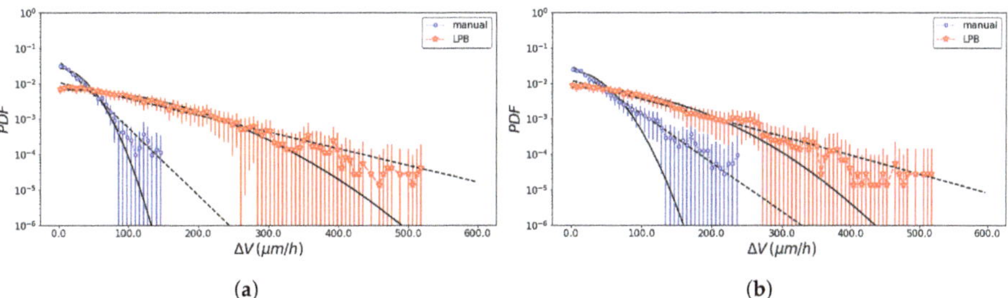

Figure 10. Comparison of the velocity increments (absolute value) obtained with the professional tool ImageJ (blue) and the texture analysis (red) method for the left wound edge (**a**) and the right wound edge (**b**); the best OLS fit of the frequencies of a exponential (dashed line) and a normal distribution (bold line) is shown for the two methods (see Table 2 for details).

Table 2. Estimated parameters with their corresponding standard error obtained through OLS regression of the corresponding model and goodness of fit (Adj. R-squared).

Model	Front	Method	$\lambda \pm$ SD (μm)	$\sigma \pm$ SD (μm^2)	Adj. R-Squared
$P(\lvert\Delta V\rvert) = 2 \cdot \mathcal{G}(\lvert\Delta V\rvert; 0, \sigma^2)$	L	ImageJ		29 ± 1	0.982
	L	texture analysis		117 ± 2	0.980
	R	ImageJ		36 ± 1	0.976
	R	texture analysis		103 ± 2	0.982
$P(\lvert\Delta V\rvert) = \frac{1}{\lambda} e^{-\lvert\Delta V\rvert/\lambda}$	L	ImageJ	23 ± 1		0.976
	L	texture analysis	92 ± 2		0.948
	R	ImageJ	31 ± 1		0.987
	R	texture analysis	82 ± 2		0.949

Table 3. Pearson correlation of the collection of the pseudo-particles' coordinates, x and y, the average position $\langle X \rangle$, the average velocity $\langle V \rangle$, and the average velocity increments $\langle A \rangle$ estimated by ImageJ with the one estimated by texture analysis for $N = 10^3$.

Front	Variable	Pearson's Coeff.	*p*-Value
L	x-coords	0.942	0.0
	y-coords	0.985	0.0
	$\langle X \rangle$	0.997	1×10^{-50}
	$\langle V \rangle$	0.547	1×10^{-4}
	$\langle A \rangle$	0.275	0.08
R	x-coords	0.913	0.0
	y-coords	0.986	0.0
	$\langle X \rangle$	0.997	1×10^{-48}
	$\langle V \rangle$	0.630	1×10^{-6}
	$\langle A \rangle$	0.156	0.19

4. Conclusions

We present an original method to extract a 2D discrete representation of the wound edge in phase contrast images acquired in time-lapse by texture analysis, and we compare the results with the ones obtained by using the professional tool ImageJ and by Otsu thresholding (see the Supplementary Material for edges derived by thresholding). The dynamics of the wound edges is defined by the SPT of N pseudo-particles uniformly distributed along the length of the fronts.

Thus, discrete SPT statistics can be performed directly on the SPT N (pseudo-particles) $\times M$ (frames) matrix dataset for the x coordinate (crossing the wound gap).

We compare SPT statistics of the data obtained by hand drawing with the texture analysis: average values, squared mean values, and the autocorrelation function of position and velocity. The two approaches lead to consistent results in terms of the trends of the dynamics (qualitative analysis) and in terms of Pearson's correlation (Table 3). By a visual check, the texture analysis appears more capable of recognizing lamellipodium protrusions than the professional tool ImageJ, because such tiny structures could be occasionally missed by human recognition (Figure 2). On the other side, the automatized procedure may also produce artifacts in the front profile, for example it would consider as part of the cell layer pieces of dead cells remaining in the middle of the wound when the cells at the wound edges get close to them. For these reasons, the wound edges detected by texture analysis are associated with larger fluctuations between different frames than the ones detected manually. For the same reasons, the pseudo-particles position fluctuates more in the texture analysis dataset between different frames, generating larger tails in the distribution of increments along the x coordinate for the velocity (and position) of the pseudo-particle (Figure 10) and larger mean squared velocity, in comparison to the one obtained by the professional tool ImageJ method. Despite such discrepancy, the average drift velocities (Figure 5), which correspond to the mean of the distribution of the position increments and the average velocity increments (Figure 10), are comparable.

Finally, by studying the SPT statistics, we are able to identify an intermediate regime characterized by a constant average of the cellular front velocity and by exponential tails for the velocity increments' distribution (Table 2). We leave the full characterization of the stochastic process and the biological meaning, which are beyond the scope of the present paper, to future research with an enlarged cohort of experiments, in order to increase the statistics, but also to characterize the inherent variability of the phenomena.

Supplementary Materials: The following are available online at https://www.mdpi.com/1099-4300/23/3/284/s1.

Author Contributions: Conceptualization, S.V., G.P. and I.Z.; software, R.S. and E.G.; investigation, I.Z.; resources, I.Z.; formal analysis, R.S. and S.V.; methodology, S.V. and G.P.; visualization, R.S. and S.V.; writing—original draft preparation, all authors; supervision, S.V., E.G. and G.P.; funding acquisition, G.P. and I.Z. All authors have read and agreed to the published version of the manuscript.

Funding: This research received no external funding.

Acknowledgments: S.V. and G.P. are supported by the Basque Government through the BERC 2018–2021 program and also funded by the Spanish Ministry of Economy and Competitiveness MINECO via the BCAM Severo Ochoa SEV-2017-0718 accreditation.

Data Availability Statement: not applicable.

Conflicts of Interest: The authors declare no conflict of interest.

References

1. Friedl, P.; Wolf, K. Tumour-cell invasion and migration: Diversity and escape mechanisms. *Nat. Rev. Cancer* **2003**, *3*, 362–374. [CrossRef] [PubMed]
2. Rorth, P. Collective cell migration. *Annu. Rev. Cell Dev. Biol.* **2009**, *25*, 407–429. [CrossRef] [PubMed]
3. Ladoux, B.; Mège, R.M. Mechanobiology of collective cell behaviours. *Nat. Rev. Mol. Cell Biol.* **2017**, *18*, 743–757. [CrossRef] [PubMed]
4. Lintz, M.; Muñoz, A.; Reinhart-King, C.A. The Mechanics of Single Cell and Collective Migration of Tumor Cells. *J. Biomech. Eng.* **2017**, *139*, 0210051–0210059. [CrossRef] [PubMed]
5. Hakim, V.; Silberzan, P. Collective cell migration: A physics perspective. *Rep. Prog. Phys.* **2017**, *80*, 076601. [CrossRef] [PubMed]
6. Ascione, F.; Vasaturo, A.; Caserta, S.; D'Esposito, V.; Formisano, P.; Guido, S. Comparison between fibroblast wound healing and cell random migration assays in vitro. *Exp. Cell Res.* **2008**, *347*, 123–132. [CrossRef] [PubMed]
7. Te, Boekhorst, V.; Preziosi, L.; Friedl, P. Plasticity of cell migration in vivo and in silico. *Annu. Rev. Cell Dev. Biol.* **2016**, *32*, 491–526. [CrossRef] [PubMed]
8. Thuroff, F.; Goychuk, A.; Reiter, M.; Frey, E. Bridging the gap between single-cell migration and collective dynamics. *eLife* **2019**, *8*, e46842. [CrossRef] [PubMed]
9. Zheng, C.; Yu, Z.; Zhou, Y.; Tao, L.; Pang, Y.; Chen, T.; Zhang, X.; Qiu, H.; Zhou, H.; Chen, Z.; Huang, Y. Live cell imaging analysis of the epigenetic regulation of the human endothelial cell migration at single-cell resolution. *Lab Chip* **2012**, *12*, 3063–3072. [CrossRef]
10. Dieterich, P.; Klages, R.; Preuss, R.; Schwab, A. Anomalous dynamics of cell migration. *Proc. Natl. Acad. Sci. USA* **2008**, *105*, 459–463. [CrossRef] [PubMed]
11. Souza, Vilela, Podestá, T.; Venzel, Rosembach, T.; Aparecida dos Santos, A.; Lobato, Martins, M. Anomalous diffusion and q-Weibull velocity distributions in epithelial cell migration. *PLoS ONE* **2017**, 12, e0180777. [CrossRef] [PubMed]
12. Luzhansky, I.D.; Schwartz, A.D.; Cohen, J.D.; MacMunn, J.P.; Barney, L.E.; Jansen, L.E.; Peyton, S.R. Anomalously diffusing and persistently migrating cells in 2D and 3D culture environments. *APL Bioeng.* **2018**, *2*, 026112. [CrossRef] [PubMed]
13. Otsu, N. A Threshold Selection Method from Gray-Level Histograms. *IEEE Trans. Syst. Man Cybern.* **1979**, *9*, 62–66.
14. Rasband, W.S. *ImageJ*; U.S. National Institutes of Health: Bethesda, MD, USA, 1997–2018. Available online: https://imagej.nih.gov/ij/ (accessed on 1 November 2020).
15. Zuiderveld, K. Contrast limited adaptive histogram equalization. In *Graphics Gems IV*; Heckbert, P., Ed.; Academic Press: Cambridge, MA, USA, 1994; pp. 474–485.
16. Python Library OpenCV-Python. Available online: https://OpenCV-python-tutroals.readthedocs.io/en/ (accessed on 1 November 2020).
17. Python Library Scikit-Image. Available online: https://scikit-image.org (accessed on 1 November 2020).

18. Jolliffe, I.T.; Cadima, J. Principal component analysis: A review and recent developments. *Philos. Trans. R. Soc. A* **2016**, *374*, 20150202. [CrossRef]
19. Python Library Scikit-Learn. Available online: https://scikit-learn.org (accessed on 1 November 2020).
20. Python Library Scipy. Available online: https://www.scipy.org/ (accessed on 1 November 2020).

MDPI

Article

Single-Particle Tracking Reveals Anti-Persistent Subdiffusion in Cell Extracts

Konstantin Speckner and Matthias Weiss *

Experimental Physics I, University of Bayreuth, Universitätsstr. 30, D-95447 Bayreuth, Germany; konstantin.speckner@uni-bayreuth.de
* Correspondence: matthias.weiss@uni-bayreuth.de

Abstract: Single-particle tracking (SPT) has become a powerful tool to quantify transport phenomena in complex media with unprecedented detail. Based on the reconstruction of individual trajectories, a wealth of informative measures become available for each particle, allowing for a detailed comparison with theoretical predictions. While SPT has been used frequently to explore diffusive transport in artificial fluids and inside living cells, intermediate systems, i.e., biochemically active cell extracts, have been studied only sparsely. Extracts derived from the eggs of the clawfrog *Xenopus laevis*, for example, are known for their ability to support and mimic vital processes of cells, emphasizing the need to explore also the transport phenomena of nano-sized particles in such extracts. Here, we have performed extensive SPT on beads with 20 nm radius in native and chemically treated Xenopus extracts. By analyzing a variety of distinct measures, we show that these beads feature an anti-persistent subdiffusion that is consistent with fractional Brownian motion. Chemical treatments did not grossly alter this finding, suggesting that the high degree of macromolecular crowding in Xenopus extracts equips the fluid with a viscoelastic modulus, hence enforcing particles to perform random walks with a significant anti-persistent memory kernel.

Keywords: anomalous diffusion; random walk; single-particle tracking

Citation: Speckner, K.; Weiss, M. Single-Particle Tracking Reveals Anti-Persistent Subdiffusion in Cell Extracts. *Entropy* **2021**, *23*, 892. https://doi.org/10.3390/e23070892

Academic Editors: Janusz Szwabiński and Aleksander Weron

Received: 15 June 2021
Accepted: 9 July 2021
Published: 13 July 2021

1. Introduction

Quantifying transport phenomena in soft and living matter on mesoscopic scales virtually always involves optical microscopy techniques, due to their high spatiotemporal resolution. Presumably, the most informative approach in this context is single-particle tracking (SPT). In SPT experiments, the rapid imaging of a sparse set of (fluorescent) particles, e.g., molecules, beads, quantum dots, or even whole organelles, allows for retrieving individual particle positions over time, eventually providing complete trajectories (see, for example, Refs. [1–3] for reviews and [4] for a quantitative comparison of SPT to other techniques). Direct access to particle trajectories facilitates the application of refined analysis approaches [5], with the mean square displacement (MSD) supposedly being the easiest and most familiar measure.

Having SPT data at hand, one can calculate, for example, the time-averaged MSD (TA-MSD), $\langle r^2(\tau)\rangle_t$, for each trajectory and compare these to their ensemble-average, $\langle r^2(\tau)\rangle_{t,E}$. A commonly observed feature is a power-law scaling of both MSDs $\langle r^2(\tau)\rangle_{t,E} \sim \langle r^2(\tau)\rangle_t \sim \tau^\alpha$, with normal Brownian diffusion being indicated by $\alpha = 1$. Scaling exponents $\alpha < 1$ are commonly referred to as 'subdiffusion', whereas values $1 < \alpha < 2$ are termed 'superdiffusion'. Subdiffusion with scaling exponents $0.3 < \alpha < 0.9$ has been observed very frequently, at least on short and intermediate time scales, for tracer particles in complex media, e.g., in equilibrated biomimetic crowded fluids [6–10], in the cytoplasm [6,11–15] and in the nucleoplasm [11,16–18] of living cells, as well as on biomembranes [19–22]; see also [23,24] for extensive reviews.

Cell extracts, which are basically only the cytosol of an ensemble of cells without larger organelle compartments, constitute an intermediate between artificial equilibrated media and nonequilibrium fluids inside living cells. Although extracts support vital processes, such as gene transcription [25] or even the formation of a mitotic spindle [26], diffusional transport in these fluids has so far been explored only sparsely. Biochemically active extracts may be derived from different sources, with the eggs of the clawfrog *Xenopus laevis* supposedly being the most popular: here, unfertilized oocytes are collected from the spawn, cooled down and crushed by centrifugation (see [27] for details). Due to different buoyancies, most membranes and the yolk can be separated from the aqueous cytosol, which eventually is obtained as extract. This extract not only includes all necessary biomolecules at physiological concentrations, but also allows for native interactions of proteins and/or nucleic acids and/or sugars. Due to its amphibian origin, the Xenopus extract provides full functionality, e.g., a dynamic cytoskeleton and even functional spindles, already at temperatures around 20 °C [27]. Given the crowdedness of these fluids and their inherent nonequilibrium background noise due to active proteins, it is unclear how particles explore such a (nearly) living fluid. While an early study, which was focused on the rheology of Xenopus extracts, revealed already that particles with sizes around 1 µm move subdiffusively in these fluids [28], a study on smaller particles but also a detailed investigation of the random walk process associated with the observed subdiffusion has been lacking so far.

As a stochastic process, subdiffusion may arise when the accessible space has a fractal geometry [29], e.g., imposed by a sufficiently dense set of immobile obstacles that form a random percolation cluster. In most experimentally relevant cases, however, obstacles are too mobile to induce an obstructed random walk in a fractal environment (see [20] for a discussion). Power-law distributed waiting times between successive steps can also induce subdiffusion [30], yet at the cost of weak ergodicity breaking [31,32], i.e., the scaling of TA-MSDs is that of normal Brownian motion ($\alpha = 1$), whereas a simple ensemble-averaged MSD of many trajectories shows subdiffusive scaling. The difference between both measures, at least in the unconfined case (see [33,34] for results in confined geometries), is related to a successive aging of the system. While such processes were sometimes observed [13,19], most experimental reports are more consistent with fully ergodic random walk processes, albeit often with a marked anti-correlation of successive steps instead of a purely Markovian random walk. For an abstract description of such a subdiffusive mode of motion, fractional Brownian motion (FBM) [35] with a Hurst coefficient $H = \alpha/2 \leq 1/2$ was used in many cases (see [5,23] for an extensive discussion). In a nutshell, FBM is a Gaussian process with stationary increments whose anti-persistent memory kernel may encode the viscoelastic characteristics of the surrounding medium. Using SPT, experimental data can be tested directly for FBM features not only via MSDs, but also via the power-spectral density (PSD) and correlation functions that report on the memory kernel [5,23].

Here, we show via SPT that beads with 20 nm radius move subdiffusively in native and chemically treated Xenopus extracts. A sublinear scaling of MSDs with an average scaling of $\langle\alpha\rangle \approx 0.9$ is found, accompanied by a significant anti-persistence peak in the velocity autocorrelation function (VACF). The VACF shows excellent agreement with the FBM prediction, and the distribution of step increments is Gaussian, suggesting that the particles perform a subdiffusive random walk of the FBM type. Further support of this notion is given by the PSD and the associated coefficient of variation, both of which also agree very well with the FBM predictions. Chemical treatments of the extract, e.g., depolymerizing the cytoskeleton, do not grossly alter the results, suggesting that the high degree of crowding in Xenopus extracts equips the fluid with a viscoelastic modulus that forces particles to perform FBM-like random walks.

2. Materials and Methods

2.1. Microscopy and Single-Particle Tracking

Fluorescence images were taken with a customized spinning-disk confocal microscope, consisting of a Leica DMI 6000 microscope stand (Leica Microsystems, Wetzlar, Germany) equipped with a CSU-X1 spinning disk unit (Yokogawa Electric, Tokyo, Japan). Samples were illuminated by a 491/561 nm dual-combined DPSS laser (Cobolt, Stockholm, Sweden), and fluorescence was detected in the range of 500–550 nm or 575–625 nm, respectively. The setup was controlled by a custom written LabView software (National Instruments, Austin, TX, USA). Time series of images were recorded at room temperature (about 19 °C) with a Hamamatsu Orca Flash 4V2.0 sCMOS camera (Hamamatsu Photonics, Hamamatsu City, Japan), using an HCPL APO 63x/1.4 oil immersion objective (Leica Microsystems). With a 2×2 hardware camera binning, the size of the squared pixels was determined as 112.4 nm.

Rhodamine-labeled microtubules in Xenopus extracts were imaged with an exposure time of 250 ms, using the 561 nm excitation channel. To improve the contrast between microtubules and unbound fluorescent tubulin monomers, images were post-processed in Fiji: the images were filtered with a median filter (radius set to 0.7 pixels), and background fluorescence was removed using a rolling-ball algorithm with a 10-pixel radius (built-in function 'subtract background'). Subsequently, the colormap *mpl-viridis* was assigned to all fluorescence images.

In our SPT experiments, fluorescent polystyrene microspheres with a diameter of 40 nm (FluoSpheres NeutrAvidin-Labeled, F8771, Thermo Fisher Scientific, Dreieich, Germany) were used. In contrast to carboxylate surface coupling, neutravidin minimizes unspecific interactions with DNA and RNA complexes or negatively charged surfaces. For calibration measurements, 1:100 stock solutions in DNase/RNase-Free Distilled water (Invitrogen) or 1:20 stock solutions (for Xenopus egg extract experiments) were prepared. On average, about 200–400 fluorescent particles were observed in the field of view of the camera sensor ($110 \times 70\ \mu m^2$). For tracking, 2000 images were recorded with an exposure time of $\Delta t = 25$ ms per frame, using the 491 nm excitation channel.

Particle positions were detected and linked to trajectories by the ImageJ/FIJI plug-in TrackMate [36]. As an input parameter for TrackMate, the diameter of fluorescent particles was estimated via the intensity profiles of 119 particles embedded in pure glycerol, yielding an average full width half maximum of the point-spread function of 2.9 ± 0.5 pixels. Particle tracking was performed using the Laplacian-of-Gaussian detection algorithm (diameter set to four pixels, threshold set to 50 ± 15 grey values and using sub-pixel localization). No additional filters were applied to the detected spots. Identified particle positions were linked with the simple linear assignment problem tracker adopted from [37]. Here, a maximum linking distance of three pixels was used and frame gaps were not allowed. The minimum trajectory length was set to $N \geq 50$ positions and trajectories with a total displacement of less than one pixel were discarded.

Non-assigned detections were cleaned from the time series experiments; particle trajectories were exported as XML and converted to ASCII files for further processing in Matlab (Matlab 2018b, The Mathworks Inc., Natick, MA, USA). All statistical analyses of particle trajectories were performed with custom-written codes in Matlab that were prior checked for proper function by random walk simulation data. In our analyses, particle trajectories were clipped exactly to lengths of $N = 70$ or $N = 150$ time steps for better comparability within the ensemble (i.e., all shorter trajectories were discarded for the analysis). In total, our ensemble (=the number M of trajectories of a given length for a given condition) consisted of 1000–3000 trajectories ($N = 70$) and 150–600 ($N = 150$) for Xenopus extract experiments. For varying glycerol water mixtures, 1000–5000 ($N = 70$) and 75–1000 ($N = 150$) trajectories were available. The specific ensemble sizes are given in Tables 1 and 2.

2.2. Xenopus Extract Preparation and Modification

Cytostatic factor-arrested (CSF) cytoplasmic extracts were prepared from freshly laid *Xenopus laevis* eggs based on standard protocols [26,27,38]. In brief, eggs in the metaphase stage of meiosis II were collected, dejellied and packed into a centrifugation tube. The packed eggs were crushed and fractioned into three distinctive layers by centrifugation. The mid cytoplasmic layer was carefully isolated and supplemented with 10 μg/mL of protease inhibitors, leupeptin, pepstatin and chymostatin (diluted in DMSO), 10 μg/mL of cytochalasin D, and ATP regeneration mix (190 mM creatine phosphate, 25 mM adenosine triphosphate, 25 mM $MgCl_2$ and 2.5 mM K-EGTA at pH 7.7) that was diluted 1:50 to the extract. Finally, 0.35 μL from the 1:20 stock solution of fluorescent particles was added to 20 μL of the extract; optionally, pharmaceuticals for affecting microtubule structures were added. Microtubules were labeled by fluorescent tubulin (TL 331M, Cytoskeleton Inc., Denver, CO, USA) according to the manufacturer's protocol: 10 mg/mL stock solution of fluorescent tubulin dissolved in general tubulin buffer (80 mM PIPES, 0.5 mM EGTA and 2 mM $MgCl_2$ at pH 6.9) with 1 mM GTP that was held on ice and added to Xenopus extracts to a working concentration of 50 μg/mL. The CSF extract was chilled immediately on ice and used within a maximum of two hours for the experiments.

Different chemicals were optionally added to Xenopus extracts to affect the microtubule integrity. Here, attention was paid to dilute the extract as little as possible. Preliminary investigations have shown that a total dilution of Xenopus extracts by up to 10% (due to the addition of beads or drugs) appears unproblematic and does not affect the microtubule structures observed. For all experimental conditions, the total volume added for modifying microtubules was balanced by the addition of distilled water to the otherwise untreated extract.

Nocodazole (Sigma-Aldrich, Munich, Germany) was used to depolymerize microtubules [18]. To this end, a stock solution of 10 mM dissolved in dimethylsulfoxide was diluted to a final concentration of 33.3 μM in Xenopus extract that was kept on ice for 10 min. Afterward, the extract was incubated for 15 min at 25 °C and then transferred to microscopy. Paclitaxel (in the remainder referred to as 'taxol', Sigma Aldrich, Germany) was used to stabilize the microtubules. It was added to the extract at a working concentration of 25 μM. Non-hydrolizable analogues of ATP (ATPγS, Merck, Darmstadt, Germany) or GTP (GTPγS, Merck, Darmstadt, Germany) were used to affect the turnover of chemical energy in the extract. To this end, ATPγS and GTPγS (stored as stock solutions of 25 mM for ATPγS and 12.5 mM for GTPγS in distilled water) were added to final concentrations of 500 μM and 250 μM to the Xenopus extracts.

3. Results and Discussion

3.1. Calibration Experiments in Viscous Media

To arrive at a proper baseline for our SPT experiments, we first tracked fluorescent beads with a radius of $R = 20$ nm in purely viscous media with varying viscosity (see Section 2 for details). In particular, we used here glycerol–water mixtures in the range of 70–90% (per weight) for which the viscosity values are known from the literature. Adding 8% of the bead stock solution to the total volume of these mixtures, viscosities in the range $\eta \in [0.016, 0.096]$ Pa · s were probed by our SPT experiments. For consistency among the trajectories and for better comparison to subsequent experiments in Xenopus cell extracts, we fixed the trajectory length to $N = 70$ or alternatively to $N = 150$, bearing in mind that short trajectories may suffer from statistical fluctuations [39] while longer trajectories might show a bias for picking slower particles [15] that are easier to track without losing them (e.g., due to leaving the focal plane).

Two-dimensional trajectories $\mathbf{r}(t) = \mathbf{r}_1, \ldots, \mathbf{r}_N$ acquired at discrete times $t = \Delta t, 2\Delta t, \ldots, N\Delta t$ with frame time $\Delta t = 25$ ms and a total measurement time $T = N\Delta t$ were first evaluated with their individual TA-MSDs, defined via the following:

$$\langle r^2(\tau)\rangle_t = \frac{1}{N-k}\sum_{j=1}^{N-k}\left(\mathbf{r}_{j+k} - \mathbf{r}_j\right)^2, \tag{1}$$

where $\tau = k\Delta t$ denotes the lag time. The resulting TA-MSDs of trajectories were fitted with a simple power-law as follows:

$$\langle r^2(\tau)\rangle = 4K\tau^\alpha \tag{2}$$

in the range $\tau \in [0.05, 0.3]$ s by linear regression of $\log[\tau]$ versus $\log[\langle r^2(\tau)\rangle_t]$. Therefore, the generalized diffusion coefficient K becomes equivalent to the familiar diffusion constant, D, for normal Brownian motion ($\alpha = 1$). In this case, the Einstein–Stokes relation predicts the diffusion constant to depend on particle radius R and medium viscosity η as follows:

$$D = \frac{k_B T}{6\pi\eta R} \tag{3}$$

yielding predictions for diffusion constants in the range $D \in [0.11, 0.67]$ $\mu m^2/s$ for our SPT experiments in glycerol–water mixtures.

In line with our expectations for purely viscous fluids, we observed on average normal diffusion, i.e., $\langle\alpha\rangle = 1$, for all viscosities and trajectory lengths (see summary in Table 1), albeit the individual TA-MSD scaling exponents showing marked fluctuations in the range $\alpha \in [0.8, 1.2]$ (see Figure 1 for representative MSDs and the probability density function of scaling exponents, $p(\alpha)$) due to the limiting statistics in TA-MSDs (see [39] for a detailed discussion). Therefore, the extracted diffusion coefficients K also showed marked fluctuations, and, yet again the average overall trajectories revealed a value $\langle K\rangle$ that compared favorably to the predicted values of D (cf. Table 1). This finding also indicates that particle radii are, on average, near to their declared and expected value, in line with previous findings [4]. Please note the slight but visible bias toward lower values of K for longer trajectories, indicating an unwanted bias toward slower particles that could be tracked over longer periods. Moreover, the amount of trajectories available for the analysis clearly increases for increasing viscosity, since slower-moving particles can be tracked easier. Overall, these calibration experiments demonstrate that normal diffusion with the anticipated mobility is found via SPT in purely viscous fluids. Deviations from unity in the (mean) scaling exponent of MSDs can therefore be taken as a clear signature of a significant anomalous diffusion.

Table 1. Summary of glycerol concentrations (weight percent) in glycerol–water mixtures, along with the respective viscosities η, and predicted diffusion constants D. Average scaling exponents $\langle\alpha\rangle$ (found via fitting all TA-MSDs for trajectories of 20 nm radius particles as described in the main text, followed by averaging the individual values of α) are near to unity and mean diffusion coefficients $\langle K\rangle$ (also obtained by averaging the results for individual TA-MSDs) compare favorably to the predicted values of D. Result for trajectories with length $N = 70$ and $N = 150$ are given in upper and lower lines, respectively. The ensemble size of evaluated trajectories for the respective condition is given by M.

glyc.	η [Pas]	D [$\mu m^2/s$]	$\langle\alpha\rangle$	$\langle K\rangle$ [$\mu m^2/s^\alpha$]	M
70%	0.016	0.67	1.00	0.56	929
			0.99	0.45	67
75%	0.023	0.46	1.01	0.42	1513
			1.03	0.35	185
80%	0.035	0.30	1.01	0.28	2158
			1.03	0.24	250
85%	0.055	0.19	1.01	0.20	4426
			1.02	0.18	749
90%	0.096	0.11	1.00	0.13	4668
			1.02	0.12	980

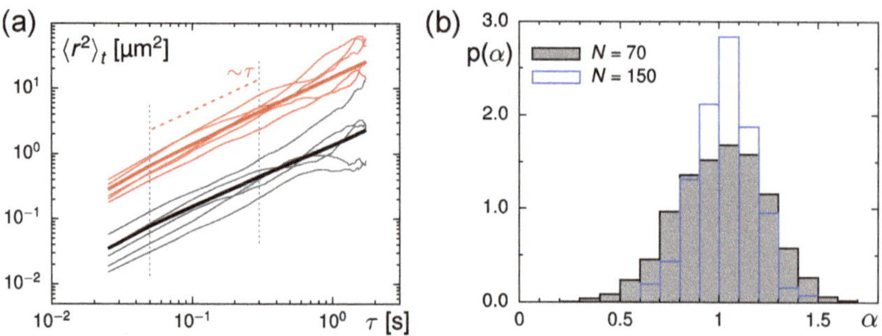

Figure 1. (**a**) Representative TA-MSDs for trajectories with length $N = 70$ (randomly chosen from the ensemble) from experiments in glycerol–water mixtures (red thin lines) and in Xenopus extract (black thin lines), together with the respective ensemble-averaged TA-MSDs (colored thick lines). For better visibility, data for calibration experiments have been shifted upward tenfold. The scaling for normal diffusion ($\langle r^2(\tau) \rangle_t \sim \tau$) is indicated by a red dashed line; vertical grey dashed lines indicate the fit region used to analyze individual TA-MSDs. (**b**) The PDF of anomaly exponents, $p(\alpha)$, as obtained from fitting TA-MSDs in glycerol–water mixtures features a mean $\langle \alpha \rangle \approx 1$, irrespective of the trajectory length (black-grey histogram: $N = 70$, blue histogram: $N = 150$).

3.2. *Evaluation of Tracer Motion in Native Xenopus Extract*

As a next step, we explored the diffusive motion of the same fluorescent particles in Xenopus extracts (see Materials and Methods for details). Given that these extracts are complex and crowded fluids with an active biochemistry, we anticipated considerable differences to the simple glycerol–water mixtures. As before, we restricted ourselves to trajectory lengths $N = 70$ and $N = 150$, and used the TA-MSD and its ensemble average for a first characterization. Representative TA-MSDs are shown together with the ensemble average in Figure 1a.

Next, we fitted all TA-MSDs with Equation (2) to extract the respective scaling exponents, α, and generalized diffusion coefficients, K. Here, we tacitly assume that static and dynamic localization errors are negligible for our data; we will confirm this assumption below. Still, to soften any remnant influence of localization errors, especially for retrieving the scaling exponent α, we did not take the first point of TA-MSDs (at $\tau = \Delta t$) into account for fitting. Fitting was performed as in the calibration measurements.

Evaluating all TA-MSDs yielded probability density functions (PDFs) for the scaling exponent, $p(\alpha)$, and the generalized diffusion coefficient, $p(K)$. Inspecting $p(\alpha)$ (shown in Figure 2a) reveals that the ensemble of trajectories shows, on average, a slight subdiffusive scaling of TA-MSDs with a mean $\langle \alpha \rangle = 0.89$. The width of the PDF (about ± 0.2 around the mean) highlights, again, marked fluctuations between individual trajectories. In fact, similar fluctuations in the trajectory-wise values of α are expected already from mere statistical fluctuations due to fairly short trajectories (see [39] for discussion). Part of the width in $p(\alpha)$, however, may also reflect the spatially varying properties of the extract that is explored and reported on by different particles. Remarkably, trajectories of length $N = 70$ and $N = 150$ resulted in comparable PDFs and the same mean, i.e., longer trajectories were not biased toward lower scaling exponents. Notably, our observation of a subdiffusive motion of beads with 20 nm radius in Xenopus extracts is consistent with an earlier report [28] that reported scaling exponents in the range $\alpha \in [0.7, 0.95]$, with lower values emerging for larger particles (radii in the range 0.1–1 μm).

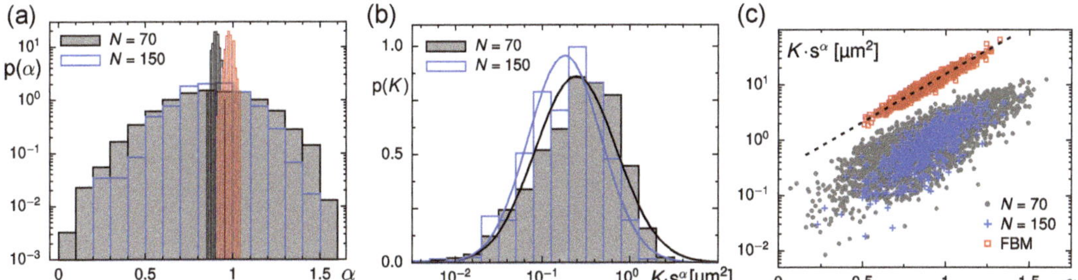

Figure 2. (**a**) The PDF of anomaly exponents, $p(\alpha)$, as obtained from fitting TA-MSDs in the interval $\tau \in [0.05, 0.3]$ s, features a mean $\langle\alpha\rangle \approx 0.9$, irrespective of the trajectory length (black-grey histogram: $N = 70$, blue histogram: $N = 150$). The considerable width of the PDF may not only reflect statistical fluctuations but is likely to also report on spatially varying material properties of the Xenopus extract. Performing a bootstrapping approach with geometric averaging (black-open histogram) confirms the slightly subdiffusive motion of particles, while an arithmetic averaging (red histogram) overestimates the mean scaling exponent; see also main text for discussion. Please note the logarithmic y-axis. (**b**) The PDF of generalized diffusion coefficients, $p(K)$, shown here versus the average area covered in one second, $K \times 1s^{\alpha}$, features an almost lognormal shape (indicated by full lines) for trajectory lengths $N = 70$ (grey/black) and $N = 150$ (blue), with a slight tendency for lower mobilities in longer trajectories. Please see the main text for discussion. (**c**) A scatter plot of trajectory-wise values of α and K (blue and grey symbols) highlights a correlation between these two quantities, in good agreement with results on simulated FBM trajectories with a Hurst coefficient $H = \alpha/2 = 0.45$ (red symbols). The black dashed line is an empiric guide for the eye. FBM simulation data have been shifted upward fivefold for better visibility.

To further confirm and validate the significance of the mean exponent $\langle\alpha\rangle \approx 0.9$, we exploited a bootstrapping approach [15]. Based on a total set of several hundred TA-MSDs, we randomly selected 100 trajectories and averaged these to a single, sub-ensemble averaged MSD from which we determined the scaling exponent α. This random drawing from the total set of TA-MSDs and the subsequent averaging was repeated 200 times to obtain a PDF for these scaling exponents. By construction, the width of this PDF can be expected to be much smaller [15], yielding a better estimate for the mean. Averaging over the subensemble (SE) of TA-MSDs was either done arithmetically ($\langle\langle r^2(\tau)\rangle_t\rangle_{\mathrm{SE}}$), or by geometric averaging ($\exp[\langle\log(\langle r^2(\tau)\rangle_t)\rangle_{\mathrm{SE}}]$). As a result, we found that the mean scaling exponent found by bootstrapping with arithmetic averaging was $\langle\alpha\rangle_{\mathrm{SE},a} = 0.98$, which is consistent with normal diffusion. In contrast, geometric averaging yielded $\langle\alpha\rangle_{\mathrm{SE},g} = 0.90$, in agreement with the mean of the raw PDF $p(\alpha)$ shown in Figure 2a. Since geometric averaging boils down to an arithmetic averaging of individual scaling exponents (due to the invoked logarithm), this result is, in fact, the more trustworthy approach for retrieving the average scaling exponent. Arithmetic averaging rather averages TA-MSDs with respect to their individual (and strongly fluctuating) diffusion coefficients, hence obscuring the mean scaling law and overestimating $\langle\alpha\rangle$ (see [15] for another example). The difference in scaling obtained by the two averaging procedures may also be understood as a consequence of Jensen's inequality. The mapping $\varphi : \alpha \mapsto t^{\alpha}$ is convex for any choice of a real number t. Jensen's inequality then states that $t^{\langle\alpha\rangle} = \varphi(\langle\alpha\rangle) \leq \langle\varphi(\alpha)\rangle = \langle t^{\alpha}\rangle$, in line with our observation.

To check for the influence of localization errors when retrieving the scaling exponent from individual TA-MSDs, we took the following approach: using immobilized beads, we determined the contribution of the static localization offset to be a positive additive constant of about 8×10^{-4} μm^2 (i.e., 20 nm accuracy of positions), whereas the dynamic localization error for our frame time and diffusion coefficients amounts to a negative constant with modulus 0.008 μm^2 or lower. Considering either of these two extreme values as additive constants to the power law Equation (2) while fitting TA-MSDs resulted in minor deviations of ± 0.03 from the previously found value for $\langle\alpha\rangle$. We therefore conclude that our SPT data show mild yet significant subdiffusion. In the following paragraphs, we further corroborate this conclusion by additional measures.

Let us now focus on the PDF of generalized diffusion coefficients, $p(K)$, retrieved from individual TA-MSDs (Figure 2b). To remove the ambiguity of units for varying scaling exponents (units of K are $\mu m^2/s^\alpha$), we report these PDFs as a function of $K \times 1\ s^\alpha$, which represents the typical area covered by the particle within a second. As indicated already in the discussion of the previous paragraph, individual values of K varied considerably between trajectories, resulting in a very broad, almost lognormal-shaped PDF. Again, statistical fluctuations as well as locus-specific mobilities are likely to contribute to the width of $p(K)$. In contrast to the anomaly exponents, a slight bias for smaller diffusion coefficients was visible for longer trajectories, seen as a slight shift of the peak in $p(K)$ between $N = 70$ and $N = 150$ (cf. Figure 2b). Still, in both cases, the typical area explored within a second matches roughly with our calibration experiments in highly viscous glycerol–water mixtures (cf. Table 1), albeit the mean scaling exponent is clearly different. Furthermore, a scatter plot of trajectory-wise values for α and K (Figure 2c) highlights a correlation of these two quantities. This finding is in line with results of FBM simulations ($M = 2000$ or $M = 1000$ two-dimensional trajectories of length $N = 70$ or $N = 150$, obtained via the circulant method [40] with $H = \alpha/2 = 0.45$ and $K = 0.4\ \mu m^2/s^\alpha$; also shown in Figure 2c), which gives a first hint that an anti-persistent stochastic process underlies the acquired trajectories.

Let us briefly insert here an intuitive explanation of why a lognormal-like shape of $p(K)$ is seen here and has also been reported frequently in the literature for other experiments (see [18,22,41–48] for examples). For the simplicity of the argument, we restrict ourselves to standard Brownian motion with Gaussian increment statistics and no memory kernel. Then, the diffusion constant $D = \langle\langle r^2(\tau)\rangle_t/(4\tau)\rangle$, as retrieved from a TA-MSD fit, is the finite mean of positive, squared Gaussian random numbers, Δr^2 (cf. Equation (1)). The associated PDF of $\vartheta = \Delta r^2/\langle\Delta r^2\rangle$ follows a special variant of the gamma distribution, known as Porter–Thomas distribution [49], $p(\vartheta) = \exp\{-\vartheta/2\}/\sqrt{2\pi\vartheta}$. Similar PDFs, featuring a power law that is cut by an exponential, have been observed, for example, for blinking quantum dots [50]. For power-law PDFs with a cutoff, it is known that a finite sum (or average) of random numbers will only slowly approach a Gaussian PDF, the hallmark of the central limit theorem. Indeed, numerically drawing N random numbers from the Porter–Thomas PDF and averaging (summing) them as $x = \sum_{i=1}^{N} \vartheta_i/N$ yields a PDF $p(x)$ with a nonzero skewness $\sim 1/\sqrt{N}$ that resembles a lognormal PDF. Hence, even mere statistical reasons can lead already to an apparently lognormal PDF of diffusion constants if trajectories are short enough. Varying mobilities, encountered by particles in different spatial positions of a heterogeneous sample, broaden the PDF even further. Please also note that applying a logarithmic transformation is basically only one variant of the more general class of Box–Cox transformations $x \mapsto (x^\lambda - 1)/\lambda$ [51], namely the one for $\lambda = 0$. These transformations were introduced as a means to symmetrize a given data set, eventually yielding a Gaussian-shaped PDF of the transformed data when choosing the optimal λ. Restricting the choice to $\lambda = 0$, i.e., simply logarithmizing the data, therefore will reduce the skewness in virtually all practical cases but may not yet completely symmetrize the data, leaving a residual non-zero skewness. As a consequence, any skewed data set will assume a more Gaussian shape upon applying a logarithmic transformation, although this mere statistical approach does, in general, not yet reveal the reason for the skewness and apparent lognormal PDF of the original data. In particular, inferring from an apparent lognormal shape of the PDF that the underlying data is the product of independent, identically-distributed variables may not be a compelling conjecture.

Coming back to our experimental data, we next explored whether the underlying random walk process is Gaussian. To this end, we considered the step increments $(\delta x_i, \delta y_i) = (x_{i+n} - x_i, y_{i+n} - y_i)$ taken within a period $\delta t = n\Delta t$. These follow a Gaussian PDF if the process is a simple FBM process. Recently, however, deviations from a Gaussian PDF were reported [14,15,52,53], highlighting heterogenous diffusion characteristics that could be rationalized by random walks with spatiotemporally fluctuating transport coefficients [54,55] and/or by systems with spatial disorders [56,57]. To account for the strongly fluctuating diffusion coefficients, we normalized the step increments δx and δy of each trajectory by the respective root-mean-square values. The resulting set of normalized steps did not exhibit significant differences between x- and y-coordinates. We therefore combined both into a single set of normalized increments, χ, which resulted in a symmetric PDF so that inspecting $p(|\chi|)$ was sufficient.

Data for different δt show overall a very good agreement with a standard Gaussian and are incompatible with a simple exponential (Figure 3), indicating that no major diffusion heterogeneity is present in our SPT data from Xenopus extracts. For $|\chi| > 3$ and $\delta t \geq 5\Delta t$, the experimental PDF falls slightly below the Gaussian benchmark for unknown reasons. Despite this slight deviation, it appears fair to conclude that the trajectories emerged from a mildly subdiffusive Gaussian process, suggesting that FBM is the most likely model that describes our experimental data.

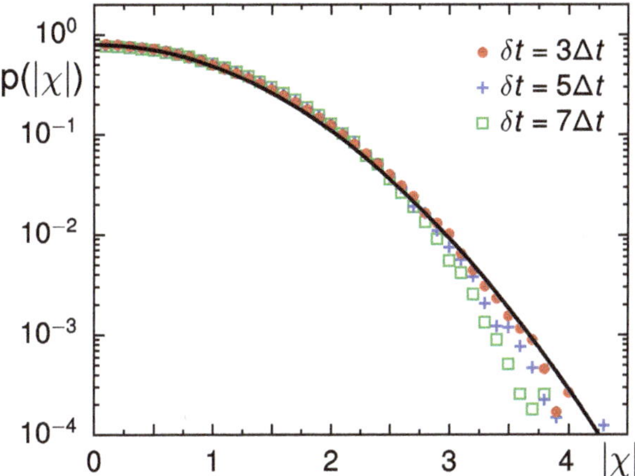

Figure 3. The PDF of normalized increments taken within a period δt, shown here as $p(|\chi|)$, complies well with a standard Gaussian (black full line) for different choices of δt (color-coded symbols). For $\delta t \geq 5\Delta t$ and $|\chi| > 3$, consistently lower probabilities than the Gaussian benchmark are observed for unknown reasons.

To follow up on this hypothesis and probe the existence of a non-trivial memory kernel in our experimental data, we employed the ensemble- and time-averaged velocity autocorrelation function (VACF) of each trajectory, defined as follows:

$$C(\tau) = \left\langle \frac{\langle \mathbf{v}(t)\mathbf{v}(t+\tau)\rangle_t}{\langle \mathbf{v}(t)^2\rangle_t} \right\rangle_E , \tag{4}$$

with $\mathbf{v}(t) = [\mathbf{r}(t+\delta t) - \mathbf{r}(t)]/\delta t$ denoting the instantaneous velocity that is simply the two-dimensional step $\mathbf{r}(t+\delta t) - \mathbf{r}(t)$ taken within integer multiples of the frame time, $\delta t = n\Delta t$.

It is convenient to rescale the lag time $\tau = k\Delta t$ with δt, yielding a dimensionless time $\xi = \tau/\delta t = k/n$. For FBM, an analytical prediction for the VACF was derived [5,35]:

$$C_{\text{FBM}}(\xi) = \{(\xi+1)^\alpha + |\xi - 1|^\alpha - 2\xi^\alpha\}/2 . \tag{5}$$

The fact that the VACF does not depend on n and k but only on the ratio $\xi = k/n$ reflects the self-similarity of FBM processes. Localization errors in SPT experiments can break this self-similarity [58,59], i.e., using different δt for rescaling τ to ξ leads to progressive deviations from Equation (5). In fact, very recently, the VACF was shown to be a sensitive reporter for detecting localization errors for FBM from the sub- to the superdiffusive regime [60], as even small localization errors lead to significant changes of $C(\xi = 1)$ at different choices of δt. In our case, however, rescaling with different δt lead to an almost perfect collapse of all data to the master curve predicted by Equation (5), see Figure 4. In fact, this finding is in favorable agreement with the earlier rheology results on Xenopus extracts that revealed a significant viscoelastic response [11,28], linking the anti-persistent dip in the VACF to a viscoelastic memory kernel of the medium. Moreover, the good agreement with Equation (5) for all choices of δt confirms our previous notion that localization offsets (to which VACFs are very sensitive) are negligible for our data.

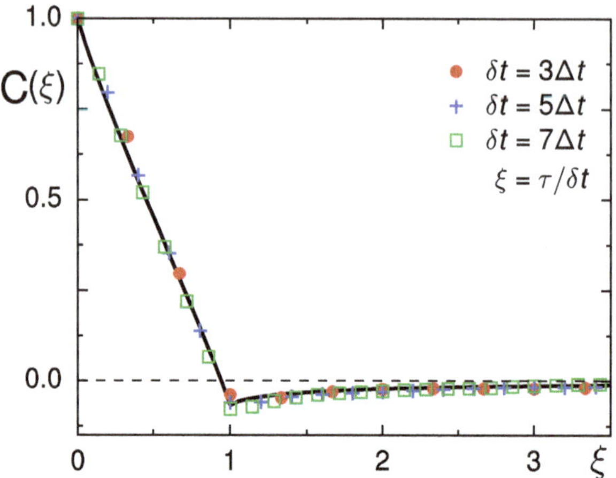

Figure 4. The normalized VACF, $C(\xi)$, for different choices of δt (color-coded symbols) shows excellent agreement with the FBM prediction [Equation (5)] when inserting the mean scaling exponent $\langle\alpha\rangle = 0.9$ (full black line). In particular, a clearly negative value of $C(\xi = 1)$ confirms an antipersistent random walk, most likely of the FBM type. No significant changes of the VACF minimum are seen for different δt, confirming that trajectories are not plagued by localization errors.

As a further piece of evidence that the mild subdiffusion seen for particle motion in Xenopus extracts is due to an antipersistent FBM process, we probed the power-spectral density (PSD) of individual trajectories:

$$S(f) = \frac{1}{T}\left|\int_0^T e^{ift}x(t)dt\right|^2 + \left|\int_0^T e^{ift}y(t)dt\right|^2, \tag{6}$$

and the corresponding ensemble average, $\langle S(f)\rangle_E$. A wealth of analytical information is available for PSDs and their trajectory-wise fluctuations [61,62]. Alerted by the observations that arithmetic and geometric averaging can perturb power-law effects in the ensemble of trajectories (cf. above) we aimed at softening the influence of the grossly varying diffusion coefficients K in the subsequent analysis. Therefore, we normalized all trajectories by their respective root-mean-square step length within successive frames, in line with the approach taken when probing the Gaussian shape of the statistics of increments (cf. context of Figure 3).

As expected, the ensemble-averaged PSD of these normalized trajectories (for $N = 70$ and $N = 150$) followed the analytical prediction $S(f) \sim 1/f^{1+\langle\alpha\rangle}$ around which PSDs of individual trajectories fluctuated to a considerable extension (Figure 5). These fluctuations encode another important hallmark of FBM via the coefficient of variation, defined as $\gamma(f) = \sigma/\langle S(f)\rangle_E$ with $\sigma(f)$ denoting the standard deviation of trajectory-wise PSDs. For FBM, asymptotic values $\gamma = 1$ for subdiffusion and $\gamma = \sqrt{5}/2$ for normal diffusion were predicted and verified before [61,62]. To calculate the coefficient of variation for our data, we randomly drew 1000 curves from the ensemble of one-dimensional TA-PSDs for the x- and y-direction and removed those 5% of TA-PSDs with the largest deviations from the ensemble-averaged PSD. The resulting values for γ fully complied with the FBM predictions (Figure 6). Normally diffusive trajectories from calibration experiments converge toward the predicted value $\gamma = \sqrt{5}/2$, whereas subdiffusive trajectories clearly assume lower values that eventually converge to the universal unity value for large frequencies.

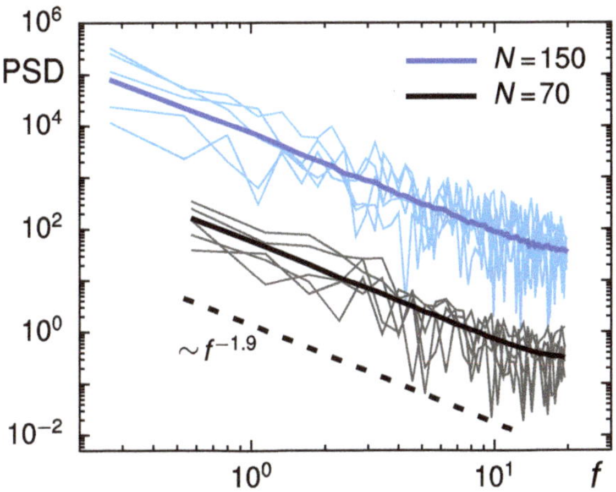

Figure 5. The PSD of individual trajectories (black and blue thin lines, representing trajectories with length $N = 70$ and $N = 150$, respectively) fluctuate around the ensemble-averaged PSD (thick colored lines). In both cases, the FBM prediction for a scaling $S(f) \sim 1/f^{1+\langle\alpha\rangle}$ (with $\langle\alpha\rangle = 0.9$, dashed line) are nicely met. For better visibility, data for $N = 150$ have been shifted upward 100-fold.

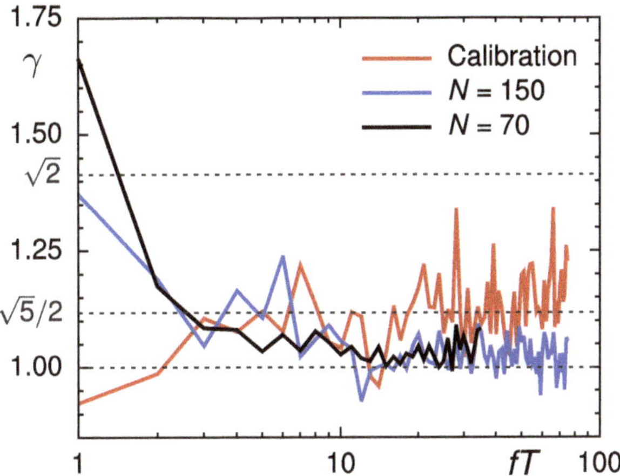

Figure 6. The coefficient of variation of individual PSDs with respect to the ensemble mean, $\gamma(f)$, for normally diffusive trajectories from calibration experiments (red line) clearly assumes higher values than those for trajectories from the Xenopus extract (blue and black lines), irrespective of the trajectory length, N. As predicted for FBM, these subdiffusive SPT data converge toward $\gamma = 1$, whereas normally diffusive data from calibration experiments converge to the predicted value $\gamma = \sqrt{5}/2$. Both are clearly distinct from the prediction for superdiffusive FBM motion, $\gamma = \sqrt{2}$. For convenience, frequencies f were made dimensionless by multiplication with the total time $T = N\Delta t$ covered in each trajectory.

Altogether, we conclude from the analyses of our SPT data that beads with 20 nm radius feature a mild antipersistent subdiffusion in Xenopus extract with a mean scaling exponent $\langle\alpha\rangle = 0.9$. Typical signatures of an FBM with a Hurst coefficient $H = \langle\alpha\rangle/2 = 0.45$ are clearly visible, suggesting that the viscoelasticity of the extract determines these distinct random-walk properties.

3.3. From Native to Pharmaceutically Treated Xenopus Extracts

To explore to what extent the observed subdiffusion is altered when challenging biochemical processes in the extract, we also performed single-particle tracking experiments after applying pharmaceuticals to the extract. In particular, we applied taxol or nocodazole to either stabilize or completely disrupt the microtubule filaments (see Materials and Methods for details). Typical fluorescence images of the extract (stained for the beads and the microtubule filaments) highlight the strong differences between untreated and chemically challenged extracts (Figure 7). While untreated extracts feature some microtubule filaments that might obstruct the free diffusion of beads, the addition of nocodazole completely erradicates these structures, potentially resulting in a decreased obstruction of bead motion. In contrast, the addition of taxol stabilizes microtubules and, therefore, even enhances the gel-like geometry within the extract.

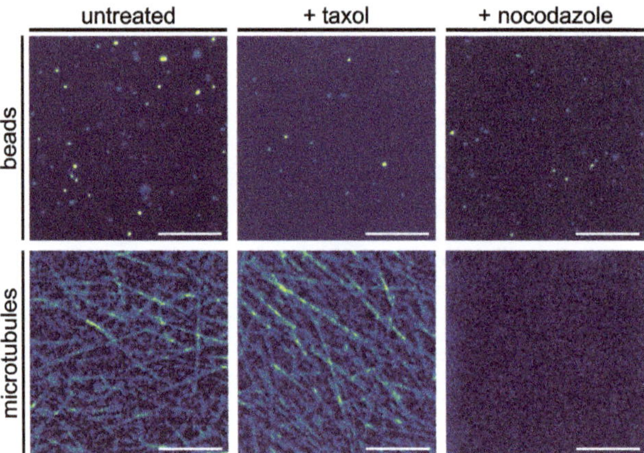

Figure 7. Representative fluorescence images of beads (upper panel) and microtubules (lower panel) in native and pharmaceutically treated Xenopus extracts (see Materials and Methods for details); scale bars indicate 10 µm. While native extracts feature a significant amount of microtubule filaments (left column), the addition of nocodazole completely eradicates these higher-order structures (right column). In contrast, stabilizing microtubules by taxol further enhances the 'filament jungle' (middle column).

Somewhat unexpectedly, however, altering the microtubule filament array had, on average, only minor effects (Table 2). While disrupting microtubules had, on average, no significant effect at all (with $\langle\alpha\rangle$ and $\langle K\rangle$ being almost unchanged), the addition of taxol induced a slight enhancement of the subdiffusion, i.e., a lower $\langle\alpha\rangle$, in line with the notion that increased density of filaments may further hamper the beads' free diffusion. Still, the effect is fairly small when bearing in mind that scaling exponents around and below $\alpha \approx 0.5$ were reported already for similar sized particles in the comparable cytoplasm of living cells [6,11,15]. Our findings, therefore, indicate that mainly macromolecular crowding of the fluid on length scales $\ll 1$ µm, which induces a viscoelastic memory kernel, underlies the observed subdiffusion. Higher-order structures, such as cytoskeletal assemblies or endomembranes, appear to be less important for the observed (sub)diffusion of beads in Xenopus extract.

Table 2. Summary of results found for different conditions of Xenopus extracts. Data for trajectories with length $N = 70$ and $N = 150$ are given in upper and lower lines, respectively. The ensemble size for the respective condition is given by M.

cond.	$\langle\alpha\rangle$	$\langle K\rangle$ [$\mu m^2/s^\alpha$]	M
untreated	0.89	0.39	3085
	0.88	0.27	582
+taxol	0.83	0.33	2548
	0.81	0.25	606
+nocodazol	0.89	0.37	1126
	0.81	0.17	130
+ATPγS +GTPγS	0.85	0.31	915
	0.81	0.10	189
+ATPγS +GTPγS +noc.	0.88	0.35	2646
	0.83	0.23	471

To complement these insights, we also applied non-hydrolizable analogues of ATP and GTP to prevent non-equilibrium processes that are fueled by these nucleotides (see Materials and Methods for details). In this case, the extract became very heterogeneous with tracking results from separated loci differing strongly (from total immobilization up to normal diffusion). Removing the immobilized tracks, the average behavior was in reasonable agreement with our findings for untreated and taxol-treated extracts. This finding suggests that (apart from immobilization loci) the ATP- and GTP-dependent processes are also of little importance for the motion of small beads. As a caveat, we would like to emphasize, however, that all of these findings might be subject to change when larger particles or different surface properties are considered, as these might interact differently with macromolecules and higher-order structures. In support of this statement, we would like to refer the reader to previous measurements on the diffusion of quantum dots in the cytoplasm of mammalian cells, where strong variations in the mobility and apparent diffusion anomaly were observed upon varying the particles' surface chemistry [63]. In fact, understanding anomalous diffusion in complex media at non-equilibrium conditions, e.g., in the crowded interior of living cells, is still a major challenge (see [64] for a recent study).

4. Conclusions

In summary, we have shown here that beads with 20 nm radius explore cell extracts from eggs of *Xenopus laevis* by mild subdiffusion that bears all properties of a FBM random walk. This mode of motion is largely conserved when treating the extract with pharmaceuticals that alter microtubule filaments or ATP/GTP-dependent processes. Therefore, the emergence of subdiffusion is most likely a consequence of the extracts' viscoelasticity that is induced by a high degree of macromolecular crowding, albeit changes in the beads' surface chemistry might also enhance or lower the diffusion anomaly due to transient and unspecific interactions with larger structures in the fluid [63]. In any case, given that even mild subdiffusion was predicted and observed to significantly alter biochemical reactions (see, e.g., [65–67]) our data provide a helpful clue for a deeper understanding of self-organization and pattern formation processes in cell extracts.

Author Contributions: Experiments and analyses: K.S.; concept, design and supervision of the study: M.W. Both authors wrote the manuscript and have read and agreed to the published version of the manuscript.

Funding: VolkswagenStiftung (Az. 92738), EliteNetwork of Bavaria (Study Program Biological Physics).

Institutional Review Board Statement: Not applicable.

Informed Consent Statement: Not applicable.

Data Availability Statement: The data presented in this study are available from the corresponding author upon reasonable request.

Acknowledgments: Financial support by the VolkswagenStiftung (Az. 92738) and by the EliteNetwork of Bavaria (Study Program Biological Physics) are gratefully acknowledged. We thank B. Neumann and O. Stemmann (University of Bayreuth, Genetics) for providing Xenopus eggs and M. Thampi, S. Krauss, P.-Y. Gires, and A. Hanold for extract preparation.

Conflicts of Interest: The authors declare no conflict of interest.

References

1. Qian, H.; Sheetz, M.P.P.; Elson, E.L.L. Single particle tracking. Analysis of diffusion and flow in two-dimensional systems. *Biophys. J.* **1991**, *60*, 910–921. [CrossRef]
2. Manzo, C.; Garcia-Parajo, M.F. A review of progress in single particle tracking: From methods to biophysical insights. *Rep. Prog. Phys.* **2015**, *78*, 124601. [CrossRef]
3. Shen, H.; Tauzin, L.J.; Baiyasi, R.; Wang, W.; Moringo, N.; Shuang, B.; Landes, C.F. Single Particle Tracking: From Theory to Biophysical Applications. *Chem. Rev.* **2017**, *117*, 7331–7376. [CrossRef]
4. Struntz, P.; Weiss, M. The hitchhiker's guide to quantitative diffusion measurements. *Phys. Chem. Chem. Phys.* **2018**, *20*, 28910–28919. [CrossRef]
5. Metzler, R.; Jeon, J.-H.; Cherstvy, A.G.; Barkai, E. Anomalous diffusion models and their properties: Non-stationarity, non-ergodicity, and ageing at the centenary of single particle tracking. *Phys. Chem. Chem. Phys.* **2014**, *16*, 24128–24164. [CrossRef]
6. Weiss, M.; Elsner, M.; Kartberg, F.; Nilsson, T. Anomalous subdiffusion is a measure for cytoplasmic crowding in living cells. *Biophys. J.* **2004**, *87*, 3518–3524. [CrossRef] [PubMed]
7. Banks, D.; Fradin, C. Anomalous diffusion of proteins due to molecular crowding. *Biophys. J.* **2005**, *89*, 2960–2971. [CrossRef] [PubMed]
8. Ernst, D.; Hellmann, M.; Köhler, J.; Weiss, M. Fractional brownian motion in crowded fluids. *Soft Matter* **2012**, *8*, 4886–4889. [CrossRef]
9. Szymanski, J.; Weiss, M. Elucidating the origin of anomalous diffusion in crowded fluids. *Phys. Rev. Lett.* **2009**, *103*, 038102. [CrossRef]
10. Pan, W.; Filobelo, L.; Pham, N.D.; Galkin, O.; Uzunova, V.V.; Vekilov, P.G. Viscoelasticity in homogeneous protein solutions. *Phys. Rev. Lett.* **2009**, *102*, 1058101. [CrossRef]
11. Guigas, G.; Kalla, C.; Weiss, M. Probing the nano-scale viscoelasticity of intracellular fluids in living cells. *Biophys. J.* **2007**, *93*, 316–323. [CrossRef]
12. Weber, S.C.; Spakowitz, A.J.; Theriot, J.A. Bacterial chromosomal loci move subdiffusively through a viscoelastic cytoplasm. *Phys. Rev. Lett.* **2010**, *104*, 238102. [CrossRef]
13. Jeon, J.-H.; Tejedor, V.; Burov, S.; Barkai, E.; Selhuber-Unkel, C.; Berg-Sørensen, K.; Oddershede, L.; Metzler, R. In vivo anomalous diffusion and weak ergodicity breaking of lipid granules. *Phys. Rev. Lett.* **2011**, *106*, 048103. [CrossRef]
14. Lampo, T.J.; Stylianidou, S.; Backlund, M.P.; Wiggins, P.A.; Spakowitz, A.J. Cytoplasmic RNA-Protein Particles Exhibit Non-Gaussian Subdiffusive Behavior. *Biophys. J.* **2017**, *112*, 532–542. [CrossRef]
15. Sabri, A.; Xu, X.; Krapf, D.; Weiss, M. Elucidating the origin of heterogeneous anomalous diffusion in the cytoplasm of mammalian cells. *Phys. Rev. Lett.* **2020**, *125*, 058101. [CrossRef] [PubMed]
16. Wachsmuth, M.; Waldeck, W.; Langowski, J. Anomalous diffusion of fluorescent probes inside living cell investigated by spatially-resolved fluorescence correlation spectroscopy. *J. Mol. Biol.* **2000**, *298*, 677–689. [CrossRef] [PubMed]
17. Pawar, N.; Donth, C.; Weiss, M. Anisotropic diffusion of macromolecules in the contiguous nucleocytoplasmic fluid during eukaryotic cell division. *Curr. Biol.* **2014**, *24*, 1905–1908. [CrossRef] [PubMed]
18. Stadler, L.; Weiss, M. Non-equilibrium forces drive the anomalous diffusion of telomeres in the nucleus of mammalian cells. *New J. Phys.* **2017**, *19*, 113048. [CrossRef]
19. Weigel, A.V.; Simon, B.; Tamkun, M.M.; Krapf, D. Ergodic and nonergodic processes coexist in the plasma membrane as observed by single-molecule tracking. *Proc. Natl. Acad. Sci. USA* **2011**, *108*, 6438–6443. [CrossRef]
20. Malchus, N.; Weiss, M. Anomalous diffusion reports on the interaction of misfolded proteins with the quality control machinery in the endoplasmic reticulum. *Biophys. J.* **2010**, *99*, 1321–1328. [CrossRef] [PubMed]
21. Jeon, J.H.; Monne, H.M.S.; Javanainen, M.; Metzler, R. Anomalous diffusion of phospholipids and cholesterols in a lipid bilayer and its origins. *Phys. Rev. Lett.* **2012**, *109*, 188103. [CrossRef] [PubMed]
22. Stadler, L.; Speckner, K.; Weiss, M. Diffusion of exit sites on the endoplasmic reticulum: A random walk on a shivering backbone. *Biophys. J.* **2018**, *115*, 1552–1560. [CrossRef] [PubMed]
23. Höfling, F.; Franosch, T. Anomalous transport in the crowded world of biological cells. *Rep. Prog. Phys.* **2013**, *76*, 046602. [CrossRef]
24. Weiss, M. Crowding, diffusion, and biochemical reactions. *Int. Rev. Cell Mol. Biol.* **2014**, *307*, 383–417.

25. Sokolova, E.; Spruijt, E.; Hansen, M.M.K.; Dubuc, E.; Groen, J.; Chokkalingam, V.; Piruska, A.; Heus, H.A.; Huck, W.T.S. Enhanced transcription rates in membrane-free protocells formed by coacervation of cell lysate. *Proc. Natl. Acad. Sci. USA* **2013**, *110*, 11692–11697. [CrossRef]
26. Good, M.C.; Vahey, M.D.; Skandarajah, A.; Fletcher, D.A.; Heald, R. Cytoplasmic volume modulates spindle size during embryogenesis. *Science* **2013**, *342*, 856–860. [CrossRef]
27. Hannak, E.; Heald, R. Investigating mitotic spindle assembly and function in vitro using xenopu s laevi s egg extracts. *Nat. Protoc.* **2006**, *1*, 2305. [CrossRef]
28. Valentine, M.T.; Perlman, Z.E.; Mitchison, T.J.; Weitz, D.A. Mechanical properties of xenopus egg cytoplasmic extracts. *Biophys. J.* **2005**, *88*, 680–689. [CrossRef] [PubMed]
29. Havlin, S.; Ben-Avraham, D. Diffusion in disordered media. *Adv. Phys.* **1987**, *36*, 695–798. [CrossRef]
30. Metzler, R.; Klafter, J. The random walk's guide to anomalous diffusion: A fractional dynamics approach. *Phys. Rep.* **2000**, *339*, 1–77. [CrossRef]
31. He, Y.; Burov, S.; Metzler, R.; Barkai, E.; Random time-scale invariant diffusion and transport coefficients. *Phys. Rev. Lett.* **2008**, *101*, 058101. [CrossRef]
32. Lubelski, A.; Sokolov, I.M.; Klafter, J. Nonergodicity mimics inhomogeneity in single particle tracking. *Phys. Rev. Lett.* **2008**, *100*, 250602. [CrossRef]
33. Neusius, T.; Sokolov, I.M.; Smith, J.C. Subdiffusion in time-averaged, confined random walks. *Phys. Rev. E* **2009**, *80*, 011109. [CrossRef]
34. Burov, S.; Metzler, R.; Barkai, E. Aging and nonergodicity beyond the khinchin theorem. *Proc. Natl. Acad. Sci. USA* **2010**, *107*, 13228–13233. [CrossRef]
35. Mandelbrot, B.B.; Van Ness, J.W. Fractional Brownian Motions, Fractional Noises and Applications. *SIAM Rev.* **1968**, *10*, 422–437. [CrossRef]
36. Tinevez, J.-Y.; Perry, N.; Schindelin, J.; Hoopes, G.M.; Reynolds, G.D.; Laplantine, E.; Bednarek, S.Y.; Shorte, S.L.; Eliceiri, K.W. Trackmate: An open and extensible platform for single-particle tracking. *Methods* **2017**, *115*, 80–90. [CrossRef]
37. Jaqaman, K.; Loerke, D.; Mettlen, M.; Kuwata, H.; Grinstein, S.; Schmid, S.L.; Danuser, G. Robust single-particle tracking in live-cell time-lapse sequences. *Nat. Methods* **2008**, *5*, 695–702. [CrossRef]
38. Dumont, S.; Mitchison, T.J. Force and length in the mitotic spindle. *Curr. Biol.* **2009**, *19*, R749–R761. [CrossRef] [PubMed]
39. Weiss, M. Resampling single-particle tracking data eliminates localization errors and reveals proper diffusion anomalies. *Phys. Rev. E* **2019**, *100*, 042125. [CrossRef] [PubMed]
40. Kroese, D.P.; Botev, Z.I. Spatial Process Simulation. In *Stochastic Geometry, Spatial Statistics and Random Fields*; Schmidt, V., Ed.; Springer: Cham, Switzerland, 2015; Volume 2120, pp. 369–404.
41. Wade, W.F.; Freed, J.H.; Edidin, M. Translational diffusion of class ii major histocompatibility complex molecules is constrained by their cytoplasmic domains. *J. Cell Biol.* **1989**, *109*, 3325–3331. [CrossRef] [PubMed]
42. Saxton, M.J. Single-particle tracking: The distribution of diffusion coefficients. *Biophys. J.* **1997**, *72*, 1744–1753. [CrossRef]
43. Laurence, T.A.; Kwon, Y.; Yin, E.; Hollars, C.W.; Camarero, J.A.; Barsky, D. Correlation spectroscopy of minor fluorescent species: Signal purification and distribution analysis. *Biophys. J.* **2007**, *92*, 2184–2198. [CrossRef]
44. Flier, B.M.I.; Baier, M.; Huber, J.; Mullen, K.; Mecking, S.; Zumbusch, A.; Woll, D. Single molecule fluorescence microscopy investigations on heterogeneity of translational diffusion in thin polymer films. *Phys. Chem. Chem. Phys.* **2011**, *13*, 1770–1775. [CrossRef] [PubMed]
45. Akin, E.J.; Sole, L.; Johnson, B.; Beheiry, M.E.; Masson, J.B.; Krapf, D.; Tamkun, M.M. Single-molecule imaging of nav1.6 on the surface of hippocampal neurons reveals somatic nanoclusters. *Biophys. J.* **2016**, *111*, 1235–1247. [CrossRef]
46. Singh, A.P.; Galland, R.; Finch-Edmondson, M.L.; Grenci, G.; Sibarita, J.B.; Studer, V.; Viasnoff, V.; Saunders, T.E. 3d protein dynamics in the cell nucleus. *Biophys. J.* **2017**, *112*, 133–142. [CrossRef] [PubMed]
47. Speckner, K.; Stadler, L.; Weiss, M. Anomalous dynamics of the endoplasmic reticulum network. *Phys. Rev. E* **2018**, *98*, 012406. [CrossRef]
48. Schneider, F.; Waithe, D.; Lagerholm, B.C.; Shrestha, D.; Sezgin, E.; Eggeling, C.; Fritzsche, M. Statistical analysis of scanning fluorescence correlation spectroscopy data differentiates free from hindered diffusion. *ACS Nano* **2018**, *12*, 8540–8546. [CrossRef]
49. Porter, C.E.; Thomas, R.G. Fluctuations of nuclear reaction widths. *Phys. Rev.* **1956**, *104*, 483–491. [CrossRef]
50. Sadegh, S.; Barkai, E.; Krapf, D. 1/fnoise for intermittent quantum dots exhibits non-stationarity and critical exponents. *New J. Phys.* **2014**, *16*, 113054. [CrossRef]
51. Box, G.E.P.; Cox, D.R. An analysis of transformations. *J. R. Stat. Soc. B* **1964**, *26*, 211–252. [CrossRef]
52. Wang, B.; Kuo, J.; Bae, S.C.; Granick, S. When brownian diffusion is not gaussian. *Nat. Mat.* **2012**, *11*, 481–485. [CrossRef]
53. Hapca, S.; Crawford, J.W.; Young, I.M. Anomalous diffusion of heterogeneous populations characterized by normal diffusion at the individual level. *J. R. Soc. Interface* **2009**, *6*, 111–122. [CrossRef]
54. Chubynsky, M.V.; Slater, G.W. Diffusing diffusivity: A model for anomalous, yet brownian, diffusion. *Phys. Rev. Lett.* **2014**, *113*, 098302. [CrossRef]
55. Chechkin, A.V.; Seno, F.; Metzler, R.; Sokolov, I.M. Brownian yet non-gaussian diffusion: From superstatistics to subordination of diffusing diffusivities. *Phys. Rev. X* **2017**, *7*, 021002. [CrossRef]

56. Goychuk, I. Viscoelastic subdiffusion in a random gaussian environment. *Phys. Chem. Chem. Phys.* **2018**, *20*, 24140–24155. [CrossRef]
57. Goychuk, I.; Pöschel, T. Finite-range viscoelastic subdiffusion in disordered systems with inclusion of inertial effects. *New J. Phys.* **2020**, *22*, 113018. [CrossRef]
58. Weber, S.C.; Thompson, M.A.; Moerner, W.E.; Spakowitz, A.J.; Theriot, J.A. Analytical tools to distinguish the effects of localization error, confinement, and medium elasticity on the velocity autocorrelation function. *Biophys. J.* **2012**, *102*, 2443–2450. [CrossRef] [PubMed]
59. Backlund, M.P.; Joyner, R.; Moerner, W.E. Chromosomal locus tracking with proper accounting of static and dynamic errors. *Phys. Rev. E* **2015**, *91*, 062716. [CrossRef] [PubMed]
60. Benelli, R.; Weiss, M. From sub- to superdiffusion: Fractional brownian motion of membraneless organelles in early c. elegans embryos. *New J. Phys.* **2021**, *23*, 063072. [CrossRef]
61. Krapf, D.; Marinari, E.; Metzler, R.; Oshanin, G.; Xu, X.; Squarcini, A. Power spectral density of a single brownian trajectory: What one can and cannot learn from it. *New J. Phys.* **2018**, *20*, 023029. [CrossRef]
62. Krapf, D.; Lukat, N.; Marinari, E.; Metzler, R.; Oshanin, G.; Selhuber-Unkel, C.; Squarcini, A.; Stadler, L.; Weiss, M.; Xu, X. Spectral content of a single non-brownian trajectory. *Phys. Rev. X* **2019**, *9*, 011019. [CrossRef]
63. Etoc, F.; Balloul, E.; Vicario, C.; Normanno, D.; Lisse, D.; Sittner, A.; Piehler, J.; Dahan, M.; Coppey, M. Non-specific interactions govern cytosolic diffusion of nanosized objects in mammalian cells. *Nat. Mater.* **2018**, *17*, 740–746. [CrossRef]
64. Hartich, D.; Godec, A. Thermodynamic uncertainty relation bounds the extent of anomalous diffusion. *Phys. Rev. Lett.* **2021**, in press.
65. Berry, H. Monte carlo simulations of enzyme reactions in two dimensions: Fractal kinetics and spatial segregation. *Biophys. J.* **2002**, *83*, 1891–1901. [CrossRef]
66. Hellmann, M.; Heermann, D.W.; Weiss, M. Enhancing phosphorylation cascades by anomalous diffusion. *EPL* **2012**, *97*, 58004. [CrossRef]
67. Stiehl, O.; Weidner-Hertrampf, K.; Weiss, M. Kinetics of conformational fluctuations in dna hairpin-loops in crowded fluids. *New J. Phys.* **2013**, *15*, 113010. [CrossRef]

Article

Information-Efficient, Off-Center Sampling Results in Improved Precision in 3D Single-Particle Tracking Microscopy

Chen Zhang and Kevin Welsher *

Department of Chemistry, Duke University, Durham, NC 27708, USA; chen.zhang2@duke.edu
* Correspondence: kdw32@duke.edu

Abstract: In this work, we present a 3D single-particle tracking system that can apply tailored sampling patterns to selectively extract photons that yield the most information for particle localization. We demonstrate that off-center sampling at locations predicted by Fisher information utilizes photons most efficiently. When performing localization in a single dimension, optimized off-center sampling patterns gave doubled precision compared to uniform sampling. A ~20% increase in precision compared to uniform sampling can be achieved when a similar off-center pattern is used in 3D localization. Here, we systematically investigated the photon efficiency of different emission patterns in a diffraction-limited system and achieved higher precision than uniform sampling. The ability to maximize information from the limited number of photons demonstrated here is critical for particle tracking applications in biological samples, where photons may be limited.

Keywords: 3D single-particle tracking; Fisher information; non-uniform illumination

Citation: Zhang, C.; Welsher, K. Information-Efficient, Off-Center Sampling Results in Improved Precision in 3D Single-Particle Tracking Microscopy. *Entropy* **2021**, 23, 498. https://doi.org/10.3390/e23050498

Academic Editors: Janusz Szwabiński and Aleksander Weron

Received: 21 March 2021
Accepted: 19 April 2021
Published: 22 April 2021

1. Introduction

Single-particle tracking (SPT) [1] has led to numerous advances in unveiling sophisticated intracellular biophysical events, including diffusion of membrane proteins [2], transportation of intracellular vesicles [3], and viral internalization events [4,5]. Despite the tremendous progress made in the field, there are limits to what can be gleaned from single-particle trajectories by the intrinsic localization precision. In a conventional microscope, this limit is given by $\sigma/\sqrt{(N)}$, where σ is the size of the microscope's point-spread function (PSF) and N is the number of photons collected [6,7]. The diffraction limit dictates the size of the PSF, so increasing the number of photons collected per localization is typically the only method available for increasing precision. While the use of artificial particles [8] can produce a higher flux of photons and improve localization precision, conventional organic fluorophores and fluorescent proteins remain essential in biophysical studies. These probes can only yield a finite number of photons before undergoing irreversible photobleaching, so it is crucial to maximize the information available from this limited number of photons. In a typical particle tracking experiment, the particle is uniformly illuminated as, typically, the emitter's position is not known at the outset of the experiment. An under-explored avenue for increasing precision is adjusting the excitation pattern around the emitter to get beyond the localization limit described above. This type of advance is only possible if the particle position is known a priori, at least to some degree of certainty. Recent efforts have focused on improving localization precision through non-uniform illumination in the context of a super-resolution technique [9]. Gallatin et al. proposed a globally optimized strategy in which the particle should be sampled at the maximum of the first-order derivative of the square-root of the intensity [10]. This theory indicated that the optimized sampling pattern is non-uniform, and the particle position should not be directly sampled, but did not provide experimental support. Both of these works suggest that non-uniform illumination shows promise for improved localization with a limited number of photons.

This study builds upon 3D single-molecule active real-time tracking microscopy (3D-SMART) [11], a previously introduced real-time 3D single-particle tracking (RT-3D-SPT)

system [12], by utilizing the emerging concept of non-uniform illumination to improve localization precision significantly. A 2-fold increase in precision was observed in both 2D (XY-plane) and 1D (Z-axis) localization, as has previously been noticed in several super-resolution-based methods [13,14]. Unlike existing methods that require sophisticated PSF engineering or specific materials, here, we have achieved higher photon efficiency in 3D localization with a diffraction-limited point-scanning confocal microscope.

In this work, we investigate the potential of non-uniform illumination in the context of real-time 3D single-particle tracking (RT-3D-SPT) [1]. Developed by several groups over the past decades [15–22], RT-3D-SPT uses active feedback to keep a single particle at the center of the microscope objective's focal volume. In 3D-SMART, the laser spot is guided to sample the XY-plane following a Knight's Tour pattern, while simultaneously sampling along the Z-axis following a sine wave, creating a scanning volume (Figure 1a–c) that samples the vicinity of the particle of interest in an approximately uniform manner (Figure 1d). Since the particle is held stationary in the lab frame, a priori information about the particle position is available during the measurement. The question then occurs: If a priori information is available regarding the particle's position, can the above-described precision limit be surpassed? We explore this potential in the following work. By examining the expected information extracted from photons collected at various locations relative to the particle center, we show that off-center sampling leads to dramatically increased precision compared to uniform illumination schemes. We then demonstrate that this photon-efficient sampling can be applied in practice using a 3D patterned laser spot. The sampling density of an off-center, information-efficient pattern in 3D is shown in Figure 1e.

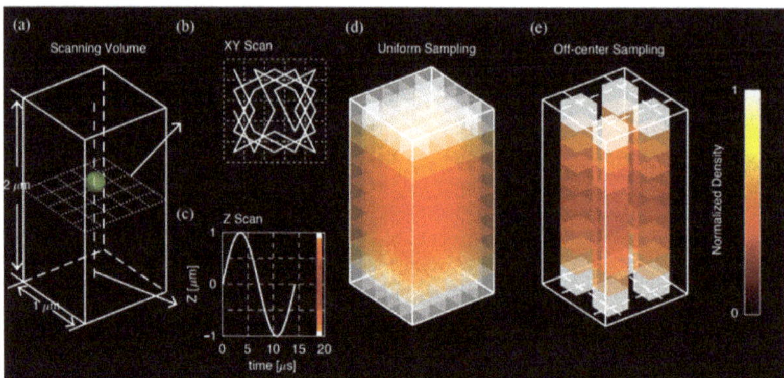

Figure 1. (**a**) Complete laser scanning volume with a dimension of 1 × 1 × 2 μm implemented in 3D-SMART; (**b**) The laser is scanned in a Knight's Tour pattern in the XY-plane using an EOD; (**c**) Along the Z-axis, a TAG lens is used to drive the focus in a sinusoidal pattern. The color bar indicates photon arrival density along the Z-axis; (**d**) Scanning density in the volume when sampling in the default pattern shown in (**b**,**c**); (**e**) Scanning density of a 3D information-efficient pattern, with the XY-plane, scanned in an off-center, information-efficient pattern. The Z-axis is scanned with a sine wave with the laser power unmodulated.

2. Theory

The probability density of observing a photon from an emitter in a microscope in one dimension is approximated by:

$$f(x|\mu,\sigma) = \frac{1}{\sqrt{2\pi\sigma^2}} e^{\frac{-(x-\mu)^2}{2\sigma^2}} \tag{1}$$

where μ is the particle position and σ is related to the width of the PSF. The Gaussian distribution is a good approximation for the actual diffraction pattern of a point-source, which is an Airy function [23]. The size of the PSF is ultimately limited by diffraction and is not user-adjustable. The prefactor $1/\sqrt{2\pi\sigma^2}$ is a normalization factor and will be neglected for the rest of this discussion. The Fisher Information (FI) is a statistical measure employed to quantify the amount of information expected when estimating a parameter of the underlying distribution [24]. The FI (J) is inversely related to the precision and is given by:

$$J(\theta) = E_\theta\left[\left(\frac{\partial}{\partial\theta} lnf(x|\theta)\right)^2\right] \tag{2}$$

Here, θ is a parameter (or vector of parameters) of the underlying distribution, and E_θ is the expectation value. The FI for Equation (1) above is:

$$J(\mu) = \int_x f(x|\mu)\frac{(x-\mu)^2}{\sigma^4}dx \tag{3}$$

Taken over all space, the expectation value above yields a constant value of $J(\mu) = \frac{1}{\sigma^2}$, the average amount of FI contributed by each observed photon. The expectation is replaced by the integral in this expression. It is also noticeable that σ will be the constant prefactor upon integration and does not affect the solution. For a total of N photons, the FI is simply $J(\mu) = \frac{N}{\sigma^2}$. The FI is the inverse of the expected variance, so it is straightforward to see that this is simply the limit of localization precision ($\sigma/\sqrt{N}$) described above. However, this is only the average value, and photons collected from certain parts of the distribution contribute more to the overall FI than others. To see this, we can more closely examine the integrand above.

$$f(x|\mu)\frac{(x-\mu)^2}{\sigma^4} \tag{4}$$

This integrand, which we will refer to as the "information density", is plotted in Figure 2 to show the contribution of observed photons versus $x-\mu$ (so the origin is the particle position). From Figure 2, it is easily seen that there are three critical points. There is a minimum in the information density at $x = \mu$, and two maxima at $x = \mu \pm \sigma\sqrt{2}$. The minimum at $x = \mu$ contributes zero FI, meaning that photons collected precisely from the particle center yield no information on the particle position. The maxima indicate that photons collected from the off-center positions yield the most FI. In a typical imaging experiment, which employs uniform illumination, the above analysis is not applicable. First, the particle's location is unknown, so it is impossible to collect photons only from specific areas around the particle's position. Second, while the photons collected exactly from the particle center yield zero FI, there is no downside to collecting them if unlimited photons are available. However, there are experimental conditions under which it is possible, and even desirable, to tailor the excitation pattern. For single-molecule tracking or super-resolution imaging, a finite number of photons can be extracted from each molecule before irreversible photobleaching occurs. In these experiments, it is therefore beneficial to collect photons from high information areas only, if possible. That being said, this approach is typically impossible because there is no a priori information of the particle position. This is where the a priori information regarding the particle position, available from active-feedback single-molecule tracking, comes in.

In the following section, we survey three-dimensional laser scanning patterns to identify the most information-efficient sampling patterns. We start with a discussion of applying this sampling along the XY-plane and Z-axis separately, followed by a discussion of achieving isotropic information-efficient sampling in all three dimensions.

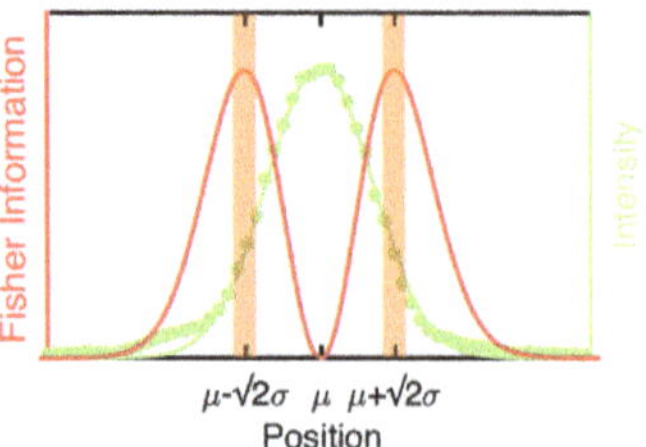

Figure 2. Demonstration of the proposed information-efficient sampling in a single dimension. The green dots represent the experimentally observed intensity of a bead along the X-axis, and the green line shows the Gaussian fit of the PSF. The red line shows the information density of the PSF. The orange-shaded areas indicate photons with high information density.

3. Results

3.1. Identification of a Photon-Efficient Sampling Pattern in the XY-Plane

In the RT-3D-SPT system reported by Hou et al. [12], a focused laser spot is guided by an electro-optic deflector (EOD) to scan a 5 × 5 grid with 250 nm between adjacent pixels in the XY-plane (Figure 3a). Each pixel is sampled for 20 μs in a scanning cycle. The 5 × 5 grid, typically scanned in a Knight's Tour pattern, ensures a uniform illumination pattern. However, the FI-based analysis above suggests that off-center positions should be selectively sampled to achieve the highest photon efficiency. To obtain an unbiased search for photon-efficient patterns, we evaluated subsets of the default 5 × 5 pattern. The 25 pixels were divided into six different groups of inequivalent pixels based on their distance to the scan center (Figure 3b). Immobilized fluorescent beads distributed in PBS buffer were then scanned using the default 5 × 5 EOD pattern in the XY-plane. A piezoelectric stage was then used to step the particle position in 20 nm increments through the center of the scan area. All 62 possible combinations of inequivalent patterns (Figure S1) were tested to identify the most photon-efficient sampling pattern. In each experiment, photons were collected from each of the 25 pixels, but only photons obtained from pixels in a given pattern were used to perform data analysis.

For each different sampling pattern, the particle position was estimated using maximum likelihood estimates (MLEs). In MLEs, the likelihood (L) for a position estimate (μ) from an arbitrary number of photons (N) is defined as the product of probability density of photon arrival positions based on a given model f which is a function of μ:

$$L(\mu X) = \prod_{i=1}^{N} f(x_i \mu) \tag{5}$$

Here the model f is a Gaussian distribution described in (1), X refers to the set of arrival positions $x_i (i = 1, 2 \ldots N)$ of photons used for estimation. The best estimate for the particle position μ is obtained when L is maximized. A detailed example of MLE is shown in Figure S2. Localization precision at different numbers of photons of each estimation is shown in Figure S3. In this study, 2000 consecutively collected photons were used for each position estimate unless otherwise stated.

MLE analysis was performed on immobilized 190 nm fluorescent beads that were stepped in 20 nm intervals over a 100 nm range for the XY-plane. The average standard deviation of MLE positions at different stage positions was used to quantify the precision (Figure 3d–f). It was observed that the MLE positions were proportional, but not exactly equal, to the expected particle positions. Off-center sampling generally results in an overestimation of the actual particle motion, compared to uniform sampling which generally yields realistic position estimation. An underestimation of the actual particle motion is associated with sampling patterns in which the center pixel is oversampled compared to uniform sampling. This observation was validated by a simulation of sampling a 2D

Gaussian emitter (Figure S4). Calibration was performed to account for these differences in response to the same changes in particle position (Figure S5).

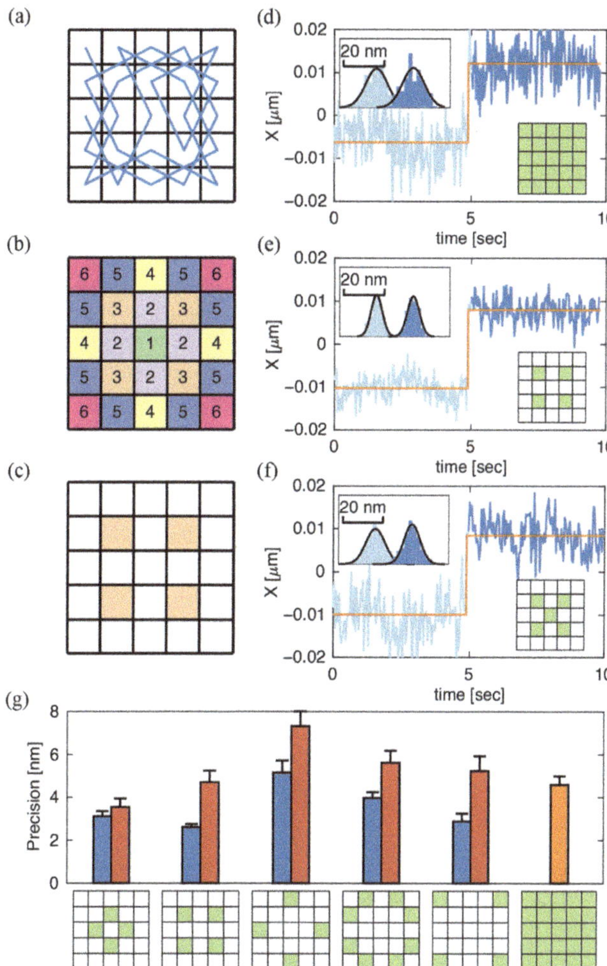

Figure 3. (**a**) 5 × 5 Knight's Tour scanning pattern; (**b**) Inequivalent pixels based on distance from the scan center; (**c**) 4-pixel 4-Corners scan, which uses only pixels from pattern 3 in (**b**), that was found to have the highest precision; (**d**–**f**) MLE positions versus stage positions obtained with the default 5 × 5, 4-Corners, and 4-Corners plus center pixel scan patterns. The average standard deviation of the estimated positions was measured to be 5.5, 2.8, and 4.4 nm, respectively. The orange line shows the relative stage position. The dark and light blue lines show the estimated position calculated for every 2000 photons. The different shades of blue indicate estimation based on two different stage steps, 20 nm apart; (**g**) Precision obtained from selectively using photons obtained from pixels in different inequivalent patterns (pattern 2–6 in 4b) alone (blue), specific pattern plus center pixel (red), and default 5 × 5 pattern (orange).

Interestingly, though we did not propose an a priori model based on Fisher information, the unbiased search resulted in an off-center, FI-efficient pattern. One 4-pixel pattern, which we refer to as the 4-Corners pattern (Figure 3c), yielded the highest precision of 2.6 ± 0.3 nm, compared with the default pattern (where all 25 pixels are sampled) which gave 4.5 ± 0.3 nm precision (Figure 3d,e). In both cases, localization was performed using 2000 photons. Additional data were obtained with the laser excitation matching the 4-Corners pattern to validate that selectively using specific photons in data post-processing was equivalent to sampling with the designated pattern. The resulting precision was 2.8 ± 0.4 nm, in excellent agreement with the post-acquisition processed data above. The effect of the number of photons on the relative advantage of the 4-Corners pattern was also investigated, showing a doubling in precision for various values of N (Figure S6). Notably, when the center pixel is added back to the 4-Corners pattern, the precision is nearly two-fold worse than the 4-Corners pattern alone and comparable to the full 25-pixel scan (4.7 ± 1.1 nm vs. 4.6 ± 0.8 nm, Figure 3f). Complete 2D trajectories of Figure 3d–f are shown in Figure S7. We note here that different sampling patterns yield different emission rates for the same laser power. For example, the default 5 × 5 pattern gives a roughly 1.5-fold increase in intensity compared to the 4-Corners pattern, but the 4-Corners pattern still exhibits doubled precision when sampling with equal bin time (Figure S8). A more thorough investigation of patterns consisting of inequivalent pixels No. 2–6 (Figure 3b) with or without the center pixel showed that precision obtained with the center pixel was always worse than without (Figure 3g). The poor precision upon sampling the center pixel shows the importance of not sampling low FI areas around the particle.

3.2. Laser Modulation in Z-Axis

We then proceeded to achieve photon-efficient sampling along the Z-axis. A Tunable Acoustic Gradient (TAG) lens [25,26] was used to create custom illumination patterns along the axial direction. The TAG lens deflects a focused laser spot in a sine wave with an amplitude of ~1 μm around the focal plane at a frequency of ~70 kHz (Figure 1c). The following mathematical relation describes the probability density of this sine wave with an amplitude of 1:

$$f(z) = \frac{1}{\pi\sqrt{1-z^2}} \tag{6}$$

The probability density has a distribution where most probability is piled up at the edges (top and bottom) of the scanning volume (Figure 1d). Photon arrivals obtained by scanning an immobilized fluorescent bead with a non-modulated, continuous wave (CW) laser spot confirm this distribution (Figure 4a). According to the theory described above, photons collected at a certain distance away from the center contain the highest information density and lead to the most efficient sampling. However, using CW laser modulation, many photons are still collected from the center of the scanning volume (where the particle spends most of its time). Real-time modulation of the laser intensity was applied to shift the photon arrival distribution away from this low information density area. To do so, the frequency and phase of the TAG lens were captured by a field-programmable gate array (FPGA, NI-7852R). A digital signal with the same frequency and adjustable phase delay was sent to a lock-in amplifier (SR850, Stanford Research Systems) and then to a multiplier circuit to double the original frequency. The frequency-doubled signal was then used to modulate the laser's power, creating custom illumination patterns along the Z-axis. A detailed illustration of how the signal was processed in the system is shown in Figure S9. Phase delays between 0° and 90° were tested. Figure 4 shows photon arrival distribution from immobilized 190 nm fluorescent beads sampled at each of the various conditions (CW, in-phase modulation, out-of-phase modulation). At 0° phase delay, photon arrivals occurred at the imaging volume center (in-phase modulation, Figure 4b). When the delay was 90°, (out-of-phase modulation) photon arrivals were clustered at the edges, with a minimal number of photons at the center (Figure 4c). Complete trajectories of data shown in Figure 4d–f are shown in Figure S10.

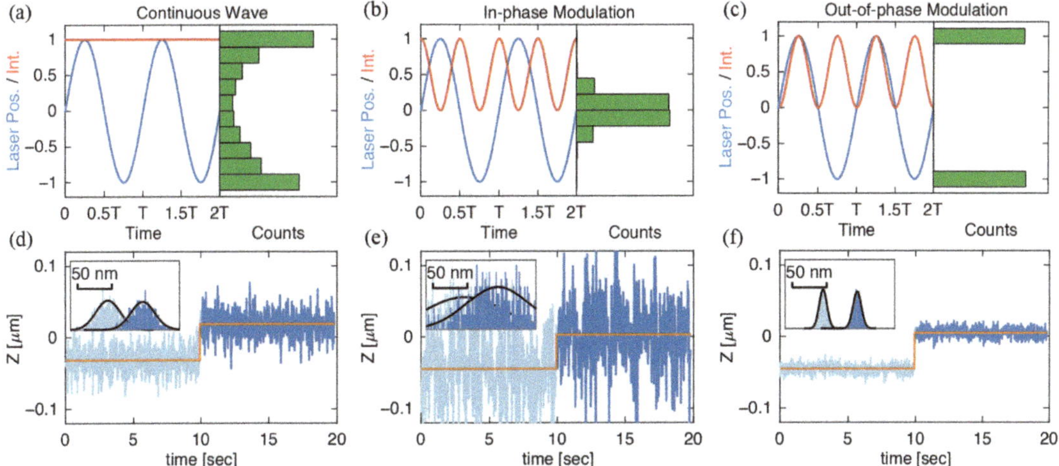

Figure 4. (**a–c**) Laser intensity versus laser position in Z-axis with photon arrival distribution of 10,000 photons from imaging a 190 nm fluorescent bead with TAG lens scanning in unmodulated (continuous wave) or modulated power (no XY scanning). Note that the scanning rate of the TAG lens was held constant at 70 kHz; (**d–f**) Estimated particle position versus stage position obtained from step tests. Each position estimation was based on 2000 photons. The average standard deviation of estimated positions at each stage position in (**d**) CW, (**e**) in-phase modulation, and (**f**) out-of-phase modulation from the partial trajectories shown in the figures was measured to be 17.6, 50.3, and 5.9 nm, respectively.

Upon performing step tests and data calibration similar to the XY scanning above, the average precision of particles scanned with CW, in-phase modulation, and out-of-phase modulation were 18.2 $\pm$ 2.3, 52.0 $\pm$ 24.2, and 9.2 $\pm$ 1.3 nm, respectively (Figure 4d–f). These results reaffirmed that off-center sampling (out-of-phase modulation) is the most information-efficient along the Z-axis, as precision was nearly doubled compared with CW power (9.2 vs. 18.2 nm). In-phase modulation, which only samples near the particle center, yielded very poor position estimates.

3.3. Determination of Optimized Sampling Parameters in 3D

Photon-efficient sampling patterns with the ability to localize in all three dimensions were determined by step tests similar to those described above. Step tests with the full 25 XY pixels and CW laser modulation along Z were first performed as a reference. These yielded precisions of 9.3 $\pm$ 0.8 nm in X and 22.2 $\pm$ 1.6 nm in Z (n = 5). Step tests were then performed with the 4-Corners EOD pattern and CW laser, which gave average precisions of 6.8 $\pm$ 0.6 nm in X and 19.6 $\pm$ 1.7 nm in Z, confirming the advantage of information-efficient off-center sampling. A comparison of step tests in X obtained with the default 5 $\times$ 5 pattern and the 4-Corners EOD pattern with TAG lens operating in CW power is shown in Figure 5. These results again confirm the importance of not collecting photons from the center of the 3D volume. The magnitude of the EOD scale (size of each pixel) and amplitude of TAG lens scan that gave the highest precision in X and Z were found to be 200 nm and 30% TAG lens amplitude (FWHM = 1.93 μm), respectively (Figure S11). Apart from the amplitude of the TAG lens, the laser power modulation pattern also altered the precision. When imaging with the 4-Corners pattern and out-of-phase modulation, the Z precision was high (12.6 $\pm$ 0.4 nm), but the X precision was low (22.8 $\pm$ 3.1 nm). Inversely, the 4-Corners pattern and in-phase modulation resulted in high X precision (4.7 $\pm$ 0.2 nm) and low precision in Z (36.0 $\pm$ 5.0 nm). Photon arrival distribution along the Z-axis of 190 nm beads sampled with the 4-Corners pattern and modulated power at different phase delay is shown in Figure S12. It is noticeable that out-of-phase modulation in axial-only scanning resulted in a "bi-plane" distribution (Figure 4c), similar to previously reported work [27].

It should be noted that when XY scanning is enabled, photons from non-edge positions are not necessarily excluded (Figure S12h).

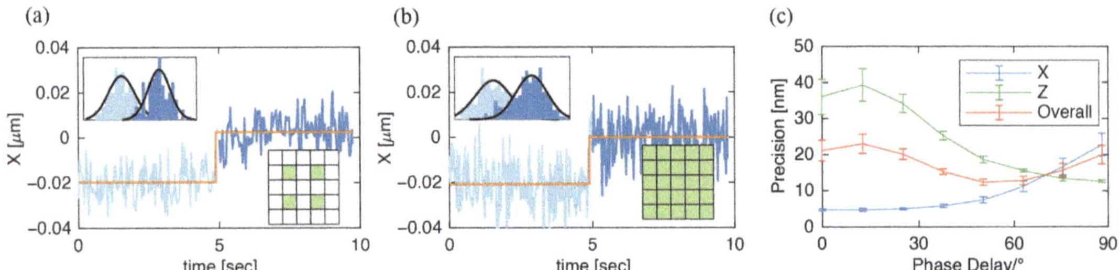

Figure 5. (**a**,**b**) Estimated position versus stage position sampled with the 4-Corners and default patterns. Both used CW laser power modulation along the axial direction. The standard deviation of estimated positions (based on 2000 photons) was (**a**) 7.4 and (**b**) 10.5 nm from the partial trajectories shown above; (**c**) X, Z, and 3D precision of five different immobilized 190 nm fluorescent beads sampled with the 4-Corners pattern in the XY-plane and modulated power at different phase delay along the Z-axis.

To get a comprehensive standard of overall precision in all dimensions in each condition, we define an overall 3D precision $= \sqrt{\frac{P_x^2+P_y^2+P_z^2}{3}} = \sqrt{\frac{2P_x^2+P_z^2}{3}}$. The 3D precision is derived from X and Z precision here since X and Y were sampled via the same mechanism. Step test trajectories obtained with the 4-Corners pattern and modulated laser power with a phase delay of 50.4° yielded the highest 3D precision (12.3 ± 0.9 nm). This result is comparable to the 3D precision found for the 4-Corners pattern and CW laser power (12.6 ± 1.1 nm). It is also noteworthy that it is possible to achieve equivalent X and Z precision when a phase delay of ~70° is applied, as is shown in Figure 5c, where the X and Z precisions intersect. The estimated precision at this point is $P = P_x = P_z = 14.3$ nm. This condition makes it possible to conduct isotropic sampling, despite having non-uniform point spread function scales in different dimensions (Figure S13c–d).

4. Discussion and Conclusions

This study showed that implementing information-efficient laser scanning patterns led to dramatic improvements in precision in all three dimensions. A 4-pixel, 4-Corners pattern yielded the highest precision of 2.6 ± 0.3 nm in the XY-plane compared to 4.6 ± 0.8 nm given by a default 5 × 5 pattern (43.5% increase). When sampling the Z-axis only, out-of-phase modulation of the laser power relative to the TAG lens phase (which gave a bi-modal distribution of photon arrivals) gave the highest precision of 9.2 ± 2.6 nm, compared to 18.2 ± 4.6 nm given by CW power modulation (49.5% increase in precision). In 3D scanning, the 4-Corners pattern with laser power modulated with a 50.3° phase delay gave a 3D precision of 12.3 ± 0.9 nm, compared to 14.9 ± 1.1 nm given by the default 5 × 5 pattern with CW power modulation (17.5% increase). These results are consistent with the hypothesis that sampling at higher Fisher information regions leads to the best precision. Moreover, it is also shown that sampling directly at the center of the particle is inefficient, with those photons carrying little or no information. Achieving high photon efficiency is extremely important as typical fluorophores give out only a limited number of photons. The off-center, information-efficient imaging proposed in this work is the first to achieve higher photon efficiency by merely scanning a focused laser spot without requiring any PSF engineering.

It is noticeable that RT-3D-SPT methods might have used similar off-center excitation patterns. For example, orbital tracking methods, developed by Gratton [28], Mabuchi [29,30], Lamb [19,31], and others, utilize a circular scanning pattern in the XY-plane, which minimizes sampling of the particle center. The motivation for such an approach was to sense changes in the particle position by modulating the particle's intensity. Others have utilized a method where the laser is scanned across the vertices of a tetrahedron [32,33]. All of these methods benefit from the off-center sampling that we demonstrate above. Unlike previous works [15,20] that utilized scan-free localization, using the EOD and TAG lens here is advantageous in defining a custom excitation pattern due to their highly tunable nature and the intrinsically information-efficient pattern caused by the TAG lens' sinusoidal motion.

Our work revealed the importance of information-efficient excitation patterns in particle localization and tracking using non-uniform illumination. The information-efficient patterns investigated in this work could shed light on an emerging field in biology: slowly moving particles. Recent studies have shown that even the previously considered "static" intracellular vesicles could still undergo small-scale motions of different rates, which could profoundly influence the fate of these particles [34]. Information-efficient sampling should optimize combined spatiotemporal precision in such demanding experiments, where the diffusive step sizes are extremely small and the fluorophores are short lived.

Supplementary Materials: The following are available online at https://www.mdpi.com/article/10.3390/e23050498/s1, Figures S1–S10, and detailed methods.

Author Contributions: Conceptualization, C.Z. and K.W.; methodology, C.Z. and K.W.; data collection; C.Z.; formal analysis, C.Z. and K.W.; writing—original draft preparation, C.Z. and K.W.; writing—review and editing, C.Z. and K.W. All authors have read and agreed to the published version of the manuscript.

Funding: The authors acknowledge financial support from the National Institute of General Medical Sciences of the National Institutes of Health under award number R35GM124868, the National Science Foundation under Grant No. 1847899, and Duke University.

Institutional Review Board Statement: Not applicable.

Informed Consent Statement: Not applicable.

Data Availability Statement: Data from this study are available from the corresponding author upon request.

Conflicts of Interest: The authors declare no conflict of interest.

References

1. Hou, S.; Johnson, C.; Welsher, K. Real-time 3D single particle tracking: Towards active feedback single molecule spectroscopy in live cells. *Molecules* **2019**, *24*, 2826. [CrossRef]
2. de Wit, G.; Albrecht, D.; Ewers, H.; Kukura, P. Revealing compartmentalized diffusion in living cells with interferometric scattering microscopy. *Biophys. J.* **2018**, *114*, 2945–2950. [CrossRef] [PubMed]
3. Wehnekamp, F.; Plucińska, G.; Thong, R.; Misgeld, T.; Lamb, D.C. Nanoresolution real-time 3D orbital tracking for studying mitochondrial trafficking in vertebrate axons in vivo. *eLife* **2019**, *8*, e46059. [CrossRef]
4. Hou, S.; Welsher, K. An adaptive real time 3D single particle tracking method for monitoring viral first contacts. *Small* **2019**, *15*, 1903039. [CrossRef]
5. Zheng, L.L.; Li, C.M.; Zhen, S.J.; Li, Y.F.; Huang, C.Z. A dynamic cell entry pathway of respiratory syncytial virus revealed by tracking the quantum dot-labeled single virus. *Nanoscale* **2017**, *9*, 7880–7887. [CrossRef]
6. Ober, R.J.; Ram, S.; Ward, E.S. Localization accuracy in single-molecule microscopy. *Biophys. J.* **2004**, *86*, 1185–1200. [CrossRef]
7. Ram, S.; Ward, E.S.; Ober, R.J. Beyond Rayleigh's criterion: A resolution measure with application to single-molecule microscopy. *Proc. Natl. Acad. Sci. USA* **2006**, *103*, 4457–4462. [CrossRef]
8. Li, Q.; Yin, W.; Li, W.; Zhang, Z.; Zhang, X.; Zhang, X.E.; Cui, Z. Encapsulating quantum dots within HIV-1 Virions through site-specific decoration of the matrix protein enables single virus tracking in live primary macrophages. *Nano Lett.* **2018**, *18*, 7457–7468. [CrossRef]
9. Balzarotti, F.; Eilers, Y.; Gwosch, K.C.; Gynnå, A.H.; Westphal, V.; Stefani, F.D.; Elf, J.; Hell, S.W. Nanometer resolution imaging and tracking of fluorescent molecules with minimal photon fluxes. *Science* **2017**, *355*, 606–612. [CrossRef]
10. Gallatin, G.M.; Berglund, A.J. Optimal laser scan path for localizing a fluorescent particle in two or three dimensions. *Opt. Express* **2012**, *20*, 16381–16393. [CrossRef]
11. Hou, S.; Exell, J.; Welsher, K. Real-time 3D single molecule tracking. *Nat. Commun.* **2020**, *11*, 1–10. [CrossRef]
12. Hou, S.; Lang, X.; Welsher, K. Robust real-time 3D single-particle tracking using a dynamically moving laser spot. *Opt. Lett.* **2017**, *42*, 2390–2393. [CrossRef]
13. Masullo, L.A.; Steiner, F.; Zähringer, J.; Lopez, L.F.; Bohlen, J.; Richter, L.; Cole, F.; Tinnefeld, P.; Stefani, F.D. Pulsed interleaved minflux. *Nano Lett.* **2020**, *21*, 840–846. [CrossRef] [PubMed]

14. Reymond, L.; Huser, T.; Ruprecht, V.; Wieser, S. Modulation-enhanced localization microscopy. *J. Phys. Photonics* **2020**, *2*, 041001. [CrossRef]
15. Wells, N.P.; Lessard, G.A.; Goodwin, P.M.; Phipps, M.E.; Cutler, P.J.; Lidke, D.S.; Wilson, B.S.; Werner, J.H. Time-resolved three-dimensional molecular tracking in live cells. *Nano Lett.* **2010**, *10*, 4732–4737. [CrossRef]
16. Welsher, K.; Yang, H. Multi-resolution 3D visualization of the early stages of cellular uptake of peptide-coated nanoparticles. *Nat. Nanotechnol* **2014**, *9*, 198–203. [CrossRef]
17. Liu, C.; Obliosca, J.M.; Liu, Y.L.; Chen, Y.A.; Jiang, N.; Yeh, H.C. 3D single-molecule tracking enables direct hybridization kinetics measurement in solution. *Nanoscale* **2017**, *9*, 5664–5670. [CrossRef] [PubMed]
18. Levi, V.; Ruan, Q.; Kis-Petikova, K.; Gratton, E. *Scanning FCS, a Novel Method for Three-Dimensional Particle Tracking*; Portland Press Ltd.: London, UK, 2003.
19. Katayama, Y.; Burkacky, O.; Meyer, M.; Bräuchle, C.; Gratton, E.; Lamb, D.C. Real-Time nanomicroscopy via three-dimensional single-particle tracking. *ChemPhysChem* **2009**, *10*, 2458–2464. [CrossRef] [PubMed]
20. Shechtman, Y.; Weiss, L.E.; Backer, A.S.; Sahl, S.J.; Moerner, W. Precise three-dimensional scan-free multiple-particle tracking over large axial ranges with tetrapod point spread functions. *Nano Lett.* **2015**, *15*, 4194–4199. [CrossRef]
21. Deich, J.; Judd, E.; McAdams, H.; Moerner, W. Visualization of the movement of single histidine kinase molecules in live Caulobacter cells. *Proc. Natl. Acad. Sci. USA* **2004**, *101*, 15921–15926. [CrossRef]
22. Estrada, L.C.; Gratton, E. Spectroscopic properties of gold nanoparticles at the single particle level in biological environments. *ChemPhysChem* **2012**, *13*, 1087–1092. [CrossRef] [PubMed]
23. Zhang, B.; Zerubia, J.; Olivo-Marin, J.C. Gaussian approximations of fluorescence microscope point-spread function models. *Appl. Opt.* **2007**, *46*, 1819–1829. [CrossRef] [PubMed]
24. Cover, T.M.; Thomas, J.A. *Elements of Information Theory (Wiley Series in Telecommunications and Signal Processing)*; Wiley-Interscience: Hoboken, NJ, USA, 2006.
25. Duocastella, M.; Sun, B.; Arnold, C.B. Simultaneous imaging of multiple focal planes for three-dimensional microscopy using ultra-high-speed adaptive optics. *J. Biomed. Opt.* **2012**, *17*, 050505. [CrossRef] [PubMed]
26. Duocastella, M.; Theriault, C.; Arnold, C.B. Three-dimensional particle tracking via tunable color-encoded multiplexing. *Opt. Lett.* **2016**, *41*, 863–866. [CrossRef]
27. Prabhat, P.; Ram, S.; Ward, E.S.; Ober, R.J. Simultaneous imaging of different focal planes in fluorescence microscopy for the study of cellular dynamics in three dimensions. *IEEE Trans. NanoBiosci.* **2004**, *3*, 237–242. [CrossRef]
28. Levi, V.; Ruan, Q.; Gratton, E. 3-D particle tracking in a two-photon microscope: Application to the study of molecular dynamics in cells. *Biophys. J.* **2005**, *88*, 2919–2928. [CrossRef]
29. McHale, K.; Mabuchi, H. Precise characterization of the conformation fluctuations of freely diffusing DNA: Beyond Rouse and zimm. *J. Am. Chem. Soc.* **2009**, *131*, 17901–17907. [CrossRef]
30. McHale, K.; Berglund, A.J.; Mabuchi, H. Quantum dot photon statistics measured by three-dimensional particle tracking. *Nano Lett.* **2007**, *7*, 3535–3539. [CrossRef]
31. Ruthardt, N.; Lamb, D.C.; Brauchle, C. Single-particle tracking as a quantitative microscopy-based approach to unravel cell entry mechanisms of viruses and pharmaceutical nanoparticles. *Mol. Ther.* **2011**, *19*, 1199–1211. [CrossRef]
32. Germann, J.A.; Davis, L.M. Three-dimensional tracking of a single fluorescent nanoparticle using four-focus excitation in a confocal microscope. *Opt. Express* **2014**, *22*, 5641–5650. [CrossRef]
33. Perillo, E.P.; Liu, Y.L.; Huynh, K.; Liu, C.; Chou, C.K.; Hung, M.C.; Yeh, H.C.; Dunn, A.K. Deep and high-resolution three-dimensional tracking of single particles using nonlinear and multiplexed illumination. *Nat. Commun.* **2015**, *6*, 1–8. [CrossRef] [PubMed]
34. Guo, Y.; Li, D.; Zhang, S.; Yang, Y.; Liu, J.-J.; Wang, X.; Liu, C.; Milkie, D.E.; Moore, R.P.; Tulu, U.S. Visualizing intracellular organelle and cytoskeletal interactions at nanoscale resolution on millisecond timescales. *Cell* **2018**, *175*, 1430–1442.e17. [CrossRef] [PubMed]

Article

Testing of Multifractional Brownian Motion

Michał Balcerek *,† and Krzysztof Burnecki †

Faculty of Pure and Applied Mathematics, Hugo Steinhaus Center, Wroclaw University of Science and Technology, Wyspianskiego 27, 50-370 Wroclaw, Poland; krzysztof.burnecki@pwr.edu.pl

* Correspondence: michal.balcerek@pwr.edu.pl

† These authors contributed equally to this work.

Received: 18 November 2020; Accepted: 10 December 2020; Published: 12 December 2020

Abstract: Fractional Brownian motion (FBM) is a generalization of the classical Brownian motion. Most of its statistical properties are characterized by the self-similarity (Hurst) index $0 < H < 1$. In nature one often observes changes in the dynamics of a system over time. For example, this is true in single-particle tracking experiments where a transient behavior is revealed. The stationarity of increments of FBM restricts substantially its applicability to model such phenomena. Several generalizations of FBM have been proposed in the literature. One of these is called multifractional Brownian motion (MFBM) where the Hurst index becomes a function of time. In this paper, we introduce a rigorous statistical test on MFBM based on its covariance function. We consider three examples of the functions of the Hurst parameter: linear, logistic, and periodic. We study the power of the test for alternatives being MFBMs with different linear, logistic, and periodic Hurst exponent functions by utilizing Monte Carlo simulations. We also analyze mean-squared displacement (MSD) for the three cases of MFBM by comparing the ensemble average MSD and ensemble average time average MSD, which is related to the notion of ergodicity breaking. We believe that the presented results will be helpful in the analysis of various anomalous diffusion phenomena.

Keywords: multifractional Brownian motion; autocovariance function; power of the statistical test; Monte Carlo simulations

---

1. Introduction

Over the last decades, massive advances in single-particle tracking (SPT), partially based on superresolution microscopy of fluorescently tagged tracers, or fluorescence correlation spectroscopy allow experimentalists to obtain insight into the motion of submicron tracer particles or even single molecules in complex environments, such as living biological cells, down to nanometer precision and at submillisecond time resolution [1,2].

The observed data obtained by SPT experiments often show pronounced deviations from Brownian motion, namely, anomalous diffusion of the power-law form

$$\mathrm{E}X(t)^2 \simeq K_\alpha t^\alpha \tag{1}$$

of the mean-squared displacement (MSD) is observed [3–8]. K_α is the anomalous diffusion coefficient. Depending on the magnitude of the anomalous diffusion exponent α we distinguish subdiffusion for $0 < \alpha < 1$ from superdiffusion with $\alpha > 1$ [5,8]. Subdiffusion is typically observed for submicron particles in both bacterial and eukaryotic cells [6,9–13], in artificially crowded [14,15] and structured [16–19] liquids, in pure and protein-crowded lipid bilayer systems [12,20–25], as well as in groundwater systems [26]. Superdiffusion occurs in the presence of active motion, for instance, in living biological cells [27–29] or due to bulk-surface exchange [30,31]. While typically anomalous diffusion refers to the power-law behavior (1) with a fixed α, an increasing number of systems are reported in which the local scaling exponent of the MSD (1) is an explicit function of time, $\alpha(t)$. Such transient behavior has, for instance, been observed for green fluorescent proteins in cells or for the motion of lipid molecules in protein-crowded bilayer membranes [25,32].

Fractional Brownian motion (FBM) introduced by Kolmogorov in 1940 and rediscovered by Mandelbrot and van Ness in 1968 [33–35] is a generalization of the classical Brownian motion (BM). Most of its statistical properties are characterized by the self-similarity (Hurst) index $0 < H < 1$. FBM is H-self-similar, namely for every $c > 0$ we have $B_H(ct) \stackrel{D}{=} c^H B_H(t)$ in the sense of all finite dimensional distributions, and has stationary increments. It is the only Gaussian process satisfying these properties. FBM is the overdamped description for viscoelastic motion and thus intimately connected to the fractional Langevin processes, an attractive framework for many physical systems [36], for instance, of lipid molecules in bilayer membranes [22,25,37]. The second moment of FBM reads $EB_H^2(t) = \sigma^2 t^{2H}$, where $EB_H^2(1) = \sigma^2 > 0$. As a consequence, for $H < 1/2$ we obtain subdiffusive dynamics with persistent motion, whereas for $H > 1/2$, the process is superdiffusive and antipersistent. Since FBM is the classical model for power-law dependence a number of statistical tests have been already introduced for this process in the literature. Let us mention here the tests based on the autocovariance function (ACVF), MSD, and detrending moving average statistics [38–40].

FBM has stationary increments that do not allow us to model processes whose regularity of paths and "memory depth" change in time [41]. Several generalizations of FBM have been proposed recently. One of these, called multifractional Brownian motion (MFBM), was proposed by Peltier and Véhel [42] with time-varying Hurst exponent $H(t)$ which is a Hölder function. The variance at time t of the MFBM $B_{H(t)}(t)$ is given by $\mathrm{Var}(B_{H(t)}(t)) = \sigma^2 t^{2H(t)}$ [43]. The time-varying Hurst exponent $H(t)$ characterizes the path regularity of the process at time t: sample paths near t with small H_t, close to 0, are space-filling and highly irregular, while paths with large H_t, close to 1, are very smooth. The variance constant σ^2 determines the "energy level" of the process. This natural extension of FBM results in some loss of some of FBMs basic properties, in particular, the increments of MFBM are non-stationary and the process is no longer self-similar.

Other, similar generalizations are limited to a piecewise constant H [44] but, what is important from a data analysis point of view, is that they lead to continuous Gaussian processes with stationary increments. Let us also mention an idea involving an appropriate class of covariance functions. Ryvkina [45] uses such covariance functions to define Gaussian processes to extend FBM and MFBM to a class of fractional Brownian motions with a variable Hurst parameter parameterized by a set of all measurable functions with values in $(1/2, 1)$, and different from MFBMs. However, from a biological data point of view, such a range for H values is not practical since it only corresponds to a superdiffusive (long-range dependent) case.

MFBMs have become popular as flexible models in describing real-life signals of high-frequency features in geoscience, microeconomics, and turbulence, to name a few [43]. They are closely related to the notion of transient diffusion dynamics observed in biological experiments. The article is structured as follows. In Section 2, the MFBM is defined and its basic properties are presented. We also recall formulas for the ensemble average MSD, time average MSD, and present three Hurst exponent functions that will be analyzed in the sequel. In Section 3, the main results are presented. We introduce a statistical test on MFBM based on its ACVF which is presented as a quadratic form. Next, the power of the test is studied for the three cases corresponding to different Hurst exponent functions. We show the areas where the test is very strong in distinguishing between the processes and the cases when it fails in this respect. Finally, Section 4 summarizes and concludes our work.

2. Model and Methods

Let us start with a definition of the MFBM.

Definition 1. *(Multifractional Brownian motion). Process* $\left(B_{H(t)}(t)\right)_{t\geq 0}$ *is called a multifractional Brownian motion (MFBM) if it is a centered Gaussian process with covariance function*

$$\operatorname{Cov}(B_{H(t)}(t), B_{H(s)}(s)) = D(H(t), H(s))\left(t^{H(t)+H(s)} + s^{H(t)+H(s)} - |t-s|^{H(t)+H(s)}\right), \tag{2}$$

where

$$D(x,y) = \frac{\sigma^2\sqrt{\Gamma(2x+1)\Gamma(2y+1)\sin(\pi x)\sin(\pi y)}}{2\Gamma(x+y+1)\sin\left(\pi\frac{x+y}{2}\right)} \tag{3}$$

for some $\sigma > 0$ *and Hölder function* $H : [0,\infty) \to [a,b] \subset (0,1)$ *of some exponent* $\beta > 0$ *[46].*

The second moment of MFBM scales as $\mathrm{E}\left(B^2_{H(t)}(t)\right) = \sigma^2 t^{2H(t)}$. Hence, we will call $H(t)$ the Hurst exponent function. Furthermore, for $H(t) \equiv H \in (0,1)$ MFBM becomes standard FBM. In general, MFBM has non-stationary increments. Its increment process $Y(t) \stackrel{def}{=} B_{H(t+1)}(t+1) - B_{H(t)}(t)$ possesses the long-range dependence property, in the sense that

$$\forall\delta > 0, \forall s \geq 0 \quad \sum_{k=0}^{\infty} |\operatorname{Corr}(Y(s), Y(s+k\delta))| = +\infty,$$

where Corr is the correlation function, i.e., $\operatorname{Corr}(Y(t), Y(s)) = \frac{\operatorname{Cov}(Y(t),Y(s))}{\sqrt{\mathrm{E}^2 Y(t)\mathrm{E}^2 Y(s)}}$ [46].

Furthermore, the function $H(t)$ can be pointwise interpreted as a local self-similarity parameter, i.e.,

$$\lim_{\varepsilon\to 0^+}\left(\frac{B_{H(u+\varepsilon t)}(u+\varepsilon t) - B_{H(u)}(u)}{\varepsilon^{H(u)}}\right)_{t\in\mathbb{R}_+} = s(u)\left(B_H(t)\right)_{t\in\mathbb{R}_+},$$

where B_H is a fractional Brownian motion with index $H \equiv H(u)$, $s(u)$ is a scaling function [47] and the convergence is on the space of continuous functions endowed with the topology of the uniform convergence on compact sets.

2.1. Mean-Squared Displacement

Let us now recall different estimators of the MSD for a sample of n trajectories $\mathbf{X_1}, \mathbf{X_2}, \ldots \mathbf{X_n}$, each with N observations, that is, $\mathbf{X_i}$ consists of $X_i(t_1), X_i(t_2), \ldots, X_i(t_N)$ equally spaced in time, $i = 1, 2, \ldots, n$. Ensemble average MSD (EAMSD) is defined as follows:

$$\operatorname{EAMSD}(\tau = m\Delta t) = \frac{1}{n}\sum_{k=1}^{n}\left(X_k(t_1+\tau) - X_k(t_1)\right)^2, \tag{4}$$

where $m = 1, \ldots, N$ and $\Delta t = t_2 - t_1$. Time average MSD (TAMSD) is defined for each trajectory as

$$\operatorname{TAMSD}(\tau = m\Delta t, k) = \frac{1}{N-m}\sum_{j=1}^{N-m}\left(X_k(t_j + m\Delta t) - X_k(t_j)\right)^2. \tag{5}$$

Finally, we consider ensemble and time average MSD (EATAMSD) which is an average of TAMSDs:

$$\operatorname{EATAMSD}(\tau) = \frac{1}{n}\sum_{k=1}^{n}\operatorname{TAMSD}(\tau, k). \tag{6}$$

Physicists often observe systems where EA and EATAMSD are different. Such behavior is called weak ergodicity breaking. In mathematics, the notion of ergodicity is restricted to stationary processes. Since the increments of MFBM lack stationarity, when analyzing the results, one has to be exceedingly meticulous.

2.2. Three Cases of the Hurst Exponent Function

Following [48], we consider three basic families of the function $H(t)$, namely

$$\begin{aligned} \text{linear} \quad & H(t) = at + b, \quad t \in [0, T], \\ \text{logistic} \quad & H(t) = \frac{c - b}{1 + \exp\left\{-d\frac{t - t_0}{T}\right\}} + b, \quad t \in [0, T], \\ \text{periodic} \quad & H(t) = a \sin\left(4\pi\frac{t}{T}\right) + b, \quad t \in [0, T] \end{aligned}$$

for some time horizon $T > 0$. Furthermore, in the sequel we consider only case with parameter $\sigma^2 = 1$ in (2). Such functions are continuous and as a consequence satisfy the Hölder condition, so in order to MFBM be properly defined we only require $H(t) \in (0, 1)$, for all $t \in [0, T]$ [42]. Such choice of considered functions can be interpreted as follows. In the linear case, MFBM can switch steadily from short- to long-range dependence or vice versa, whereas in the logistic case such change is quite rapid and it happens between two levels. The latter case closely resembles instantaneous change in dependence or jump-type regime switching (such cases would lead to non-Hölder function). An alternative function to the logistic which is also considered is the literature is the arctan function [49]. Finally, the periodic case represents a situation where such changes are gradual and repetitive.

In the paper, we focus on the following special cases with specified parameters:

$$\begin{aligned} \text{linear function}: \quad & H^{(1)}(t) = \frac{0.3}{1000}t + 0.3, \quad t \in [0, 1000], \\ \text{logistic function}: \quad & H^{(2)}(t) = \frac{0.3}{1 + \exp\left\{-100\frac{t - 500}{1000}\right\}} + 0.3, \quad t \in [0, 1000], \\ \text{periodic case}: \quad & H^{(3)}(t) = 0.15 \sin\left(4\pi\frac{t}{1000}\right) + 0.45. \quad t \in [0, 1000]. \end{aligned}$$

We choose those specific parameters so that all of the cases have a similar "average" behavior of the function $H(t)$, i.e., its mean is close to 0.45, and the function itself has values in the interval $[0.3, 0.6]$. We illustrate those cases in Figures 1–3. On the top left panel of each of these figures, we can see three simulated trajectories. The function $H(t)$ is presented on the top right panel, whereas on the bottom panel we can see a behavior of the corresponding MSDs. It is important to note that EAMSD (blue line) is directly related to the variance of the model at time τ, i.e., $\text{EAMSD}(\tau) = \widehat{\text{Var}}(X(\tau))$, thus, from (2), it should behave like $\tau^{2H(\tau)}$.

For the linear case, see Figure 1, since $H^{(1)}(t)$ increases steadily from 0.3 to 0.6 we can see trajectories exhibit more variability for bigger times. This is also related to the EAMSD dynamics. In addition, such a model exhibits weak ergodicity breaking behavior (i.e., lack of equality between EATAMSD and EAMSD), which can be inferred from the bottom panel.

Next, for the logistic case, see Figure 2, $H^{(2)}(t)$ increases quite rapidly from 0.3 to 0.6 near $t = 500$. As a consequence, we can see a switch in the behavior of simulated trajectories: for times $t < 500$ they exhibit far less variability than for $t > 500$. Intuitively, for times $t < 500$ trajectories locally exhibit short-range dependence, whereas for $t > 500$ they locally exhibit long-range dependence. Again, we can see weak ergodicity breaking on the bottom panel.

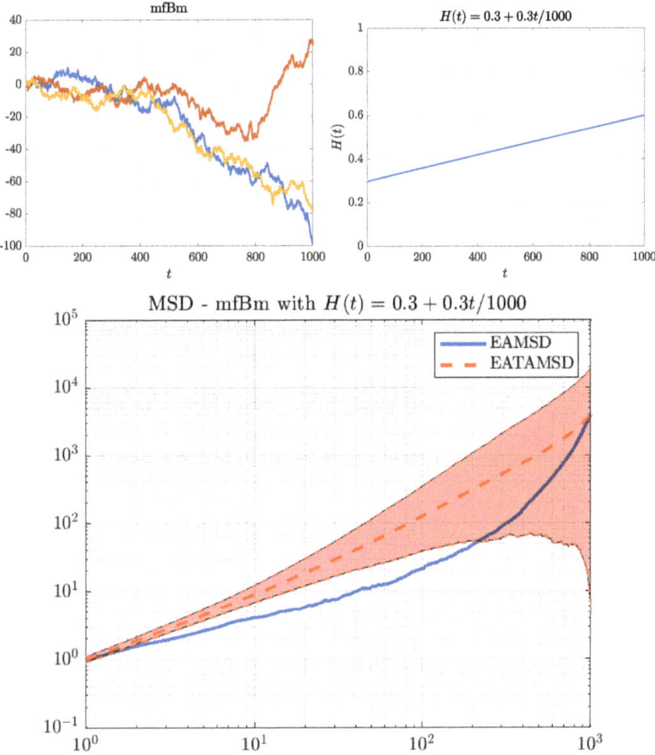

Figure 1. MFBM with the linear Hurst exponent function. Top left panel: three simulated trajectories. Top right panel: illustration of the function $H(t)$ used in simulations. Bottom panel: comparison of EAMSD (solid blue line) with EATAMSD (dashed red line) and its 95% confidence interval (red shaded area). EAMSD and ETAMSD with confidence interval were calculated on the basis of 1000 simulated trajectories of MFBM.

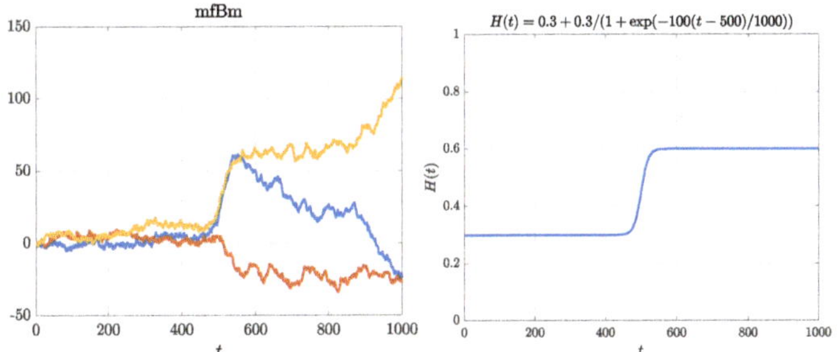

Figure 2. *Cont.*

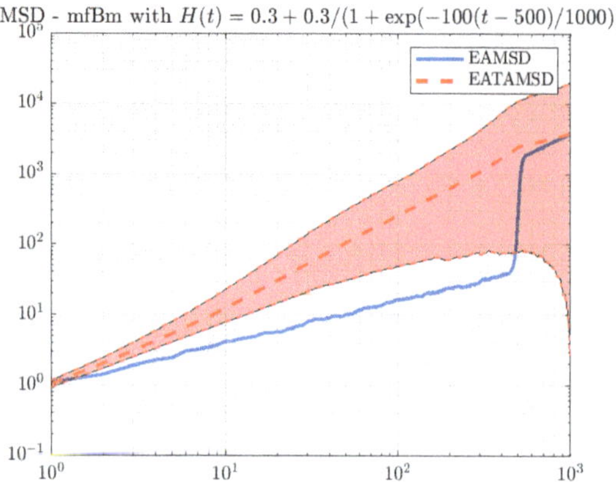

Figure 2. MFBM with the logistic Hurst exponent function. Top left panel: three simulated trajectories. Top right panel: illustration of the function $H(t)$ used in simulations. Bottom panel: comparison of EAMSD (solid blue line) with EATAMSD (dashed red line) and its 95% confidence interval (red shaded area). EAMSD and ETAMSD with confidence interval were calculated on the basis of 1000 simulated trajectories of MFBM.

Finally, for the periodic case, see Figure 3, $H^{(3)}(t)$ varies between 0.3 and 0.6. We can clearly see two different regimes of behavior: for times when $H^{(3)}(t)$ is bigger, trajectories are smoother and generally have larger values, in contrast to times when $H^{(3)}(t)$ is smaller. Despite lack of stationarity, here, EAMSD almost always lies in the confidence region of EATAMSD, which could suggest there is no weak ergodicity breaking.

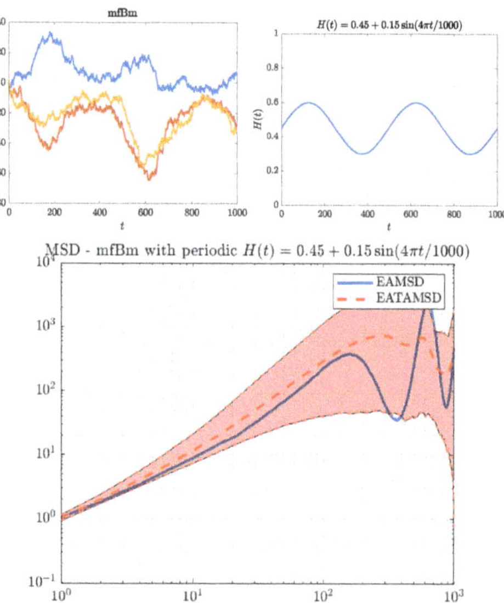

Figure 3. MFBM with the periodic Hurst exponent function function. Top left panel: three simulated trajectories. Top right panel: illustration of the function $H(t)$ used in simulations. Bottom panel: comparison of EAMSD (solid blue line) with EATAMSD (dashed red line) and its 95% confidence interval (red shaded area). EAMSD and ETAMSD with confidence interval were calculated on the basis of 1000 simulated trajectories of MFBM.

3. Results

In applications, it is crucial to be able to check whether a stochastic model describes empirical data well. Despite dedicated identification methods for the MFBM [50–53], to the best of the authors' knowledge, there is no rigorous statistical test designed for such process. Here, we propose an approach using a simple test statistic which also contains useful information about the process itself.

3.1. Test

For the testing purposes, we follow an approach based on the ACVF which was introduced by Balcerek and Burnecki [38]. ACVF is a very popular statistic and it is also one of the simplest quadratic forms. For a random sample $\mathbf{X}_N = \{X(1), X(2), \ldots, X(N)\}$ and $\tau \in \{1, 2, \cdots, N-1\}$, it is defined as follows:

$$\mathrm{ACVF}_N(\tau) = \frac{1}{N-\tau} \sum_{i=1}^{N-\tau} X(i+\tau)X(i). \tag{7}$$

Here, we only consider a version of ACVF without subtracting the sample mean as it does not influence performance of tests based on this statistic for a centered process [38] and it makes the formulas much simpler.

Let us now introduce a matrix $\mathbf{A}(\tau) = \{a(\tau; i, j)\}_{i,j=1}^{N}$, where

$$a(\tau; i, j) = \begin{cases} \frac{1}{N}\mathbb{I}(i = j) & \text{if } \tau = 0 \\ \frac{1}{2}\frac{1}{N-\tau}\mathbb{I}(|i-j| = \tau) & \text{if } \tau = 1, 2, \ldots, N-1 \\ 0 & \text{otherwise,} \end{cases} \tag{8}$$

and $\mathbb{I}$ is the indicator. To summarize, the matrix $\mathbf{A}(\tau)$ is either diagonal (for $\tau = 0$) with elements $\frac{1}{N}$ on diagonal or Toeplitz, with only two nonzero subdiagonals (starting at $(1+\tau)$th row and $(1+\tau)$th column) with elements $\frac{1}{2}\frac{1}{N-\tau}$. The statistic ACVF_N can be now expressed as a quadratic form (as shown in [38]) as a generalized χ^2 distribution, that is

$$\mathrm{ACVF}_N(\tau) = \sum_{i=1}^{N} \lambda_i(\tau) Z_i^2, \tag{9}$$

where Z_is are i.i.d standard normal variables (so Z_i^2 has a χ_1^2 distribution) and $\lambda_k(\tau)$ are eigenvalues of the matrix $\mathbf{\Sigma}_N(\tau) = \mathbf{\Sigma}^{1/2}\mathbf{A}(\tau)\mathbf{\Sigma}^{1/2}$ with $\mathbf{\Sigma}$ being the (theoretical) autocovariance matrix of our trajectory $\mathbf{X}_N$. It is important to note that this result is true regardless of whether the considered model is stationary or not.

Let us now formulate a test for checking whether a random sample $\mathbf{X}_N$ comes from the MFBM with function $H : [0, T] \to (0, 1)$, where T is the time horizon:

H_0 : sample comes from the model with function $H(t)$

versus

H_1 : sample comes from the model with function different than $H(t)$.

We will use ACVF_N as a test statistic with its distribution given by Equation (9) to calculate critical regions of such test for a given significance level. Naturally, eigenvalues $\lambda_i(\tau)$ depend on the matrix $\mathbf{A}(\tau)$ as well as on the matrix $\mathbf{\Sigma}$. Elements of $\mathbf{\Sigma}$ are given by ACVF (2) and are calculated using the function $H(t)$ from the null hypothesis.

3.2. Three Power Case Studies

The power of the test is the probability to reject the null hypothesis when the alternative is true. The power is an important characteristic of any statistical test. We consider the following null hypotheses, which correspond to the examples presented in Figures 1–3.

$$\begin{aligned} \text{linear function null hypothesis } H_0: &\quad H(t) = H^{(1)}(t) \quad t \in [0, 1000], \\ \text{logistic function null hypothesis } H_0: &\quad H(t) = H^{(2)}(t), \quad t \in [0, 1000], \\ \text{periodic case null hypothesis } H_0: &\quad H(t) = H^{(3)}(t), \quad t \in [0, 1000]. \end{aligned}$$

In our studies, for all considered cases, we calculate the power of the test by using Monte Carlo simulations. We assume that the significance level is equal to 5%. In our Monte Carlo simulations we consider the time horizon $T = 1000$ and equally spaced time points $t = 1, 2, \ldots, 1000$. For each set of parameters from the alternative hypothesis, we simulate $n = 1000$ trajectories, calculate test statistic (7), and check if the null hypothesis is rejected at 5% significance level. In the test statistic, we consider only $\tau = 1$ since other choices of τ lead to worse results. Finally, we estimate the power of this test for each considered case by calculating the fraction of rejected null hypotheses.

We present the results in the form of power functions with arguments being the parameters of the function $H(t)$ from the alternative hypothesis. For all of the cases, we considered the alternative coming from the same family of functions as the function $H(t)$ in the null hypothesis, i.e., linear alternative for linear null, etc.

First, let us consider testing MFBM with the linear Hurst exponent function. We can see the power function related to that case in Figure 4. The left panel presents the power function with respect to parameters a and b from the alternative hypothesis $H_1 : H(t) = at + b$, the right panel the corresponding heat map. We can see "layers" (regions on the heat map with the same color) for which our test has a very similar power. For example, the deep blue region on the right panel corresponds to the processes indistinguishable from the null hypothesis process. We believe that the shape of the region is related to the construction of our test, namely our test statistic ACVF_N takes into account all addends $X(t)X(t + \tau)$ with the same weight, thus it is not that relevant whether t is big or not. In the case of MFBM, for which neither the process nor its increments are stationary, it might be an important factor. As a consequence, we can see that the test has a difficulty in distinguishing between MFBMs with increasing and decreasing Hurst exponent functions if their means are similar. However, this conjecture is not very precise, namely the mean case $H \equiv 0.45$, which matches the alternative hypothesis with $a = 0, b = 0.45$, yields a much higher power of the test than the significance level. We also note that for parameters from the null hypothesis: $a = 0.3$, $b = 0.3$ power of such test is approximately equal to 5%, which is the assumed significance level. On the heat map, white regions represent areas for which the process MFBM is not well-defined, i.e., $H(t) \notin (0, 1)$ for some $t \in [0, 1000]$.

Let us now consider the second case, that is MFBM with the logistic function $H(t)$. We can observe the power function in Figure 5. Left panel presents the power function with respect to parameters b and c from the alternative hypothesis $H_1 : H(t) = \frac{c-b}{1+\exp\left\{-d\frac{t-t_0}{T}\right\}} + b$, the right panel the corresponding heat map. The parameter b is related to the local self-similarity parameter for $t < 500$, and c for $t > 500$. Similarly to the case with the linear function null hypothesis, here we can observe "layers" in which parameters b and c are almost symmetric (e.g., the null hypothesis case where $b = 0.3$ and $c = 0.6$ is closely related to the case $b = 0.6$ and $c = 0.3$). Moreover, we can see that the power function seems to be quite high in the cases when a tested sample has the b parameter close to the value 0.3 from the null hypothesis, but c is far from 0.6, or vice versa.

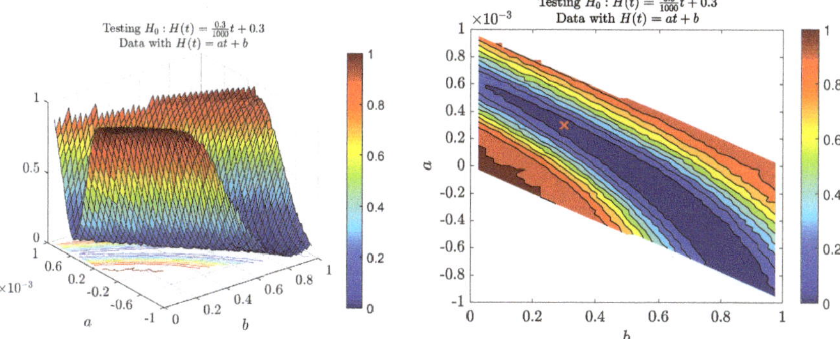

Figure 4. Power of the introduced test for the linear Hurst exponent function $H(t) = at + b$ with respect to parameters a and b. The null hypothesis is $a = \frac{0.3}{1000}$ and $b = 0.3$. The right panel depicts the results in the form of a heat map with the red 'x' sign representing parameters in the null hypothesis. White regions represent areas for which MFBM is not well defined. The powers were calculated by means of Monte Carlo simulations on the basis of simulated data from the MFBM with different *a*s and *b*s.

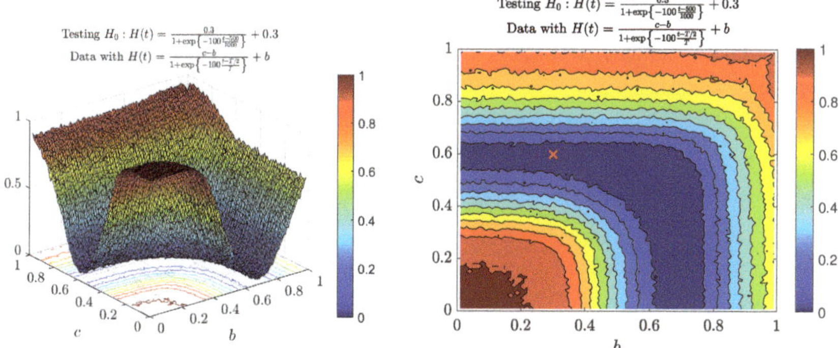

Figure 5. Power of the introduced test for the logistic Hurst exponent function $H(t) = \frac{c-b}{1+\exp\{-100\frac{t-500}{1000}\}} + b$ with respect to parameters c and b. The null hypothesis is $c = 0.6$ and $b = 0.3$. The right panel depicts the results in the form of heat map with the red 'x' sign representing parameters in the null hypothesis. White regions represent areas for which MFBM is not well defined. The powers were calculated by means of Monte Carlo simulations on the basic of simulated data from the MFBM with different *c*s and *b*s.

Lastly, let us consider the case of MFBM with the periodic function $H(t)$. We can observe the power function in Figure 6. The left panel presents the power function with respect to parameters a and b from the alternative hypothesis $H_1 : H(t) = a\sin\left(4\pi\frac{t}{T}\right) + b$, the right panel the corresponding heat map. Parameter b is related to the "mean" behavior of the function $H(t)$, whereas the parameter a corresponds to its amplitude. Again, similarly to the two previous null hypotheses, we can observe "layers" of similar power values. Those layers are symmetric with respect to $a = 0$. This means that for alternatives with opposite parameters a the test seems to return the same power. Let us note that this is not intuitive, namely such opposite *a*s are related to completely different local behaviors of the self-similarity parameter. On the heat map, white regions represent areas for which process MFBM is not well defined, i.e., $H(t) \notin (0,1)$ for some $t \in [0, 1000]$.

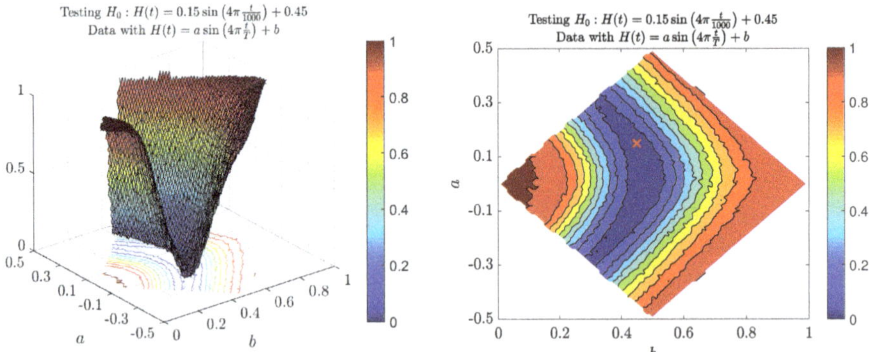

Figure 6. Power of the introduced test for the periodic Hurst exponent function $H(t) = a\sin\left(4\pi\frac{t}{1000}\right) + b$ with respect to parameters a and b. The null hypothesis is $a = 0.15$ and $b = 0.45$. The right panel depicts the results in the form of heat map with the red 'x' sign representing parameters in the null hypothesis. White regions represent areas for which MFBM is not well defined. The powers were calculated by means of Monte Carlo simulations on the basic of simulated data from the MFBM with different *a*s and *b*s.

Finally, we would like to emphasize that the introduced test requires MFBM parameters to be fixed (we test if the data follow MFBM with fixed parameters). In practice, when analyzing empirical data the parameters are often estimated. In the literature, methods for estimation of the Hurst exponent function $H(t)$ in the MFBM framework have been already introduced [50–52] and later combined to improve both the goodness of fit and the computational speed of the algorithm [53].

4. Discussion and Conclusions

For power-law anomalous diffusion of the form (1) with constant anomalous diffusion exponent α a number of models exist, including continuous-time random walks, fractional Langevin equation motion, FBM, or scaled Brownian motion [8]. These models all have different physical properties such as the PDF or their ergodic and aging properties [8].

In this paper, we concentrated on MFBM which is a generalization of FBM for Hölder continuous functions $H(t)$ that allows the Hurst exponent to vary in time. The time-varying Hurst exponent has an impact on both the statistical properties of the process and trajectory characteristics. MFBM helps to model phenomena whose regularity of paths and anomalous diffusion exponent change in time. The process has no longer stationary increments and it is not self-similar but the variance scales in a natural way as $t^{2H(t)}$.

Following the idea of testing FBM based on the ACVF statistic [38], in this paper, we introduced a rigorous statistical test on MFBM with the ACVF statistic presented as a quadratic form. We derived the distribution of the statistic which is the generalized χ^2. In order to study the efficiency of the test, we took into consideration three possible classes of the Hurst exponent function, namely linear, logistic, and periodic. For those cases, we conducted power studies with the help of Monte Carlo simulations. As alternatives, we considered MFBMs within the same class of the Hurst exponent function but with different parameters.

We found ranges of the parameters where the test is more sensitive to differences and ranges where it fails to distinguish between the models. It appears that for the linear Hurst exponent function the test is most sensitive to changes in the mean of the function. If the means are similar then the test often fails, even if the functions have completely different patterns, namely, one is increasing and the other decreasing. The latter observation may sound like a serious objection for using the test, but, in practice, an experimentalist knows whether the anomalous diffusion exponent increases or decreases in time. In the logistic case, the situation is different, namely the mean does not matter much as for the linear case. Now, the test is most sensitive to deviations from the true values of the two levels (c and b) with the exception that replacing c with b does not change the power of the test (so, again, it does not matter if the function increases or decreases). For the periodic case, we have again a different situation. The test is sensitive to the changes of the amplitude of the sine function and the value of the free term but it does not detect a sign of the parameter related to the magnitude.

Finally, we note that we checked the behavior of the test for other sets of parameters of the null hypotheses and different sample lengths, and the conclusions were similar. We only found that the range of possible H values from the null hypothesis has an influence on the width of the acceptance regions (the wider the range the wider the acceptance region, which is reasonable). We present some of the additional tests' power simulation studies in Appendix A. Figure A1 presents different cases of the null hypothesis for the linear case, Figure A2 for the logistic case, and Figure A3 for the periodic case. Tests for the linear case were performed for length $N = 1000$, whereas logistic and periodic cases were studied for length $N = 200$.

In sum, we introduced a rigorous statistical test for MFBM based on ACVF statistic presented as a quadratic form. We highlighted the weak and strong points of the test. Improving the efficiency of the test will be a subject of our future studies. We believe that the obtained results can help to understand the mechanisms underlying various anomalous diffusion phenomena.

Author Contributions: Conceptualization, M.B.; methodology, M.B. and K.B.; visualization, M.B.; writing—original draft preparation, M.B. and K.B.; writing—review and editing, M.B. and K.B.; investigation, K.B. All authors have read and agreed to the published version of the manuscript.

Funding: The second author would like to acknowledge the Beethoven Grant no.: DFG-NCN 2016/23/G/ST1/04083.

Conflicts of Interest: The authors declare no conflict of interest. The funders had no role in the design of the study; in the collection, analysis, or interpretation of data; in the writing of the manuscript, or in the decision to publish the results.

Abbreviations

The following abbreviations are used in this manuscript:

SPT	Single-particle tracking
FBM	Fractional Brownian motion
MFBM	Multifractional Brownian motion
MSD	Mean-squared displacement
TAMSD	Time average mean-squared displacement
EAMSD	Ensemble average mean-squared displacement
EATAMSD	Ensemble and time average mean-squared displacement
ACVF	Autocovariance function

Appendix A

In Figures A1–A3, we present power functions of tests related to different null hypotheses. In Figure A1, we consider $H_0 : H(t) = \frac{-0.3}{1000}t + 0.7$ (top panel), so a case in which function H begins in the superdiffusive regime and then decreases linearly to value 0.4; $H_0 : H(t) = \frac{-0.4}{1000}t + 0.5$ (middle panel), so a case in which function H begins in the diffusion regime and then decreases linearly to value 0.1; and $H_0 : H(t) = \frac{0.6}{1000}t + 0.2$ (bottom panel), so a case in which function H begins in the strong subdiffusive regime and then increases linearly to 0.8.

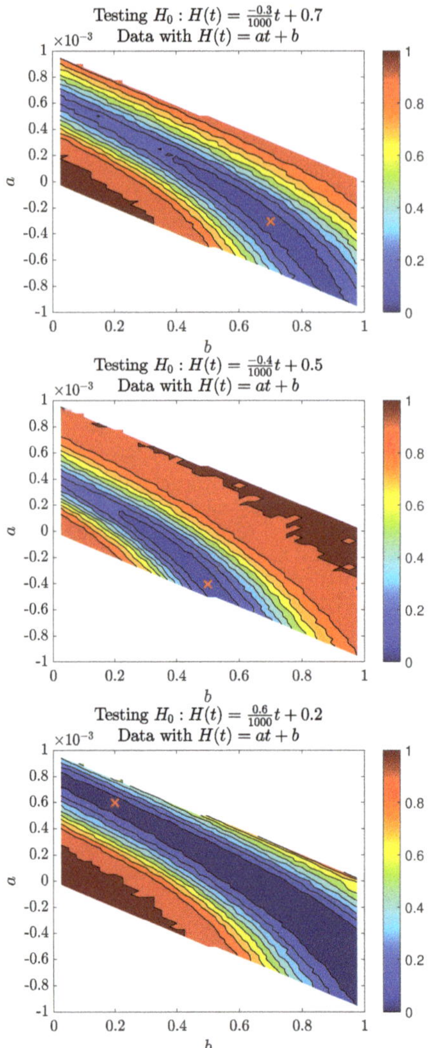

Figure A1. Power of the introduced test for the linear Hurst exponent function $H(t) = at + b$ with respect to parameters a and b. The null hypotheses are: $a = \frac{-0.3}{1000}$ and $b = 0.7$ (top panel), $a = \frac{-0.4}{1000}$ and $b = 0.5$ (middle panel), $a = \frac{0.6}{1000}$ and $b = 0.2$ (bottom panel). All of the panels depict the results in the form of a heat map with the red '**x**' sign representing parameters in the null hypothesis. White regions represent areas for which MFBM is not well defined. The powers were calculated by means of Monte Carlo simulations on the basis of simulated data from the MFBM with different as and bs.

In Figure A2, we consider $H_0 : H(t) = \frac{-0.4}{1+\exp\{-100\frac{t-500}{1000}\}} + 0.5$ (top panel), so a case in which function H begins in the diffusive regime and ends in the strong subdiffusive regime; $H_0 : H(t) = \frac{-0.3}{1+\exp\{-100\frac{t-500}{1000}\}} + 0.6$ (middle panel), so a case in which function H begins in the superdiffusive regime and ends in the subdiffusive regime; and $H_0 : H(t) = \frac{0.6}{1+\exp\{-100\frac{t-500}{1000}\}} + 0.2$ (bottom panel), so a case in which function H begins in the strong subdiffusive regime and ends in the strong superdiffusive regime.

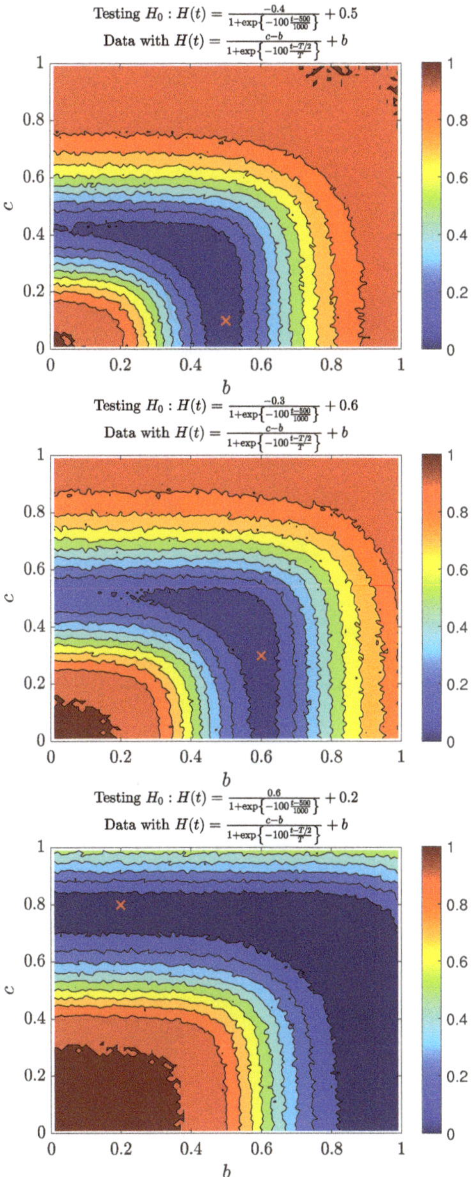

Figure A2. Power of the introduced test for the logistic Hurst exponent function $H(t) = \frac{c-b}{1+\exp\{-100\frac{t-500}{1000}\}} + b$ with respect to parameters c and b. The null hypotheses are: $c = 0.1$ and $b = 0.5$ (top panel), $c = 0.3$ and $b = 0.6$ (middle panel), $c = 0.8$ and $b = 0.2$ (bottom panel). All of the panels depict the results in the form of heat map with the red 'x' sign representing parameters in the null hypothesis. White regions represent areas for which MFBM is not well defined. The powers were calculated by means of Monte Carlo simulations on the basic of simulated data from the MFBM with different cs and bs.

In Figure A3, we consider $H_0 : H(t) = 0.1\sin\left(4\pi\frac{t}{1000}\right) + 0.8$ (top panel), so a case in which function H varies periodically between 0.7 and 0.9, i.e., in the strong superdiffusion regime; H_0 :

$H(t) = 0.2\sin\left(4\pi\frac{t}{1000}\right) + 0.7$ (middle panel), so a case in which function H varies periodically between 0.5 and 0.9, i.e., in the superdiffusion regime; and $H_0 : H(t) = 0.5\sin\left(4\pi\frac{t}{1000}\right) + 0.4$ (bottom panel), so a case in which function H varies periodically between 0.1 and 0.9, i.e., in the whole spectrum of anomalous diffusion.

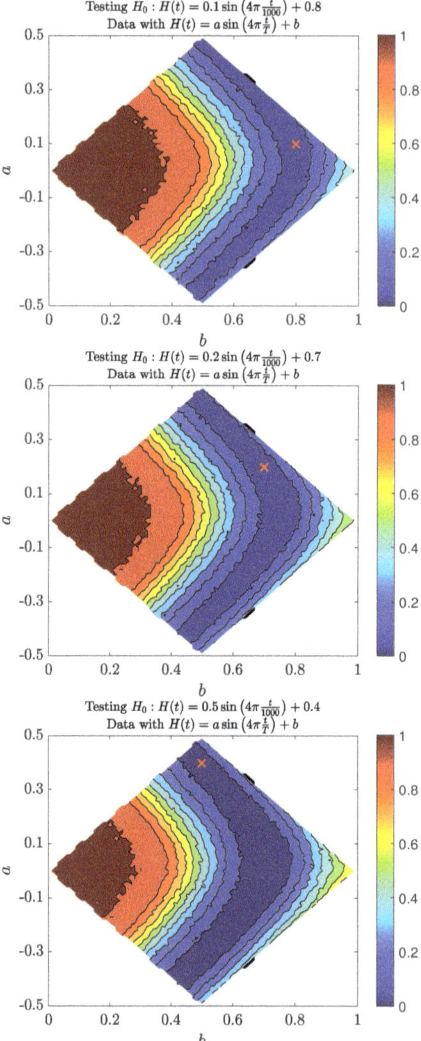

Figure A3. Power of the introduced test for the periodic Hurst exponent function $H(t) = a\sin\left(4\pi\frac{t}{1000}\right) + b$ with respect to parameters a and b. The null hypotheses are: $a = 0.1$ and $b = 0.8$ (top panel), $a = 0.2$ and $b = 0.7$ (middle panel), $a = 0.5$ and $b = 0.4$ (bottom panel). All of the panels depict the results in the form of heat map with the red 'x' sign representing parameters in the null hypothesis. White regions represent areas for which MFBM is not well defined. The powers were calculated by means of Monte Carlo simulations on the basic of simulated data from the MFBM with different as and bs.

References

1. Braüchle, C.; Lamb, D.C.; Michaelis, J. *Single Particle Tracking and Single Molecule Energy Transfer*; Wiley-VCH: Weinheim, Germany, 2010.
2. Xie, X.S.; Choi, P.J.; Li, G.W.; Lee, N.K.; Lia, G. Single-molecule approach to molecular biology in living bacterial cells. *Annu. Rev. Biophys.* **2008**, *37*, 417–444. [CrossRef] [PubMed]
3. Metzler, R.; Jeon, J.H.; Cherstvy, A. Non-Brownian diffusion in lipid membranes: Experiments and simulations. *Biochim. Biophys. Acta (BBA)-Biomembr.* **2016**, *1858*, 2451–2467. [CrossRef] [PubMed]
4. Saxton, M.J.; Jacobson, K. Single-particle tracking: Applications to membrane dynamics. *Annu. Rev. Biophys. Biomol. Struct.* **1997**, *26*, 373–399. [CrossRef] [PubMed]
5. Bouchaud, J.P.; Georges, A. Anomalous diffusion in disordered media: Statistical mechanisms, models and physical applications. *Phys. Rep.* **1990**, *195*, 127–293. [CrossRef]
6. Höfling, F.; Franosch, T. Anomalous transport in the crowded world of biological cells. *Rep. Prog. Phys.* **2013**, *76*, 046602. [CrossRef] [PubMed]
7. Barkai, E.; Garini, Y.; Metzler, R. Strange kinetics of single molecules in living cells. *Phys. Today* **2012**, *65*, 29. [CrossRef]
8. Metzler, R.; Jeon, J.H.; Cherstvy, A.G.; Barkai, E. Anomalous diffusion models and their properties: Non-stationarity, non-ergodicity, and ageing at the centenary of single particle tracking. *Phys. Chem. Chem. Phys.* **2014**, *16*, 24128–24164. [CrossRef]
9. Golding, I.; Cox, E.C. Physical nature of bacterial cytoplasm. *Phys. Rev. Lett.* **2006**, *96*, 098102. [CrossRef]
10. Bronstein, I.; Israel, Y.; Kepten, E.; Mai, S.; Shav-Tal, Y.; Barkai, E.; Garini, Y. Transient anomalous diffusion of telomeres in the nucleus of mammalian cells. *Phys. Rev. Lett.* **2009**, *103*, 018102. [CrossRef]
11. Jeon, J.H.; Tejedor, V.; Burov, S.; Barkai, E.; Selhuber-Unkel, C.; Berg-Sørensen, K.; Oddershede, L.; Metzler, R. In vivo anomalous diffusion and weak ergodicity breaking of lipid granules. *Phys. Rev. Lett.* **2011**, *106*, 048103. [CrossRef]
12. Weigel, A.V.; Simon, B.; Tamkun, M.M.; Krapf, D. Ergodic and nonergodic processes coexist in the plasma membrane as observed by single-molecule tracking. *Proc. Natl. Acad. Sci. USA* **2011**, *108*, 6438–6443. [CrossRef] [PubMed]
13. Tabei, S.A.; Burov, S.; Kim, H.Y.; Kuznetsov, A.; Huynh, T.; Jureller, J.; Philipson, L.H.; Dinner, A.R.; Scherer, N.F. Intracellular transport of insulin granules is a subordinated random walk. *Proc. Natl. Acad. Sci. USA* **2013**, *110*, 4911–4916. [CrossRef] [PubMed]
14. Szymanski, J.; Weiss, M. Elucidating the origin of anomalous diffusion in crowded fluids. *Phys. Rev. Lett.* **2009**, *103*, 038102. [CrossRef] [PubMed]
15. Jeon, J.H.; Leijnse, N.; Oddershede, L.B.; Metzler, R. Anomalous diffusion and power-law relaxation of the time averaged mean squared displacement in worm-like micellar solutions. *New J. Phys.* **2013**, *15*, 045011. [CrossRef]
16. Wong, I.; Gardel, M.; Reichman, D.; Weeks, E.R.; Valentine, M.; Bausch, A.; Weitz, D.A. Anomalous diffusion probes microstructure dynamics of entangled F-actin networks. *Phys. Rev. Lett.* **2004**, *92*, 178101. [CrossRef] [PubMed]
17. Hansing, J.; Ciemer, C.; Kim, W.K.; Zhang, X.; DeRouchey, J.E.; Netz, R.R. Nanoparticle filtering in charged hydrogels: Effects of particle size, charge asymmetry and salt concentration. *Eur. Phys. J. E* **2016**, *39*, 53. [CrossRef] [PubMed]
18. Xu, Q.; Feng, L.; Sha, R.; Seeman, N.; Chaikin, P. Subdiffusion of a sticky particle on a surface. *Phys. Rev. Lett.* **2011**, *106*, 228102. [CrossRef]
19. Godec, A.; Bauer, M.; Metzler, R. Collective dynamics effect transient subdiffusion of inert tracers in flexible gel networks. *New J. Phys.* **2014**, *16*, 092002. [CrossRef]
20. Weiss, M.; Hashimoto, H.; Nilsson, T. Anomalous protein diffusion in living cells as seen by fluorescence correlation spectroscopy. *Biophys. J.* **2003**, *84*, 4043–4052. [CrossRef]
21. Kneller, G.R.; Baczynski, K.; Pasenkiewicz-Gierula, M. Communication: Consistent picture of lateral subdiffusion in lipid bilayers: Molecular dynamics simulation and exact results. *J. Chem. Phys.* **2011**, *135*, 141105. [CrossRef]
22. Jeon, J.H.; Monne, H.M.S.; Javanainen, M.; Metzler, R. Anomalous diffusion of phospholipids and cholesterols in a lipid bilayer and its origins. *Phys. Rev. Lett.* **2012**, *109*, 188103. [CrossRef] [PubMed]

23. Yamamoto, E.; Akimoto, T.; Yasui, M.; Yasuoka, K. Origin of subdiffusion of water molecules on cell membrane surfaces. *Sci. Rep.* **2014**, *4*, 4720. [CrossRef] [PubMed]
24. Manzo, C.; Torreno-Pina, J.A.; Massignan, P.; Lapeyre, G.J., Jr.; Lewenstein, M.; Parajo, M.F.G. Weak ergodicity breaking of receptor motion in living cells stemming from random diffusivity. *Phys. Rev. X* **2015**, *5*, 011021. [CrossRef]
25. Jeon, J.H.; Javanainen, M.; Martinez-Seara, H.; Metzler, R.; Vattulainen, I. Protein crowding in lipid bilayers gives rise to non-Gaussian anomalous lateral diffusion of phospholipids and proteins. *Phys. Rev. X* **2016**, *6*, 021006. [CrossRef]
26. Berkowitz, B.; Cortis, A.; Dentz, M.; Scher, H. Modeling non-Fickian transport in geological formations as a continuous time random walk. *Rev. Geophys.* **2006**, *44*. [CrossRef]
27. Caspi, A.; Granek, R.; Elbaum, M. Enhanced diffusion in active intracellular transport. *Phys. Rev. Lett.* **2000**, *85*, 5655. [CrossRef]
28. Reverey, J.F.; Jeon, J.H.; Bao, H.; Leippe, M.; Metzler, R.; Selhuber-Unkel, C. Superdiffusion dominates intracellular particle motion in the supercrowded cytoplasm of pathogenic Acanthamoeba castellanii. *Sci. Rep.* **2015**, *5*, 1–14. [CrossRef]
29. Gal, N.; Weihs, D. Experimental evidence of strong anomalous diffusion in living cells. *Phys. Rev. E* **2010**, *81*, 020903. [CrossRef]
30. Monserud, J.H.; Schwartz, D.K. Interfacial molecular searching using forager dynamics. *Phys. Rev. Lett.* **2016**, *116*, 098303. [CrossRef]
31. Campagnola, G.; Nepal, K.; Schroder, B.W.; Peersen, O.B.; Krapf, D. Superdiffusive motion of membrane-targeting C2 domains. *Sci. Rep.* **2015**, *5*, 1–10. [CrossRef]
32. Javanainen, M.; Hammaren, H.; Monticelli, L.; Jeon, J.H.; Miettinen, M.S.; Martinez-Seara, H.; Metzler, R.; Vattulainen, I. Anomalous and normal diffusion of proteins and lipids in crowded lipid membranes. *Faraday Discuss.* **2013**, *161*, 397–417. [CrossRef] [PubMed]
33. Kolmogorov, A.N. Wienersche Spiralen und einige andere interessante Kurven in Hilbertscen Raum. *C.R. (Dokl.) Acad. Sci. URSS (NS)* **1940**, *26*, 115–118.
34. Mandelbrot, B.B.; Van Ness, J.W. Fractional Brownian motions, fractional noises and applications. *SIAM Rev.* **1968**, *10*, 422–437. [CrossRef]
35. Graves, T.; Gramacy, R.; Watkins, N.; Franzke, C. A brief history of long memory: Hurst, Mandelbrot and the road to ARFIMA, 1951–1980. *Entropy* **2017**, *19*, 437. [CrossRef]
36. Goychuk, I. Viscoelastic subdiffusion: Generalized Langevin equation approach. *Adv. Chem. Phys.* **2012**, *150*, 187.
37. Klafter, J.; Lim, S.; Metzler, R. *Fractional Dynamics: Recent Advances*; World Scientific: Singapore, 2012.
38. Balcerek, M.; Burnecki, K. Testing of fractional Brownian motion in a noisy environment. *Chaos Solitons Fractals* **2020**, *140*, 110097. [CrossRef]
39. Sikora, G.; Burnecki, K.; Wyłomańska, A. Mean-squared displacement statistical test for fractional Brownian motion. *Phys. Rev. E* **2017**, *95*, 032110. [CrossRef]
40. Sikora, G. Statistical test for fractional Brownian motion based on detrending moving average algorithm. *Chaos Solitons Fractals* **2018**, *114*, 54–62. [CrossRef]
41. Ralchenko, K.; Shevchenko, G. Path properties of multifractal Brownian motion. *Theory Probab. Math. Stat.* **2010**, *80*, 119–130. [CrossRef]
42. Peltier, R.F.; Véhel, J.L. *Multifractional Brownian Motion: Definition and Preliminary Results*; Research Report, RR-2645 INRIA, 1995. Ffinria-00074045; Inria Paris: Rocquencourt, France, 1995.
43. Lee, K. Characterization of turbulence stability through the identification of multifractional Brownian motions. *Nonlinear Process. Geophys.* **2013**, *20*, 97–106. [CrossRef]
44. Perrin, E.; Harba, R.; Iribarren, I.; Jennane, R. Piecewise fractional Brownian motion. *IEEE Trans. Signal Process.* **2005**, *53*, 1211–1215. [CrossRef]
45. Ryvkina, J. Fractional Brownian Motion with variable Hurst parameter: Definition and properties. *J. Theor. Probab.* **2015**, *28*, 866–891. [CrossRef]
46. Ayache, A.; Cohen, S.; Véhel, J.L. The covariance structure of multifractional Brownian motion, with application to long range dependence. In Proceedings of the 2000 IEEE International Conference on Acoustics, Speech, and Signal Processing, Istanbul, Turkey, 5–9 June 2000; Volume 6, pp. 3810–3813.

47. Benassi, A.; Cohen, S.; Istas, J. Identifying the multifractional function of a Gaussian process. *Stat. Probab. Lett.* **1998**, *39*, 337–345. [CrossRef]
48. Chan, G.; Wood, A.T. Simulation of multifractional Brownian motion. In *COMPSTAT*; Springer: Berlin/Heidelberg, Germany, 1998; pp. 233–238.
49. Stoev, S.A.; Taqqu, M.S. How rich is the class of multifractional Brownian motions? *Stoch. Process. Appl.* **2006**, *116*, 200–221. [CrossRef]
50. Bianchi, S. Pathwise identification of the memory function of multifractional Brownian motion with application to finance. *Int. J. Theor. Appl. Financ.* **2005**, *8*, 255–281. [CrossRef]
51. Bardet, J.M.; Bertrand, P. Identification of the multiscale fractional Brownian motion with biomechanical applications. *J. Time Ser. Anal.* **2007**, *28*, 1–52. [CrossRef]
52. Bianchi, S.; Pantanella, A.; Pianese, A. Modeling stock prices by multifractional Brownian motion: An improved estimation of the pointwise regularity. *Quant. Financ.* **2013**, *13*, 1317–1330. [CrossRef]
53. Pianese, A.; Bianchi, S.; Palazzo, A.M. Fast and unbiased estimator of the time-dependent Hurst exponent. *Chaos Interdiscip. J. Nonlinear Sci.* **2018**, *28*, 031102. [CrossRef]

MDPI

Article

Cusp of Non-Gaussian Density of Particles for a Diffusing Diffusivity Model

M. Hidalgo-Soria [1,*,†], E. Barkai [1,†] and S. Burov [2,*,†]

[1] Department of Physics, Institute of Nanotechnology and Advanced Materials, Bar-Ilan University, Ramat-Gan 5290002, Israel; Eli.Barkai@biu.ac.il

[2] Department of Physics, Bar-Ilan University, Ramat-Gan 5290002, Israel

* Correspondence: mariohidalgosoria@gmail.com (M.H.-S.); stasbur@gmail.com (S.B.)

† These authors contributed equally to this work.

Abstract: We study a two state "jumping diffusivity" model for a Brownian process alternating between two different diffusion constants, $D_+ > D_-$, with random waiting times in both states whose distribution is rather general. In the limit of long measurement times, Gaussian behavior with an effective diffusion coefficient is recovered. We show that, for equilibrium initial conditions and when the limit of the diffusion coefficient $D_- \longrightarrow 0$ is taken, the short time behavior leads to a cusp, namely a non-analytical behavior, in the distribution of the displacements $P(x,t)$ for $x \longrightarrow 0$. Visually this cusp, or tent-like shape, resembles similar behavior found in many experiments of diffusing particles in disordered environments, such as glassy systems and intracellular media. This general result depends only on the existence of finite mean values of the waiting times at the different states of the model. Gaussian statistics in the long time limit is achieved due to ergodicity and convergence of the distribution of the temporal occupation fraction in state D_+ to a δ-function. The short time behavior of the same quantity converges to a uniform distribution, which leads to the non-analyticity in $P(x,t)$. We demonstrate how super-statistical framework is a zeroth order short time expansion of $P(x,t)$, in the number of transitions, that does not yield the cusp like shape. The latter, considered as the key feature of experiments in the field, is found with the first correction in perturbation theory.

Keywords: CTRW; diffusing-diffusivity; occupation time statistics

Citation: Hidalgo-Soria, M.; Barkai, E.; Burov, S. Cusp of Non-Gaussian Density of Particles for a Diffusing Diffusivity Model. *Entropy* **2021**, *23*, 231. https://doi.org/10.3390/e23020231

Received: 20 December 2020
Accepted: 9 February 2021
Published: 17 February 2021

1. Introduction

The emergence of non-Gaussian features for the positional probability density function (PDF) of particle spreading, denoted $P(x,t)$, in a disordered environment is a common attribute that arises in many different physical and biological systems. Specifically, a tent like shape of the PDF, in the semi-log scale, together with a linear time dependence of the mean square displacement (MSD) appear for diffusion in glassy system [1], biological cells [2–7], and colloidal suspensions [8–11]. This tent shape, sometimes fitted with a Laplace distribution $P(x,t) \sim \exp(-C|x|)$ with C a constant, suggests that the decay of the PDF is exponential. This feature is becoming a more frequent observation for the spreading of molecules. Phenomenological approaches are diffusing diffusivity models, in which non-Gaussianity is obtained by coupled stochastic differential equations with random diffusion coefficients [7,12–20], and path integrals formalism for Brownian motion in the presence of a sink [21]. More recently, theoretical frameworks describing this behavior emerged from continuous time random walk (CTRW) approaches employing large deviations theory [22–24] and microscopical models like molecular dynamics of tracer particles in polymer networks [25,26] and interacting particles with fluctuating sizes [27–29], the so-called Hitchhiker model [28].

While in some of the systems the non-Gaussian behavior disappears when the measurement time is made long enough, the short time tent-like decay of the PDF seems to be a universal phenomenon [22]. It is then natural to ask if there is some sort of universality that can be deduced for the temporal limit of short times. Within the diffusing diffusivity models for the large x limit, exponentially decaying propagators have been observed by employing a dichotomous process for the diffusivity [13]. The latter model consists of a "fast" and a "slow" phases, each one with a diffusion coefficient D_+ and D_-, respectively [13,30]. Furthermore, the appearance of a cusp at small displacements also has been reported in different diffusive approaches like the Sinai model [31], employing the quenched trap model [32–37] or spatial dependence in the diffusivity [38], within the Lévy–Lorentz gas model [39] and using the fractional Fokker–Planck equation [40]. It is important to notice that the cusp found in [31–34,36,38–40] is within the context of anomalous diffusion in the MSD sense, and those presented in [35–37] are for normal diffusive systems.

It is worth mentioning that several systems in nature exhibit, or can be reduced to, a dichotomous process. Examples of two state systems include nuclear magnetic imaging to measure the diffusion of heterogeneous molecules [41], diffusion in glassy materials [1], blinking quantum dots [42,43], diffusion in single molecules tracking experiments [4,7], and protein conformational dynamics [44]. Other approaches for analyzing two state systems were also devised over the years; see heterogeneous molecular transport [41], telegraphic noise [43], Lèvy Flights [45], and CTRW models [1,22].

In this work, we deal with a two state jumping diffusivity model with equilibrium initial conditions, i.e., we assume that the process started long before the measurement began. The long measurement time behavior of the positional PDF for this model is Gaussian and is independent of the specifics of the waiting times at the different diffusive states. A rather unexpected result is achieved for the opposite temporal regime. We obtain that the behavior in the limit of the short measurement times and the shape of the positional PDF of the molecule spreading in the two state jumping diffusivity model attains a cusp or a general tent-like shape. Our result is based on the statistics of the temporal occupation fraction of the diffusivity states, the latter is defined as the time spent in state D_+ over the total measurement time. The Gaussian behavior in the long measurement time is dictated by the δ-function shape of the distribution of this temporal occupation fraction, a feature that is solely based on the ergodic properties of the system. We show that, in the limit of short measurement time, the distribution of the temporal occupation fraction attains a uniform distribution that leads to the mentioned cusp behavior of $P(x,t)$. The uniformity of the occupation fraction is a general result in the sense that it does not depend on the statistics of the waiting times in the two states, and the latter can be arbitrary. The non-Gaussian behavior of $P(x,t)$ for short measurement times is similarly general as the Gaussian behavior of the propagator for long times. We then show that our approach reproduces the results of a specific representative system with exponentially distributed waiting times.

Our manuscript is organized as follows: in Section 2.1, we introduce the jumping diffusivity model and the initial conditions utilized in this work. In Section 2.2, we develop our theory for the statistics of the occupation time in the short measurement time limit, for which the PDFs of the waiting times in states $D_\pm$ are rather general. The obtained behavior of the occupation fraction is used in order to describe the non-Gaussian features of $P(x,t)$, i.e., its cusp shape, that is observed in this model. In Section 2.3, we corroborate our previous results for a system with exponentially distributed waiting times. In Section 3.1, we discuss briefly how these theoretical results differ from those found within the super-statistical approach [12,14,21] and further how our approach may be applicable in experiments. Finally, in Section 4, we present a summary of our results, and we discuss briefly recent work of Postnikov et al. [37] who considered a model with quenched disorder, emphasizing the importance of equilibrium initial conditions. The main derivations are given in the corresponding Appendixes.

2. Results

2.1. The Model

We consider a two state renewal model, with a stochastic diffusion field $D(t)$ for a particle in a random medium. The position of the particle is following a diffusion process given by $dx(t)/dt = \sqrt{2D(t)}\xi$. $D(t) \in \{D_+, D_-\}$ is a dichotomous model, considering the case when $D_- < D_+$ and ξ is a standard white noise, i.e., with mean zero, variance one, and delta correlated. As an example of the dynamics of the model, at a given time, the particle follows a pure diffusion process with a diffusion coefficient $D_+ > 0$ during a period τ. After this time period has elapsed, the diffusion coefficient jumps and, during the next time interval, the particle diffuses with diffusion coefficient D_-. The waiting times at each state $D_\pm$ are distributed according to a general PDF $\psi_\pm(\tau)$, with mean waiting times $\langle\tau\rangle_\pm$. The subscript $\pm$ denotes whether the waiting times are defined for the D_+ or D_- states. In the following, we present the two-state model with $D_- = 0$, while the case with $D_+ > D_- > 0$ is analyzed in Appendix A. In Figure 1, we show representative trajectories for the position at time t, $x(t)$, while, in Figure 2, we present the same for $D(t)$ and we show the notation we use.

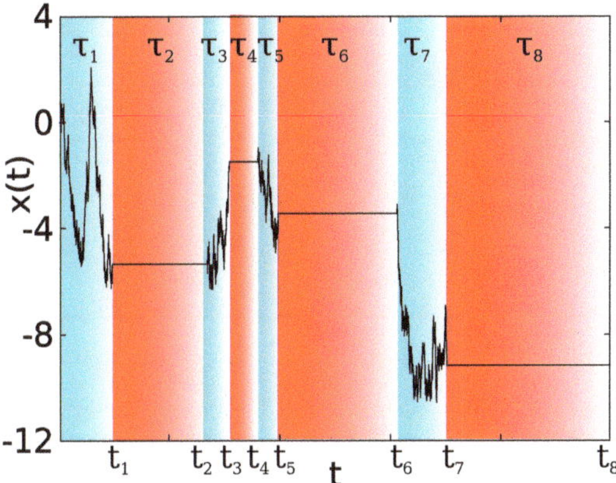

Figure 1. Typical trajectory of $x(t)$ given by Equation (2) with $D_+ = 10$ (blue regions), and $D_- = 0$ (red regions). For this trajectory, exponential waiting times with $\langle\tau\rangle_+ = 1$ and $\langle\tau\rangle_- = 5$ were used.

We define $T_\pm$ as the occupation time in state "$\pm$", namely the total amount of time that the process diffuses with D_+ or D_- during t. Jumps between states D_+ and D_- occur at random times t_1, t_2, etc., until a final measurement time t and clearly $t = T_+ + T_-$. The intervals of time between each jump are defined by $\tau_1 = t_1$, $\tau_2 = t_2 - t_1$, $\tau_3 = t_3 - t_2$, etc., see Figure 2. Then, the occupation times in each state, when started from D_+, are explicitly provided by

$$
\begin{aligned}
T_+ &= \tau_1 + \tau_3 + \ldots + \tau_N \\
T_- &= \tau_2 + \tau_4 + \ldots + \tau_{N-1} + \tau^* \quad if \quad N = 2k+1, \\
T_+ &= \tau_1 + \tau_3 + \ldots + \tau_{N-1} + \tau^*, \\
T_- &= \tau_2 + \tau_4 + \ldots + \tau_N \qquad\qquad if \quad N = 2k,
\end{aligned} \tag{1}
$$

where N is the random number of transitions that were performed between the two states during the measurement time t and k is an integer. The measurement time t and N satisfy $t \geq t_N$, with $t_N = \tau_1 + \tau_2 + \ldots + \tau_N$, i.e., the exact time when the Nth jump was performed. The backward recurrence time τ^* is defined by $\tau^* = t - t_N$ [46]. Each waiting time τ_i follows $\tau_i = t_i - t_{i-1}$ with $i \in (1, N)$. For this particular initial condition, odd values of i in τ_i refer to waiting times at D_+ and even values of i to waiting times during which the diffusion coefficient is D_- (see Figure 2). Expressions similar to Equation (1) are also obtained when the process starts from D_-, see Equation (A65) .

Since the particle is diffusing with a constant diffusion constant D_+ for time τ_1, when starting from D_+, the position $x(\tau_1)$ is simply $x(\tau_1) = \sqrt{2D_+\tau_1}\xi_1$, where ξ_1 is a zero mean Gaussian random variable with $\langle\xi_1^2\rangle = 1$. When at the state with diffusion constant $D_- = 0$, the particle is not moving, therefore $x(t_2) - x(t_1) = 0$ and $x(t_3) - x(t_2) = \sqrt{2D_+\tau_3}\xi_3$, where ξ_3 is a zero mean Gaussian random variable with $\langle\xi_3^2\rangle = 1$ independent of ξ_1. Generally, $x(t_i) - x(t_{i-1}) = \sqrt{2D_\pm\tau_i}\xi_i$, where all ξ_i are independent zero mean Gaussian random variables that satisfy $\langle\xi_i^2\rangle = 1$. By using Equation (1) and exploiting the properties of summation of independent Gaussian variables, we obtain that the position at general time t is provided by

$$x(t) = \sqrt{2D_+T_+}\xi, \tag{2}$$

when $D_+ > 0$ and $D_- = 0$. Equation (2) holds irrespective of the state at $t = 0$. We see that the particles' position is a product of two independent random variables, the square root of the time staying at the state D_+ times a standard Gaussian random variable.

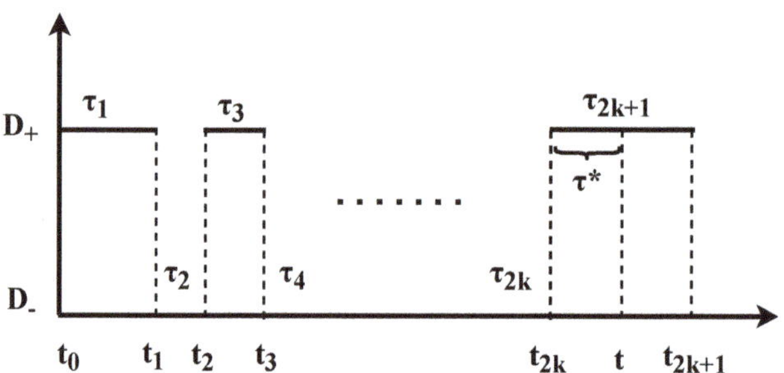

Figure 2. Alternating process for the diffusivity, starting from the state '+' and $N = 2k + 1$. For the case of equilibrium initial conditions exposed in Section 2.2, for $N = 1$, τ_1 works as the forward recurrence time with PDF Equation (10).

In the following, we consider a situation in which the process has started long before the measurement began, i.e., at $t = 0$, the process was already running for a very long time. In this way, the measurement begins from an initial condition in which the system is in equilibrium, meaning that the probability to start from D_+ is $\langle\tau\rangle_+/[\langle\tau\rangle_+ + \langle\tau\rangle_-]$ and accordingly the probability to start from D_- is $\langle\tau\rangle_-/[\langle\tau\rangle_+ + \langle\tau\rangle_-]$ (see [46,47]). For this set-up, the PDF of the occupation time T_+, $f_t(T_+)$, is determined by the contribution to start from D_+ and the contribution to start from D_-, yielding

$$f_t(T_+) = \frac{\langle\tau\rangle_+}{\langle\tau\rangle_+ + \langle\tau\rangle_-} f_t^+(T_+) + \frac{\langle\tau\rangle_-}{\langle\tau\rangle_+ + \langle\tau\rangle_-} f_t^-(T_+), \tag{3}$$

where $f_t^\pm(T_+)$ is the PDF of T_+ for measurement time t, given that the process has started from $\pm$. Since $D_- = 0$, Equation (2) dictates that the positional PDF, provided that the system has occupied the state with D_+ for a time T_+, is given by

$$P(x|T_+) = \frac{e^{-\frac{x^2}{4D_+T_+}}}{\sqrt{4\pi D_+T_+}}. \tag{4}$$

The propagator of the system is obtained via integrating over all possible values of the occupation time T_+, whose PDF is $f_t(T_+)$ Equation (3), yielding

$$P(x,t) = \int_0^t P(x|T_+)f_t(T_+)dT_+. \tag{5}$$

Likewise, we can work with the temporal occupation fraction, which is defined by $p_+ = T_+/t$ with $0 \le p_+ \le 1$. In this case, the positional PDF for a specific value of p_+ follows

$$P(x|p_+) = \frac{e^{-\frac{x^2}{4D_+tp_+}}}{\sqrt{4\pi D_+tp_+}}. \tag{6}$$

and the propagator is obtained similarly to Equation (5), but using the PDF of p_+, which we denote by $g_t(p_+)$,

$$P(x,t) = \int_0^1 P(x|p_+)g_t(p_+)dp_+. \tag{7}$$

Since the properties of $P(x|T_+)$ or $P(x|p_+)$ are known, the task of computing the propagator completely depends on our ability to calculate the PDF of T_+ or p_+. In the following section, we address this problem.

2.2. *The General Case: Arbitrary Distribution of Waiting Times*

Two regimes of the process are of special interest. The long and the short limits of the measurement time t. The two different limits involve different considerations when computing the PDFs of the occupation time (T_+) and fraction (p_+). We first handle the regime of small t and then we treat the $t \to \infty$ limit.

2.2.1. Short Time Regime

The PDF of the occupation time T_+ is defined by Equation (3). We condition on the number of transitions N, and each term $f_t^{\pm}(T_+)$ is provided by

$$f_t^{\pm}(T_+) = \sum_{N=0}^{\infty} f_t^{\pm}(T_+|N)Q_t^{\pm}(N), \tag{8}$$

where $Q_t^{\pm}(N)$ is the probability to perform exactly N transitions during t when the process started at $\pm$. $f_t^{\pm}(T_+|N)$ is the PDF of T_+ when exactly N transitions were performed (during t), and the process has started from $\pm$. This conditional probability is obtained by counting the number of trajectories of temporal span t that started from the $\pm$ state and performed exactly N transitions, out of the total number of trajectories that started from the $\pm$ state and for which the diffusion spent a total time T_+ at this state. Utilizing Equation (8), we rewrite Equation (3) as

$$f_t(T_+) = \frac{\langle\tau\rangle_+}{\langle\tau\rangle_+ + \langle\tau\rangle_-}\sum_{N=0}^{\infty} f_t^{+}(T_+|N)Q_t^{+}(N) + \frac{\langle\tau\rangle_-}{\langle\tau\rangle_+ + \langle\tau\rangle_-}\sum_{N=0}^{\infty} f_t^{-}(T_+|N)Q_t^{-}(N). \tag{9}$$

Since we consider a renewal process, the expression for $Q_t^{\pm}(N)$ is known in the Laplace space [42], as $\hat{Q}_s^{\pm}(N) = \mathcal{L}\{Q_t^{\pm}(N)\} = \int_0^{\infty} Q_t^{\pm}(N)\exp(-ts)\,dt$, for any general $\hat{\psi}_{\pm}(s) = \mathcal{L}\{\psi_{\pm}(\tau)\}$. Concretely, $Q_t^{\pm}(N)$ is obtained by taking into account all the possibilities to perform N jumps up to time $t_N < t$, and no additional jumps during the backward recurrence time τ^*. This sums up to a convolution of $N+1$ random variables. It is important to notice that, since we assume equilibrium initial conditions, τ_1, which is measured from $t = 0$, is only a part of a full renewal event and is termed the forward recurrence time. The PDF of τ_1 for the $\pm$ state, $f_{eq}^{\pm}(\tau_1)$, is provided by (see [46])

$$f_{eq}^{\pm}(\tau_1) = \left(1 - \int_0^{\tau_1} \psi_{\pm}(\tau)\, d\tau\right) / \langle\tau\rangle_{\pm} \tag{10}$$

and in the Laplace space $\mathcal{L}\{f_{eq}^{\pm}(\tau_1)\} = (1 - \hat{\psi}_{\pm}(s)) \big/ \langle\tau\rangle_{\pm} s$. This initial condition stems from the equilibrium of the underlying process, in which we do not have a jump at the initial time ($t_0 = 0$ in Figure 2). In the literature [13,42,46,48,49], the case where the renewal process starts at $t = 0$ is called ordinary or non-equilibrium, and as we will see below, by following our approach, this does not yield any universal features for $P(x,t)$, hence the assumption of an equilibrium process is important in our methodology, (see discussion about non-equilibrium initial conditions in Appendix B).

The probability of not performing any jumps during τ^* is equivalent to the probability of obtaining a waiting time $\tau_{N+1} > \tau^*$, i.e., $1 - \int_0^{\tau^*} \psi_{\pm}(\tau)\, d\tau$. Eventually, by implementing the initial equilibrium condition, we obtain

$$\begin{aligned}
\hat{Q}_s^{\pm}(0) &= \frac{1 - \frac{1-\hat{\psi}_{\pm}(s)}{\langle\tau\rangle_{\pm} s}}{s}, \\
\hat{Q}_s^{\pm}(1) &= \left(\frac{1 - \hat{\psi}_{\pm}(s)}{\langle\tau\rangle_{\pm} s}\right)\left(\frac{1 - \hat{\psi}_{\mp}(s)}{s}\right), \\
\hat{Q}_s^{\pm}(2) &= \left(\frac{1 - \hat{\psi}_{\pm}(s)}{\langle\tau\rangle_{\pm} s}\right)\hat{\psi}_{\mp}(s)\left(\frac{1 - \hat{\psi}_{\pm}(s)}{s}\right), \\
\hat{Q}_s^{\pm}(3) &= \left(\frac{1 - \hat{\psi}_{\pm}(s)}{\langle\tau\rangle_{\pm} s}\right)\hat{\psi}_{\mp}(s)\psi_{\pm}(s)\left(\frac{1 - \hat{\psi}_{\mp}(s)}{s}\right).
\end{aligned} \tag{11}$$

In all the equations above on the right-hand side, we have a multiplication of functions in the Laplace space, this implies convolutions as we transform from s to t. The first term in the multiplication on the right-hand side of Equation (11) obviously stems from the equilibrium initial condition under study. We assume that the PDF of the waiting times is analytic for $\tau \to 0$, thus we can express $\psi_{\pm}(\tau)$ as [22,23]

$$\psi_{\pm}(\tau) \sim C_{A_{\pm}}^{\pm} \tau^{A_{\pm}} + C_{A_{\pm}+1}^{\pm} \tau^{A_{\pm}+1} + \dots, \tag{12}$$

with $A_{\pm} \geq 0$ an integer number. As an example, consider the case with exponential waiting times, i.e., $\psi_{\pm}(\tau) = \psi(\tau) = \exp(-\tau/\langle\tau\rangle)/\langle\tau\rangle$, namely the waiting times at the $D_{\pm}$ states are identically distributed. Its analytic expansion is $\psi(\tau) \sim 1/\langle\tau\rangle - \tau/\langle\tau\rangle^2$, with $A_{\pm} = 0$, $C_{A_{\pm}}^{\pm} = 1/\langle\tau\rangle$ and $C_{A_{\pm}+1}^{\pm} = 1/\langle\tau\rangle^2$. The analyticity of $\psi_{\pm}(\tau)$ Equation (12) is a very mild demand that covers a wide range of sojourn times distributions. Since we are interested in the small t limit, the corresponding behavior in the Laplace space is found for $s \to \infty$, where the leading terms of $\hat{\psi}_{\pm}(s)$ are [22]

$$\hat{\psi}_{\pm}(s) \sim \frac{\Gamma(A_{\pm}+1)C_{A_{\pm}}^{\pm}}{s^{A_{\pm}+1}} + \frac{\Gamma(A_{\pm}+2)C_{A_{\pm}+1}^{\pm}}{s^{A_{\pm}+2}} + \dots, \tag{13}$$

For the mentioned example with exponential waiting times, $\hat{\psi}(s) \sim 1/[\langle\tau\rangle s]$. Using Equation (13) for $\hat{Q}_s^{\pm}(N)$, we obtain that, in the $s \to \infty$ limit, corresponding to the short time limit, which is at the focus of our interest

$$
\begin{aligned}
\hat{Q}_s^{\pm}(0) &\sim \frac{1}{s} - \frac{1}{\langle\tau\rangle_{\pm} s^2} + \frac{\Gamma(A_{\pm}+1)C_{\pm}^{\pm}}{\langle\tau\rangle_{\pm} s^{A_{\pm}+3}} + \ldots, \\
\hat{Q}_s^{\pm}(1) &\sim \frac{1}{\langle\tau\rangle_{\pm} s^2} - \frac{2C_{A_{\pm}}^{\pm}\Gamma(A_{\pm}+1)}{\langle\tau\rangle_{\pm} s^{A_{\pm}+3}} + \ldots \\
\hat{Q}_s^{\pm}(2) &\sim \frac{\Gamma(A_{\mp}+1)C_{A_{\mp}}^{\mp}}{\langle\tau\rangle_{\pm} s^{A_{\mp}+3}} + \ldots \\
\hat{Q}_s^{\pm}(3) &\sim \frac{\Gamma(A_{\pm}+1)\Gamma(A_{\mp}+1)C_{A_{\pm}}^{\pm}C_{A_{\mp}}^{\mp}}{\langle\tau\rangle_{\pm} s^{A_{\pm}+A_{\mp}+4}} + \ldots.
\end{aligned}
\tag{14}
$$

We see that the leading terms for all $\hat{Q}_s^{\pm}(N)$ with $N > 1$ are of the order $1/s^{\gamma}$ with $\gamma > 2$. Thus, in the small t limit, terms with $N > 1$ contain contributions that scale like $t^{\gamma-1}$ and are negligible with respect to the $N \in \{0, 1\}$ cases. Therefore, only the first two $Q_t^{\pm}(N)$s are taken into account, i.e.,

$$Q_t^{\pm}(0) \sim 1 - \frac{t}{\langle\tau\rangle_{\pm}}, \tag{15}$$

$$Q_t^{\pm}(1) \sim \frac{t}{\langle\tau\rangle_{\pm}}. \tag{16}$$

This is an expected result, as, for short times, only contributions from a single transition and zero transitions are important. By calculating $Q_t^{\pm}(N)$, we advanced towards obtaining the behavior of the PDF of T_+, according to Equation (9), in order to complete this mission, one needs to compute the relevant contributions of $f_t^{\pm}(T_+|N)$ in the $t \to 0$ limit. First, we see that the conditional distribution $f_t^{\pm}(T_+|0)$ depends only on the starting state. There are only two types of trajectories that have performed 0 transitions, i.e., for all the time, they have been either at D_+ or at D_-. Consequently,

$$f_t^{+}(T_+|0) = \delta(t - T_+), \tag{17}$$

$$f_t^{-}(T_+|0) = \delta(T_+). \tag{18}$$

The calculation of $f_t^{\pm}(T_+|N)$ is obtained by conditioning over the first event. If starting from the $+$ state, the process will spend a time τ_1 at this state before jumping to the $-$ state. τ_1 can attain any value $0 \le \tau_1 \le T_+$ and for the remaining time $t - \tau_1$ the process has to perform one transition less. In general, without regarding the initial conditions of the problem, an integration over all possible τ_1's provides the relation

$$f_t^{+}(T_+|N'+1) = \int_0^{T_+} \frac{1}{B_+}\psi_+(\tau_1) f_{t-\tau_1}^{-}(T_+ - \tau_1|N')\, d\tau_1 \tag{19}$$

with $N' + 1 = N$, and B_+ a normalization factor. For instance, for $N = 1$, we have that $\int_0^t \psi_+(\tau_1)/B_+\, d\tau_1 = 1$, which stems from the fact that we consider only trajectories of time span t. The corresponding formula for $f_t^{-}(T_+|N'+1)$ is

$$f_t^{-}(T_+|N'+1) = \int_0^{t-T_+} \frac{1}{B_-}\psi_-(\tau_1) f_{t-\tau_1}^{+}(T_+|N')\, d\tau_1. \tag{20}$$

Since we are assuming equilibrium initial conditions, the $\psi_{\pm}$ in the $N' + 1$ element of the iterative forms (Equations (19) and (20)) must be replaced by $f_{eq}^{\pm}$ (Equations (10)). As was already noted above, only the $N = 0$ and $N = 1$ are of interest in the small t limit, then, according to Equations (17), (20), and (10), we get for $N = 1$

$$f_t^+(T_+|1) = \frac{f_{eq}^+(T_+)}{\int_0^t f_{eq}^+(t')\,dt'}, \tag{21}$$

$$f_t^-(T_+|1) = \frac{f_{eq}^-(t-T_+)}{\int_0^t f_{eq}^-(t')\,dt'}. \tag{22}$$

Using the small time approximation of $\psi_\pm(\tau)$ Equation (12), in Equations (21) and (22), we obtain that, independently of the starting state,

$$f_t^\pm(T_+|1) \sim \frac{1}{t}. \tag{23}$$

The $1/t$ dependence comes from the integral factors in Equations (21) and (22), all the other terms in the numerator and denominator simply cancel out. See Appendix B for a complementary derivation of Equation (23) using the definition of the joint PDF of T_+ and N. Gathering Equations (15)–(18), and (23) in Equation (9), we find that

$$\begin{aligned} f_t(T_+) &\sim \frac{\langle\tau\rangle_+}{\langle\tau\rangle_+ + \langle\tau\rangle_-}\left(1-\frac{t}{\langle\tau\rangle_+}\right)\delta(t-T_+) \\ &+ \frac{\langle\tau\rangle_-}{\langle\tau\rangle_+ + \langle\tau\rangle_-}\left(1-\frac{t}{\langle\tau\rangle_-}\right)\delta(T_+) + \frac{2}{\langle\tau_+\rangle + \langle\tau_-\rangle}. \end{aligned} \tag{24}$$

The PDF of the occupation fraction is obtained trivially from Equation (24) by changing variables to $p_+ = T_+/t$

$$\begin{aligned} g_t(p_+) &\sim \frac{\langle\tau\rangle_+}{\langle\tau\rangle_+ + \langle\tau\rangle_-}\left(1-\frac{t}{\langle\tau\rangle_+}\right)\delta(1-p_+) \\ &+ \frac{\langle\tau\rangle_-}{\langle\tau\rangle_+ + \langle\tau\rangle_-}\left(1-\frac{t}{\langle\tau\rangle_-}\right)\delta(p_+) + \frac{2t}{\langle\tau\rangle_+ + \langle\tau\rangle_-}. \end{aligned} \tag{25}$$

The third term in Equations (24) and (25) is uniform, i.e., terms which are independent of T_+ or p_+, and this is the first main result of this paper. All the additional terms and contributions to the PDF of p_+ only introduce terms that depend on higher orders of t and are thus negligible in the small t limit. This means that, for equilibrium initial conditions, regardless of the exact form of $\psi_\pm(\tau)$, the PDF of p_+ (Equation (25)) is always uniform for $0 < p_+ < 1$. This general uniform behavior of the PDF of the occupation fraction is applicable for an extremely large class of waiting times PDFs $\psi_\pm(\tau)$. As a remark, the connection between the conditional PDF of T^+, $f_t^\pm(T_+|N)$, and the joint PDF of T^+ and N, $f_t^\pm(T_+, N)$ is discussed in Appendix B.1. In the following, it is shown that this uniformity leads to universal features of the propagator in the limit of small t. In Sections 2.3.1 and 2.3.2, we treat particular examples (with exponential waiting times) that are exactly tractable, without any simplifications or assumptions. The results agree perfectly with the general form in Equation (25). It is important to notice that our approximations affect only the form of $g_t(p_+)$ and do not affect $P(x|p_+)$. This allows us to obtain the behavior of $P(x,t)$ for any $-\infty < x < \infty$, as is shown below.

2.2.1.1. $P(x,t)$ for Arbitrary Waiting Times

In order to obtain the positional PDF for small t, we combine Equations (6), (7), and (25), which, after integration, gives

$$P(x,t) = \frac{\langle\tau\rangle_+}{\langle\tau\rangle_+ + \langle\tau\rangle_-}\left(1 - \frac{t}{\langle\tau\rangle_+}\right)\frac{e^{-\frac{x^2}{4D_+t}}}{\sqrt{4\pi D_+ t}} + \frac{\langle\tau\rangle_-}{\langle\tau\rangle_+ + \langle\tau\rangle_-}\left(1 - \frac{t}{\langle\tau\rangle_-}\right)\delta(x)$$
$$+ \left(\frac{2t}{\langle\tau\rangle_+ + \langle\tau\rangle_-}\right)\left\{\frac{e^{-\frac{x^2}{4D_+t}}}{\sqrt{\pi D_+ t}} - \frac{|x|}{2D_+t}\left(1 - Erf\left(\frac{|x|}{\sqrt{4D_+t}}\right)\right)\right\}. \tag{26}$$

Considering $x \neq 0$, in the limit of $x \longrightarrow 0$ when $\exp(-x^2/4D_+t) \sim 1 - x^2/4D_+t$ and $1 - Erf(|x|/\sqrt{4D_+t}) \sim 1 - 2|x|/\sqrt{4\pi D_+ t}$. After substituting in Equation (26), it turns into

$$P(x,t) \sim \frac{(3t + \langle\tau\rangle_+)}{\sqrt{4\pi D_+ t}[\langle\tau\rangle_+ + \langle\tau\rangle_-]} - \frac{|x|}{D_+[\langle\tau\rangle_+ + \langle\tau\rangle_-]} + K_1 x^2, \tag{27}$$

with $K_1 = (5t - \langle\tau\rangle_+)/[8(\langle\tau\rangle_+ + \langle\tau\rangle_-)\sqrt{\pi}(D_+t)^{\frac{3}{2}}]$. We can see that, in Equation (27), there is a linear dependence on $|x|$ in the vicinity of $x = 0$. This means that for short enough measurement times the PDF of x will always have a tent like shape, irrespective of the distributions $\psi_\pm$ that were chosen (see Figure 3 below). Only the mean sojourn times affect the shape. This is a general result for the short time regime, and it is based on the general fact that the PDF of the temporal occupation fraction is uniform for $0 < p_+ < 1$. Concretely, at short times when $|x|$ is small, the decay of $P(x,t)$ will always resemble an exponential one. For large $|x|$, the form of $P(x,t)$ must be Gaussian, due to the fact that this limit is determined by the instances when no transition to D_- was ever made and the transport is controlled by diffusion with D_+. However, if we only look at the particles that have moved, i.e., we get rid of the delta function at $x = 0$ in Equation (26). We can relate these dynamics with some experiments which condition the measurements on the movement of the particles. This procedure is called population splitting; see [50,51]. Technically, if $D_- > 0$, the cusp is not found; however, as long as $D_-/D_+ << 1$, the tent like shape will be found; for further details, see Appendix A.

2.2.2. Long Time Regime

In the limit $t \longrightarrow \infty$, the PDF of the temporal occupation fraction $g_t(p_+)$ follows a different but also a general form. As mentioned, we are focusing on the case where both $\psi_\pm$ have finite first moments, $\langle\tau\rangle_\pm > 0$. In the long time limit, ergodicity is satisfied, namely the equivalence of *ensemble* and temporal averages are attained. Particularly, in this case, the *ensemble* average of the occupation fraction at D_+ is equal to the temporal average which is defined by the fraction of average waiting times at D_+ and D_-, i.e., $\langle p_+\rangle = \langle\tau\rangle_+/[\langle\tau\rangle_+ + \langle\tau\rangle_-]$ (see Appendix F). Thus, in the long time limit, $g_t(p_+)$ converges to a δ-function,

$$g_t(p_+) \underset{t\to\infty}{\longrightarrow} \delta\left(p_+ - \frac{\langle\tau\rangle_+}{\langle\tau\rangle_+ + \langle\tau\rangle_-}\right). \tag{28}$$

Since ergodicity prevails, by using Equation (28) in Equation (7), the positional PDF gets the form

$$P(x,t) = \sqrt{\frac{\langle\tau\rangle_+ + \langle\tau\rangle_-}{4\pi D_+ t\langle\tau\rangle_+}}e^{-\frac{x^2(\langle\tau\rangle_+ + \langle\tau\rangle_-)}{4D_+t\langle\tau\rangle_+}}. \tag{29}$$

In the long time limit, the positional PDF given by Equation (29) represents a Gaussian propagator with an effective diffusion coefficient $D_+\langle\tau\rangle_+/[\langle\tau\rangle_+ + \langle\tau\rangle_-]$. Since $\langle\tau\rangle_+/[\langle\tau\rangle_+ + \langle\tau\rangle_-] < 1$, and the effective diffusion coefficient is always smaller compared with D_+. Indeed, this slow-down is an expected result due to the portion of the time that the particle spends in the state with $D_- = 0$ and basically not moving during this period.

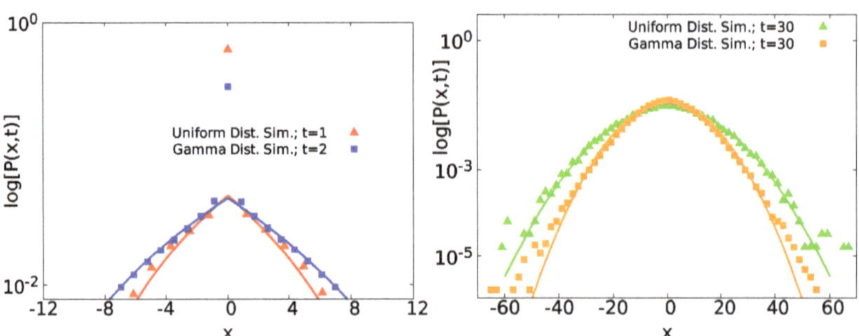

Figure 3. Distribution of displacements $P(x,t)$ in semi-log scale, obtained by simulations, for a two state system with uniform distributed waiting times and gamma distributed waiting times. The left panel presents short time results where a tent like shape is clearly visible and a non-analytical feature is obvious, while the right panel exhibits Gaussian statistics for long times. Left: $P(x,t)$ for $t=1$ for $\tau \sim U(0,5)$ at D_+ and $\tau \sim U(0,10)$ at D_- (red triangles)—with $\langle\tau\rangle_+ = 2.5 < \langle\tau\rangle_- = 5$. In addition, $t=2$ with $\tau \sim Gamma(0.5,8)$ at D_+ and $\tau \sim Gamma(0.5,12)$ at D_- (blue squares), such that $\langle\tau\rangle_+ = 4 < \langle\tau\rangle_- = 6$. Both cases fit with Equation (26) (red and blue solid lines) with a tent like shape. In both normalized histograms at $x=0$, there is a peak representing the Dirac delta function in Equation (26). Right: $P(x,t)$ for $t=30$ and waiting times uniformly distributed (green triangles) with the same parameters as above and for gamma distributed waiting times (orange squares) with $\tau \sim Gamma(2,1)$ at D_+, $\tau \sim Gamma(8,1)$ at D_-, and $\langle\tau\rangle_+ = 2 < \langle\tau\rangle_- = 8$. We employed the last set of parameters in the gamma distributed waiting times in order to avoid an overlapping between curves. $P(x,t)$ converges to the Gaussian statistics Equation (29) (green and orange solid lines). In all the presented cases, $D_+ = 10$ and $D_- = 0$ were used.

2.2.3. Simulations

The two general limits of $g_t(p_+)$ Equations (25) and (28) produce two different prevailing distributions of $P(x,t)$ Equations (26) and (29). In Figure 3, we compare analytical formulas Equations (26) and (29) (solid lines) with simulations of two different state models—one with uniform distributed waiting times $\tau \sim U(0,5)$ for D_+ and $\tau \sim U(0,10)$ for D_- and with $t=1$ (red triangles) and $t=30$ (green triangles), such that $\langle\tau\rangle_+ = 2.5 < \langle\tau\rangle_- = 5$. Here, the notation $\tau \sim U(a,b)$ means that τ has a uniform distribution with a and b the minimum and maximum values, respectively. In addition, the other with gamma distributed waiting times, such that $\tau \sim Gamma(k,\theta)$. The latter notation denotes that τ has a gamma distribution with k its shape parameter and θ the corresponding scale parameter. In this case, the PDF follows

$$\psi_{\pm}(\tau) = \frac{\tau^{k-1} e^{-\frac{\tau}{\theta}}}{\Gamma(k)\theta^k}, \tag{30}$$

particularly the PDF of the gamma distribution Equation (30) implies a cumulative distribution function $F(\tau) = \gamma(k,\tau/\theta)/\Gamma(k)$, with $\gamma(x,y)$ the incomplete gamma function and $\Gamma(x)$ the standard gamma function. For the latter case, we used $\tau \sim Gamma(0.5,8)$ at D_+ and $\tau \sim Gamma(0.5,12)$ at D_-, for $t=2$ (blue squares) and $t=30$ (orange squares), with $\langle\tau\rangle_+ = 4 < \langle\tau\rangle_- = 6$. As we can see in the short time regime, for uniform and gamma distributed waiting times (red triangles and blue squares), $P(x,t)$ has a tent shape for short displacements, and it agrees with Equation (26), joined with a peak at $x=0$ due to the Dirac delta function in Equation (26). For $t=30$ (green triangles and orange squares), each case of $P(x,t)$ converges to Gaussian statistics (Equation (29)).

The cusp we have found for small $|x|$ implies that we may approximate the distribution on a small scale with a Laplace like distribution, $P(x,t) \sim \exp(-C|x|)$. However, clearly this does not hold globally for large x, see Figure A2 in Appendix C. Still within the interval of short displacements, due to the presence of the delta peak at the origin, we expect for this span a considerable contribution on the normalization of $P(x,t)$. Particularly, we find that the area underneath the curve for the case of uniformly distributed waiting times (red line) in Figure 3 in the left panel has a value of 0.88 for $x \in (-4,4)$. In addition, the corresponding area within the same figure, but, for gamma distributed waiting times, the (blue curve) has a value of 0.89 for $x \in (-8,8)$.

2.3. Exponentially Distributed Waiting Times

In this section, we obtain $g_t(p_+)$ for a specific distribution of waiting times, but using different methods, which let us corroborate the validity of our general approach described above. We analyze the case of exponential waiting times in states with D_+ and D_-, each waiting time following a PDF given by

$$\psi_\pm(\tau) = \frac{e^{-\frac{\tau}{\langle\tau\rangle_\pm}}}{\langle\tau\rangle_\pm}. \tag{31}$$

We show first the case of a two state system with the same mean waiting times and then investigate the complimentary case. In Appendix D, we analyze both cases for non-equilibrium initial conditions, e.g., a system starting from D_+.

2.3.1. Equal Mean Waiting Times $\langle\tau\rangle_+ = \langle\tau\rangle_-$

Let us consider a system with $\langle\tau\rangle_+ = \langle\tau\rangle_- = \langle\tau\rangle$. We know that the temporal fraction occupation p_+ and T_+ can be related to the difference of occupation times defined by $S_t = T_+ - T_-$, as $S_t = 2T_+ - t = 2p_+t - t$ [46]. In this section, we analyze the double Laplace transform of the PDF of S_t, called $\phi_t(S_t)$, with Laplace pairs $S_t \Leftrightarrow v$ and $t \Leftrightarrow s$. In [46], $\phi_t(S_t)$ is provided by

$$\hat{\phi}_s(v) = \frac{s[1-\psi(s+v)\psi(s-v)] + v[\psi(s+v) - \psi(s-v)]}{(s^2-v^2)[1-\psi(s+v)\psi(s-v)]}. \tag{32}$$

The Laplace transform of $\psi(\tau)$ in Equation (31) is given by $\mathcal{L}\{\psi(\tau)\} = \hat{\psi}(s) = \frac{1}{1+\langle\tau\rangle s}$. Substituting $\hat{\psi}(s)$ in Equation (32), we obtain

$$\hat{\phi}_s(v) = \frac{s+2\langle\tau\rangle}{s^2+2\langle\tau\rangle s - v^2}. \tag{33}$$

In Appendix E, an analytical expression for the PDF of S_t is found, i.e., inverse Laplace transform of Equation (33) is performed (see Equation (A44)). Then, remember that the temporal occupation fraction in the plus state p_+ is related to the difference of occupation times as $S_t = 2p_+t - t$. We can employ Equation (A44) for obtaining the PDF of p_+, which is given by

$$\begin{aligned} g_t(p_+) &= \tfrac{1}{2}e^{-\frac{t}{\langle\tau\rangle}}\Big\{\delta(1-p_+)+\delta(p_+)\Big\} \\ &+ \frac{t\Theta(t-|2p_+t-t|)e^{-\frac{t}{\langle\tau\rangle}}}{\langle\tau\rangle}\left[I_0\left(\frac{2t\sqrt{p_+(1-p_+)}}{\langle\tau\rangle}\right) + \frac{I_1\left(\frac{2t\sqrt{p_+(1-p_+)}}{\langle\tau\rangle}\right)}{2\sqrt{p_+(1-p_+)}}\right]. \end{aligned} \tag{34}$$

A similar expression for a system with non-equilibrium initial conditions (always starting from D_+) is found in Appendix D. Expanding Equation (34) in the short time limit $t \longrightarrow 0$, i.e., $t << \langle\tau\rangle$, Equation (34) can be approximated by a uniform distribution

$$g_t(p_+) \sim \frac{e^{-\frac{t}{\langle\tau\rangle}}}{2}\Big\{\delta(1-p_+)+\delta(p_+)\Big\} + \frac{t}{\langle\tau\rangle}. \tag{35}$$

For $\langle\tau\rangle_+ = \langle\tau\rangle_- = \langle\tau\rangle$, Equation (35) agrees with Equation (25) obtained by the general approach of Secton 2.2. In the left panel of Figure 4, we show the short time approximation of $g_t(p_+)$ (Equation (35)) compared with the general formula in Equation (34); it is evident that both results agree perfectly. In the right panel of Figure 4, we show Equation (34) for short and long measurement times. $g_t(p_+)$ evolves from a uniform distribution to a peaked distribution centered at its mean value $p_+ = 1/2$ (see Appendix E for a deduction of the central moments of $g_t(p_+)$).

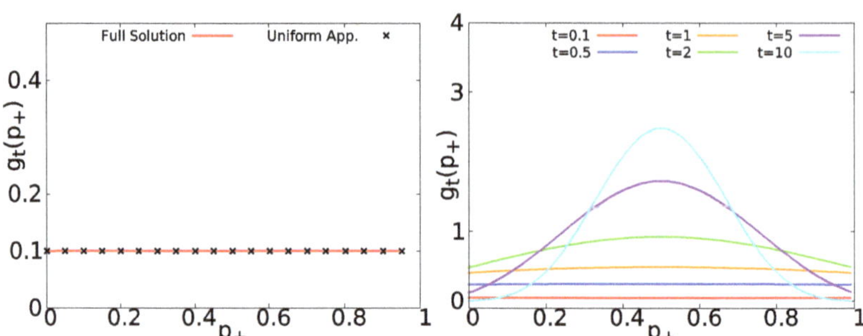

Figure 4. Left: Comparison between $g_t(p_+)$ Equation (34) (red solid line) and the short time uniform approximation Equation (35) (black asterisks) for exponentially distributed waiting times Equation (31) with $\langle\tau\rangle_\pm = \langle\tau\rangle = 1$ and $t = 0.1$. Right: $g_t(p_+)$ Equation (34) for $\langle\tau\rangle = 1$ and $t \in \{0.1, 0.5, 1, 2, 5, 10\}$.

Positional Distribution Function

An analytical expression for the positional distribution function $P(x,t)$ (given by Equation (7)), with $g_t(p_+)$ provided by Equation (34), can be deduced by using the series representation of the modified Bessel functions, $I_\nu(y) = \sum_{k=0}^{\infty} (\frac{y}{2})^{2k+\nu}/[k!\Gamma(\nu+k+1)]$. The integration in Equation (7) yields

$$P(x,t) = \frac{e^{-\frac{t}{\langle\tau\rangle}-\frac{x^2}{4Dt}}}{2\sqrt{4\pi Dt}} + \frac{\delta(x)e^{-\frac{t}{\langle\tau\rangle}}}{2} + \frac{te^{-\frac{t}{\langle\tau\rangle}-\frac{x^2}{4Dt}}}{2\langle\tau\rangle\sqrt{4\pi Dt}}\left\{\sum_{k=0}^{\infty}\frac{(-1)^k\pi}{k!}\left(\frac{t}{\langle\tau\rangle}\right)^{2k}\left[\frac{{}_1F_1(k+1;\frac{1}{2}-k;\frac{x^2}{4DT})}{\Gamma(2k+\frac{3}{2})\Gamma(\frac{1}{2}-k)} - \frac{(\frac{x^2}{4Dt})^{k+\frac{1}{2}}\,{}_1F_1(2k+\frac{3}{2};k+\frac{3}{2};\frac{x^2}{4Dt})}{\Gamma(k+1)\Gamma(k+\frac{3}{2})}\right] + \frac{1}{2}\sum_{k=0}^{\infty}\frac{(-1)^k\pi}{(k+1)!}\left(\frac{t}{\langle\tau\rangle}\right)^{2k+1}\left[\frac{{}_1F_1(k+1;\frac{1}{2}-k;\frac{x^2}{4Dt})}{\Gamma(2k+\frac{3}{2})\Gamma(\frac{1}{2}-k)} - \frac{(\frac{x^2}{4Dt})^{k+\frac{1}{2}}\,{}_1F_1(2k+\frac{3}{2};k+\frac{3}{2};\frac{x^2}{4Dt})}{\Gamma(k+1)\Gamma(k+\frac{3}{2})}\right]\right\}, \tag{36}$$

with ${}_1F_1(a;b;z)$ the confluent hypergeometric function of the first kind. Nonetheless, in a short time limit, we can use the uniform approximation of $g_t(p_+)$ (Equation (35)), and then Equation (7) provides

$$P(x,t) \sim \frac{e^{-\frac{t}{\langle\tau\rangle}-\frac{x^2}{4D_+t}}}{2\sqrt{4\pi D_+t}} + \frac{\delta(x)e^{-\frac{t}{\langle\tau\rangle}}}{2} + \frac{t}{\langle\tau\rangle}\left\{\frac{2e^{-\frac{x^2}{4D_+t}}}{\sqrt{4\pi D_+t}} - \frac{|x|}{2D_+t}\left[1 - Erf\left(\frac{|x|}{\sqrt{4D_+t}}\right)\right]\right\}, \tag{37}$$

which agrees with the results obtained above in Equation (26), when $\langle\tau\rangle_+ = \langle\tau\rangle_-$ and for $t \longrightarrow 0$, since, in that limit, $\exp(-t/\langle\tau\rangle) \sim 1 - t/\langle\tau\rangle$. Particularly for $x \neq 0$ and taking $x \longrightarrow 0$, Equation (37) yields a tent shaped propagator described by

$$P(x,t) \sim \frac{3t + \langle\tau\rangle}{4\langle\tau\rangle\sqrt{\pi D_+t}} - \frac{|x|}{2D_+\langle\tau\rangle} + K_2x^2, \tag{38}$$

with $K_2 = (5t - \langle\tau\rangle)/[16\langle\tau\rangle\sqrt{\pi}(D_+t)^{\frac{3}{2}}]$ and in concordance with Equation (27). On the other hand, within this short time limit, for large displacements $x \longrightarrow \infty$, the two terms between curly braces in Equation (37) cancel each other, and only the first term in Equation (37) is left (when $x \neq 0$). This is due to the expansion of $1 - Erf(z) \sim \exp(-z^2)/(\sqrt{\pi}z)$ for $z \longrightarrow \infty$, in our case $z = |x|/\sqrt{4D_+t}$. Then, Equation (37) can be approximated by

$$P(x,t) \underset{x\to\infty}{\sim} \frac{e^{-\frac{t}{\langle\tau\rangle}-\frac{x^2}{4D_+t}}}{2\sqrt{4\pi D_+t}}. \tag{39}$$

This Gaussian behavior of $P(x,t)$ at the tails is expected. The large $|x|$ limit is dominated by trajectories for which no transitions to D_- were performed and a pure diffusion process with D_+ occurs.

When $t >> \langle\tau\rangle$, ergodicity is satisfied and, therefore, the system on average visits the two states the same amount of time. Namely, the *ensemble* average of p_+ is equal to the corresponding fraction of the average waiting times. In this case, when $\langle\tau\rangle_+ = \langle\tau\rangle_-$, the occupation fraction is concentrated at $p_+ = 1/2$. Thus, the PDF of p_+ is represented by the delta function

$$g_t(p_+) \xrightarrow[t\to\infty]{} \delta\Big(p_+ - \frac{1}{2}\Big). \tag{40}$$

Substituting Equation (40) in Equation (7), we recover Gaussian statistics for the displacements

$$P(x,t) \sim \frac{e^{-\frac{x^2}{2D_+t}}}{\sqrt{2\pi D_+t}}. \tag{41}$$

In Figure 5, we present the two different limit distributions for $P(x,t)$ in the short time limit $t = 0.1$ (red circles) and $t = 0.5$ (blue crosses) Equation (37) and the Gaussian limit for $t = 5$ (orange circles) and $t = 10$ (green crosses) Equation (41), for the normalized variable $z = x/\sqrt{t}$. As we can see, the displacements for short times follow a tent shape (black solid line) and a Gaussian one in the long time limit (magenta solid line).

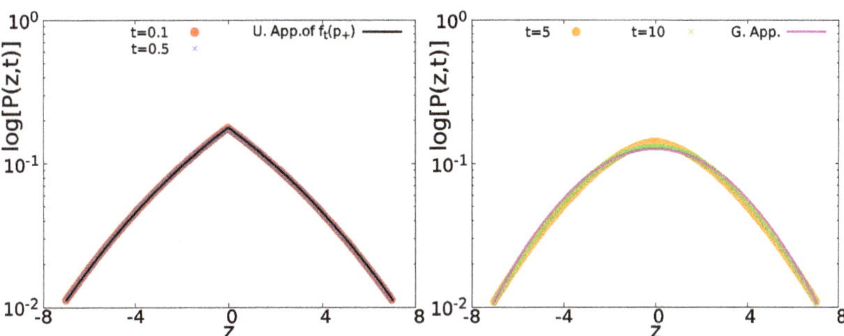

Figure 5. $P(z,t)$ in semi-log scale, with $z = x/\sqrt{t}$. Left: For short times $t = 0.1$ (red circles) and $t = 0.5$ (blue crosses), $P(z,t)$ is represented by Equation (37) (black solid line) with a tent like shape. Right: The same for large times $t = 5$ (orange circles) and $t = 10$ (green crosses), $P(z,t)$ converges to the Gaussian distribution Equation (41) (magenta solid line). In all the cases, $D_+ = 10$, $D_- = 0$, and $\langle\tau\rangle = 1$ were used.

2.3.2. Different Mean Waiting Times $\langle\tau\rangle_+ \neq \langle\tau\rangle_-$

Relaxing the assumption of equal mean waiting times for exponentially distributed sojourn times in the model, we have that $\langle\tau\rangle_+ \neq \langle\tau\rangle_-$, with waiting times following Equation (31). As mentioned, for equilibrium initial conditions, the PDF of T_+ is given by Equation (3). Let $\hat{f}_s^{\pm}(u)$ be the double Laplace transform of $f_t^{\pm}(T_+)$, defined as $\hat{f}_s^{\pm}(u) = \int_0^\infty \int_0^\infty f_t(T_+) \exp(-uT_+ - st)\, dT_+\, dt$. Then, the different terms of the PDF of T_+ in Equation (3) are provided in Laplace space, by [42,47,49]

$$\hat{f}_s^{+}(u) = \left\{\hat{\psi}_+(s+u)\left[\frac{1-\hat{\psi}_-(s)}{s}\right] + \frac{1-\hat{\psi}_+(s+u)}{s+u}\right\}\frac{1}{1-\hat{\psi}_+(s+u)\hat{\psi}_-(s)}, \quad (42)$$

$$\hat{f}_s^{-}(u) = \left\{\hat{\psi}_-(s)\left[\frac{1-\hat{\psi}_+(s+u)}{s+u}\right] + \frac{1-\hat{\psi}_-(s)}{s}\right\}\frac{1}{1-\hat{\psi}_+(s+u)\hat{\psi}_-(s)}. \quad (43)$$

Summing up Equations (42) and (43) according to Equation (3), we obtain, for exponentially distributed waiting times,

$$\hat{f}_s(u) = \frac{\langle\tau\rangle_-^2 + \langle\tau\rangle_+^2(1+\langle\tau\rangle_- s) + \langle\tau\rangle_+\langle\tau\rangle_-[2+\langle\tau\rangle_-(s+u)]}{(\langle\tau\rangle_+ + \langle\tau\rangle_-)[\langle\tau\rangle_- s + \langle\tau\rangle_+(1+\langle\tau\rangle_- s)(s+u)]}. \quad (44)$$

Taking the double inverse Laplace transform of Equation (44) with respect to $u \Leftrightarrow T_+$ and $s \Leftrightarrow t$ and changing variables to $p_+ = T_+/t$, we obtain the PDF for p_+ (see details in Appendix F)

$$g_t(p_+) = \frac{\langle\tau\rangle_- e^{-\frac{t}{\langle\tau\rangle_-}}}{\langle\tau\rangle_+ + \langle\tau\rangle_-}\delta(p_+) + \frac{\langle\tau\rangle_+ e^{-\frac{t}{\langle\tau\rangle_+}}}{\langle\tau\rangle_+ + \langle\tau\rangle_-}\delta(1-p_+) + \frac{2t}{\langle\tau\rangle_+ + \langle\tau\rangle_-}\left\{ I_0\left(2t\sqrt{\frac{p_+(1-p_+)}{\langle\tau\rangle_+\langle\tau\rangle_-}}\right) + \left[\frac{(1-p_+)\sqrt{\langle\tau\rangle_+\langle\tau\rangle_-}}{\langle\tau\rangle_+} + \frac{p_+\sqrt{\langle\tau\rangle_+\langle\tau\rangle_-}}{\langle\tau\rangle_-}\right]\frac{I_1\left(2t\sqrt{\frac{p_+(1-p_+)}{\langle\tau\rangle_+\langle\tau\rangle_-}}\right)}{2\sqrt{p_+(1-p_+)}}\right\} e^{-\frac{tp_+}{\langle\tau\rangle_+} - \frac{t(1-p_+)}{\langle\tau\rangle_-}}. \quad (45)$$

For the case when $\langle\tau\rangle_+ = \langle\tau\rangle_- = \langle\tau\rangle$, Equation (45) recovers Equation (34) obtained by the methods reported in [52,53]. The case of non-equilibrium initial conditions is shown in Appendix D.

In the short time regime, strictly speaking when $t << \langle\tau\rangle_\pm$, by expanding Equation (45) for $t \longrightarrow 0$, $g_t(p_+)$ can be approximated by the uniform distribution

$$g_t(p_+) \sim \frac{\langle\tau\rangle_- e^{-\frac{t}{\langle\tau\rangle_-}}}{\langle\tau\rangle_+ + \langle\tau\rangle_-}\delta(p_+) + \frac{\langle\tau\rangle_+ e^{-\frac{t}{\langle\tau\rangle_+}}}{\langle\tau\rangle_+ + \langle\tau\rangle_-}\delta(1-p_+) + \frac{2t}{\langle\tau\rangle_+ + \langle\tau\rangle_-}. \quad (46)$$

As mentioned above, Equation (25) that was deduced for general PDFs of waiting times, encloses the particular case of Equation (46). For the uniform approximation of $g_t(p^+)$ (Equation (46)), the positional PDF (Equation (7)) is

$$P(x,t) \sim \frac{\langle\tau\rangle_+ e^{-\frac{t}{\langle\tau\rangle_+} - \frac{x^2}{4D_+t}}}{(\langle\tau\rangle_+ + \langle\tau\rangle_-)\sqrt{4\pi D_+ t}} + \frac{\langle\tau\rangle_-}{\langle\tau\rangle_+ + \langle\tau\rangle_-} e^{-\frac{t}{\langle\tau\rangle_-}}\delta(x) + \frac{2te^{-\frac{x^2}{4D_+t}}}{(\langle\tau\rangle_+ + \langle\tau\rangle_-)\sqrt{\pi D_+ t}} - \frac{|x|}{D_+(\langle\tau\rangle_+ + \langle\tau\rangle_-)}\left[1 - Erf\left(\frac{|x|}{\sqrt{4D_+t}}\right)\right], \quad (47)$$

which agrees with the general case described by Equation (26).

Similar to Section 2.2, in the limit $t \longrightarrow \infty$, the PDF of the occupation fraction $g_t(p_+)$ follows Equation (28). In addition, the PDF of the displacements in the long time regime is given by Equations (7) and (29), recovering Gaussianity.

In Figure 6, we show $g_t(p_+)$ for exponential waiting times with $\langle\tau\rangle_+ = 1$ and $\langle\tau\rangle_- = 5$, in the left panel, we compare the uniform approximation of Equation (46) (black asterisks) with the full solution Equation (45) (red solid line), observing an excellent agreement. In the right panel of Figure 6, the behavior of $g_t(p_+)$ (as provided by Equation (45)) is displayed. As we can see, it starts with a uniform distribution for short times and then it evolves to a peaked distribution centered at $p_+ = \langle\tau\rangle_+/(\langle\tau\rangle_+ + \langle\tau\rangle_-) = 1/6$. As shown in

Appendix D, for non-equilibrium initial condition, the PDF of p_+ is still uniform within the short time regime. See also Appendix B for other similar cases.

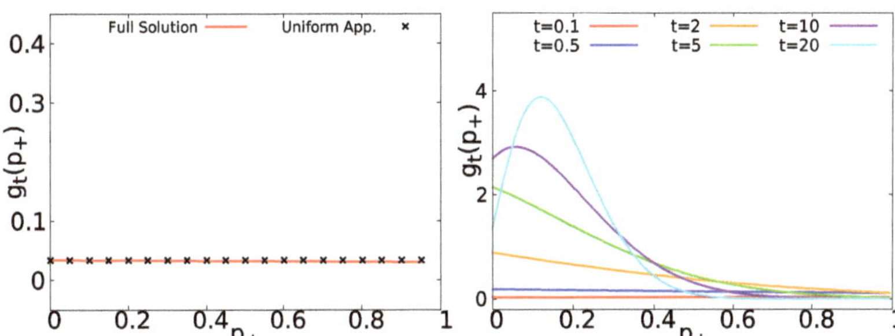

Figure 6. Left: Comparison between $g_t(p_+)$ Equation (45) (red solid line) and the uniform approximation Equation (46) (black asterisks) for $\langle\tau\rangle_+ = 1$, $\langle\tau\rangle_- = 5$ and $t = 0.1$. Right: $g_t(p_+)$ Equation (45) for $\langle\tau\rangle_+ = 1$, $\langle\tau\rangle_- = 5$ and $t \in \{0.1, 0.5, 2, 5, 10, 20\}$.

Finally, in Figure 7, we show the corresponding positional spreading for the normalized variable $z = x/\sqrt{t}$. As we can see in the short time, $t = 0.1$ (red circles) and $t = 0.5$ (blue crosses) $P(z,t)$ (given by Equation (47)) attain a tent-like shape. In the long run, $t = 20$ (orange circles) and $t = 30$ (green squares) $P(z,t)$ have a Gaussian distribution given by Equation (29).

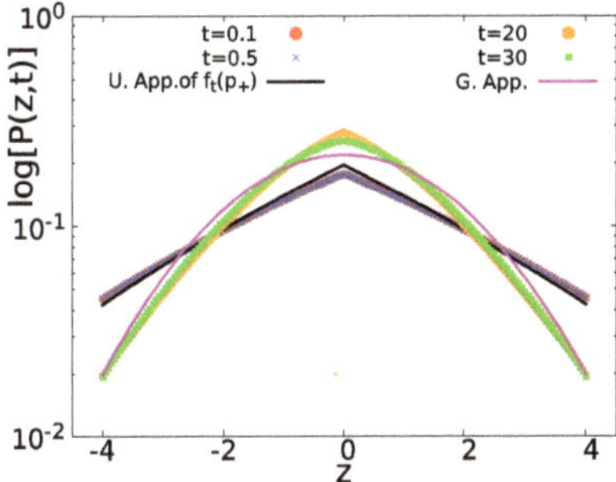

Figure 7. For a system with, $\langle\tau\rangle_+ = 1$ and $\langle\tau\rangle_- = 5$, $P(z,t)$ in semi-log scale, with $z = x/\sqrt{t}$. For short times $t = 0.1$ (red circles) and $t = 0.5$ (blue crosses), $P(z,t)$ is represented by Equation (47) (black solid line) with a tent like shape. For large times $t = 20$ (orange circles) and $t = 30$ (green diamonds), $P(z,t)$ converges to the Gaussian statistics Equation (29) (magenta solid line). In all the cases, $D_+ = 10$ and $D_- = 0$ were used. Compared with Figure 5, in this case, the Gaussian curve is above the tent curve, contrary to the case with equal mean waiting times. This is because the coefficient of the Gaussian curve Equation (29) is bigger compared with the weight of the delta peak in Equation (47). In Figure 5, we have the opposite, and the weight of the corresponding delta function in Equation (37) is bigger compared with the Gaussian Equation (41).

3. Discussion

3.1. The Histogram of the Diffusion Coefficient as Extracted from Experimental Data

3.1.1. Super-Statistics

We have found that at $x = 0$, $P(x,t)$ exhibits a cusp. A mathematically similar non-analytical behavior is found using an approach called super-statistics [12,14,21,54], which was used to explain laboratory observations. This framework postulates that the distribution of diffusion constants in the system is exponential, namely $P(D) =$

$\exp(-D/\langle D\rangle)/\langle D\rangle$ for $D > 0$ and $\langle D\rangle$ the average diffusivity. Then, the diffusion follows a Gaussian process with a random D. This approach gives

$$P(x,t) = \int_0^\infty \frac{e^{-\frac{x^2}{4Dt}}}{\sqrt{4\pi Dt}} \frac{e^{-\frac{D}{\langle D\rangle}}}{\langle D\rangle} dD = \frac{e^{-\frac{|x|}{\langle D\rangle t}}}{4\langle D\rangle t}. \tag{48}$$

Here, on the right-hand side, we have the Laplace PDF, which was used by Laplace in 1774 [55] to describe his linear law of errors [56]. In addition, as in our case, within the super-statistics method, we see in Equation (48) a non-analytical behavior since $P(x,t) \sim C_1 - C_2|x|$, for small x and with C_1, C_2 constants. Our work does not support the Laplace law, see Equations (26) and (27). However, maybe more importantly, the whole approach presented in this manuscript differs from the super-statistical approach in the following way. In our model, we have two diffusion constants, D_+ and $D_- = 0$ (see Appendix A for the case when $D_- \neq 0$). Hence, the PDF of diffusion constants is $P(D) = a\delta(D) + b\delta(D - D_+)$, with $a, b \geq 0$. It follows that the super-statistical approach predicts that the diffusing packet $P(x,t)$ is a sum of a delta function corresponding to non-moving particles and a Gaussian packet describing the movers. Thus, when the non-moving particles are excluded, we have perfect Gaussian behavior. This is actually correct, to leading order, for very short times. Thus, the super-statistical approach gives the correct $t \longrightarrow 0$ behavior but fails to predict the main issue (in our opinion), and that is the cusp on $x = 0$. To explore the non-analytical behavior, one needs to go to the next order terms in the expansion to include paths with a transition between states. Then, as we have shown, the equilibrium initial condition yields a uniform distribution of the occupation fraction Equation (25). It is this fact that brings the non-analytical behavior in the final result for $P(x,t)$ Equation (27), graphically represented by a "tent" see Figures 3, 5 and 7. It follows that the exponential conspiracy in which distribution of diffusion constants is exponential is not a necessary condition for a cusp like behavior of $P(x,t)$. We further remark that the non-analytical behavior is found also in the context of normal diffusion in [35–37] and within the anomalous one at [31–34,36,38–40].

3.1.2. Time Average MSD

We note that, in single molecule experiments, the time average mean squared displacement (TAMSD) is used in many cases to estimate the distribution of diffusion constants [2,5,6]. Since time averages are recorded over a finite measurement time, the time average fluctuates. Hence, we have naturally a distribution of the estimator for the diffusion parameters. In addition, the aforementioned two delta peak distribution of D, i.e., on D_+ and on D_-, is expected to be smeared out. This topic was extensively studied in a wide variety of models [57,58].

We now investigate the fluctuations of the time averaged diffusivities in a two state model and their implications in the distribution of diffusion coefficients obtained from real experimental data. For a further analysis of the time average diffusivity within a two state system, see [59,60].

We note that, in different single particle tracking experiments with non-Gaussian propagators, the recorded distribution of the diffusion coefficient D (obtained by means of TAMSD analysis) is relatively broad and peaked close to the origin [2,5,6]. Those experimental distributions of D are typically fitted by exponential [6] or gamma [2] distributions. Within the two state model, the diffusivity takes only two possible values D_- or D_+, but the respective TAMSD analysis gives values of D around D_- and D_+ [59]. The average D is given by $\langle D\rangle = (D_+\langle\tau\rangle_+ + D_-\langle\tau\rangle_-)/(\langle\tau\rangle_+ + \langle\tau\rangle_-)$. Thus, how different is the distribution of the diffusivities, extracted via TAMSD techniques, in a two state model compared with the one present in single molecule experiments? As we show next, this will be determined by the values of $D_\pm$ and $\langle\tau\rangle_\pm$. In Figure 8, we show the distribution of the diffusion coefficients obtained by means of TAMSD analysis for $D_+ = 10$, $D_- = 0$ and exponentially distributed waiting times. We show two different cases, the first one with the same mean waiting times $\langle\tau\rangle_+ = \langle\tau\rangle_- = 1$ (see red boxes). In addition, the second one with different mean waiting times, such that $\langle\tau\rangle_+ = 1$ and $\langle\tau\rangle_- = 5$ (see blue boxes).

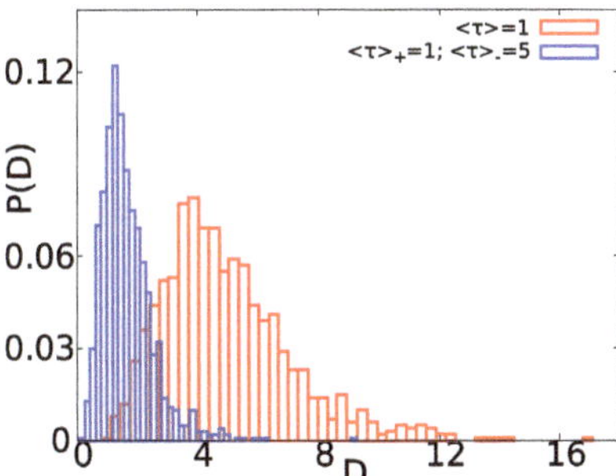

Figure 8. Distribution of diffusion coefficients $P(D)$ obtained via TAMSD analysis of simulated trajectories of a two state system with $D_+ = 10$, $D_- = 0$ and exponentially distributed waiting times. From the linear plots of the TAMSD versus the lag time estimates of D were extracted. We show two cases, the first for a system with the same mean waiting times $\langle\tau\rangle_+ = \langle\tau\rangle_- = \langle\tau\rangle = 1$ (red boxes). In addition, the PDF of D for a system with different mean waiting times with $\langle\tau\rangle_+ = 1$ and $\langle\tau\rangle_- = 5$ is also shown (blue boxes). For the system with the same mean waiting times, the average diffusivity found in the simulations is $\langle D\rangle = 4.98$, and, for the case of different mean waiting times, we have $\langle D\rangle = 1.69$. In both cases, we used $t = 1000$ and 1000 trajectories.

As we can see in Figure 8, when the difference between the diffusion coefficients is large, as in our case $D_+ = 10 > D_- = 0$, $P(D)$ is relatively broad. Nonetheless, for the case with $\langle\tau\rangle_+ = 1$ and $\langle\tau\rangle_- = 5$, the peak of $P(D)$ is closer to the origin compared to the case with $\langle\tau\rangle = 1$.

This difference between mean waiting times in each state is the second factor that determines the shape of $P(D)$. For instance, when this difference is such that $\langle\tau\rangle_+ < \langle\tau\rangle_-$, it is straightforward that the more the process spends in the state "$-$", the more the observed values of D will be closer to D_-. In this latter case, the distribution of D is peaked close to the origin since $D_- < D_+$. Thus, we can say that, when the differences between the diffusivities (and the mean waiting times) in the different states are pronounced, i.e., $D_- << D_+$ and $\langle\tau\rangle_+ << \langle\tau\rangle_-$, $P(D)$ in the two sate model resembles the distributions found in single molecule experiments [2,5,6].

4. Conclusions

From symmetry of the density of spreading particles $P(x,t) = P(-x,t)$, we expect an analytical expansion of the propagator as $P(x,t) \sim K_1 - K_2x^2 + \ldots$, with K_1, K_2 constants. Instead, in the two state model handled throughout this work, we get an expansion that is linear in $|x|$, see Equation (27). This is a non-analytical expansion graphically represented by a tent like structure, see Figures 3, 5 and 7. As mentioned above, Laplace in 1774 considered a similar non-analytical PDF, $P(x) = \exp(-|x|)/2$ for $-\infty < x < \infty$ [55,56]. However, the expression we find is clearly non-exponential, see Equation (26). Furthermore, for large x, we get a Gaussian behavior for $P(x,t)$. It should be noted that a non-analytical behavior is found only if $D_- = 0$, see Appendix A for further details. In practice, we may approach the non-analytical features of $P(x,t)$, as D_- is getting small.

Recently, a very general theory was developed for the non-Gaussian spreading of packets of particles. Using a CTRW framework, it was shown that, for any analytical PDF of waiting times, for large x limit $P(x,t) \sim \exp(-C|x|\ln|x|)$, with C a constant [22]. In the former model, we thus find exponential tails for large x, while, here, the anomaly, i.e., the cusp or tent like feature of $P(x,t)$, comes from the small x limit.

Recently, Postnikov et al. [37] investigated a model of diffusion in a quenched disordered setting, where the diffusive field is spatially varying. They showed that equilibrium initial conditions play a major role stating: "within the class of models with quenched disorder, the Itô model under equilibrium conditions is the only promising candidate for the description of Brownian Non Gaussian diffusion (BnG)." Note that here the definition of BnG means a model or system where the MSD is increasing linearly *for all times* and the propagator is non-Gaussian. Our model uses a time dependent diffusivity, and we showed that equilibrium initial conditions are indeed a key requirement. Here, we note that BnG does not imply a cusp, and vice versa. Namely, we may find a system where the MSD is increasing linearly in time, for the entire span of time, with or without a cusp for $P(x,t)$ at $x = 0$. The main focus of our work is the presence of a cusp for $P(x,t)$. Regarding the behavior of the MSD, it can be shown that, when equilibrium initial conditions are applied, $\langle T_+ \rangle = (\langle \tau \rangle_+ t)/[\langle \tau \rangle_+ + \langle \tau \rangle_-]$, for all times t (see Appendix G). Then, by Equations (A62) and (A68), the MSD is provided by

$$\langle x^2(t) \rangle = \left(\frac{D_+ \langle \tau \rangle_+ + D_- \langle \tau \rangle_-}{\langle \tau \rangle_+ + \langle \tau \rangle_-} \right) t, \tag{49}$$

for any time t. Thus, if the process starts from equilibrium, the MSD grows linearly for all times and we have BnG. Nevertheless, we would like to emphasize that our model is exhibiting BnG, but specifically $P(x,t)$ has a cusp only if $D_- = 0$, and practically when $D_- << D_+$.

To summarize, we emphasize that we have shown, by means of the statistics of the temporal occupation, that there is a universality for the PDF of the temporal occupation fraction in a two state model. For PDFs of waiting times with finite first moments, $g_t(p_+)$ can be approximated by a uniform distribution following Equation (25). This leads to tent like decaying propagators (Equation (26)) similar to those found in many experimental systems. We corroborate our results by solving analytically a two state system with exponentially distributed waiting times. We have shown that, either for short or long times, the distribution of displacements $P(x,t)$ has a general form, either a "tent" or a Gaussian bell curve. These two endpoints of the positional PDF are independent of the actual form of the distribution of waiting times. The crucial point within our framework is the generality of the behavior of the PDF of the occupation fraction p_+, being a uniform distribution for short times and a delta peak for long times. The former was found for a system with equilibrium initial conditions. We note that, for certain types of non-equilibrium initial conditions, we can still get a uniform PDF for the fraction occupation time; however, this is not generic (see details in Appendix B). Therefore, the non-Gaussian features are readily present in our model within the short time regime, and regardless of the specifics of the waiting times.

Mathematically, we presented an expansion in terms of the number of transitions from state $+$ to $-$ and backwards. Naturally, for very short times, the leading contribution to the packet comes from the paths with zero transitions, and then the packet is simply a sum of two Gaussian curves with diffusion coefficients D_+ and D_-. However, we showed that, by going to next order terms in the expansion, namely considering the paths with a single jump, we get the cusp like shape, found in the limit $D_- \to 0$. Thus, the whole effect is achieved by using a perturbation approach obtaining the leading order correction to the trivial behavior. Put differently, a widely popular super statistical approach is found to miss one of the main issues of the field, namely the cusp in $P(x,t)$. A super-statistical approach [14,54] uses a distribution of diffusivities, which in our model is a sum of two delta functions, at $D_- = 0$ and D_+. This does not give the cusp, as it is merely the zeroth order of the perturbation theory developed here.

Author Contributions: Conceptualization, E.B. and S.B.; methodology, E.B. and S.B.; software, M.H.-S.; validation, E.B., S.B. and M.H.-S.; formal analysis, M.H.-S. and S.B.; investigation, M.H.-S.; writing—original draft preparation, M.H.-S.; writing—review and editing, S.B. and E.B.; visualization, M.H.-S.; supervision, E.B. All authors have read and agreed to the published version of the manuscript.

Funding: E.B. and M.H.-S. are thankful for the support of the Israel Science Foundation Grant No. 1898/17. S.B. is grateful for the support of the Pazy foundation Grant No. 61139927 and the Israel Science Foundation Grant No. 2796/20.

Institutional Review Board Statement: Not applicable.

Informed Consent Statement: Not applicable.

Data Availability Statement: All the data sets obtained by numerical simulations or data analysis are available from the corresponding authors upon request.

Conflicts of Interest: The authors declare no conflict of interest.

Abbreviations

The following abbreviations are used in this manuscript:

CTRW	Continuous Time Random Walk
TAMSD	Time Average Mean Squared Displacement

Appendix A. A Two State Model with $D_+ > D_- > 0$

When $D_+ > D_- > 0$, the process for the displacements becomes

$$x(t) = \sqrt{2D_+T_+}\xi_1 + \sqrt{2D_-(t-T_+)}\xi_2, \tag{A1}$$

with ξ_1 and ξ_2 each i.i.d. Gaussian variables. In this case, the form of the conditioned PDF is given by

$$P(x,t|T_+) = \frac{e^{-\frac{x^2}{4[D_+T_+ + D_-(t-T_+)]}}}{\sqrt{4\pi[D_+T_+ + D_-(t-T_+)]}}. \tag{A2}$$

Then, the marginal distribution for the displacements follows

$$P(x,t) = \int_0^1 \frac{e^{-\frac{x^2}{4t[D_+p_+ + D_-(1-p_+)]}}}{\sqrt{4\pi t[D_+p_+ + D_-(1-p_+)]}} g_t(p_+)dp_+. \tag{A3}$$

Appendix A.1. $P(x,t)$ for Arbitrary Waiting Times

As we did in Section 2.2.1.1, using the general forms obtained above, i.e., Equations (9), (17), (18), (23), and (28), we can analyze $P(x,t)$ in the short and long time limits.

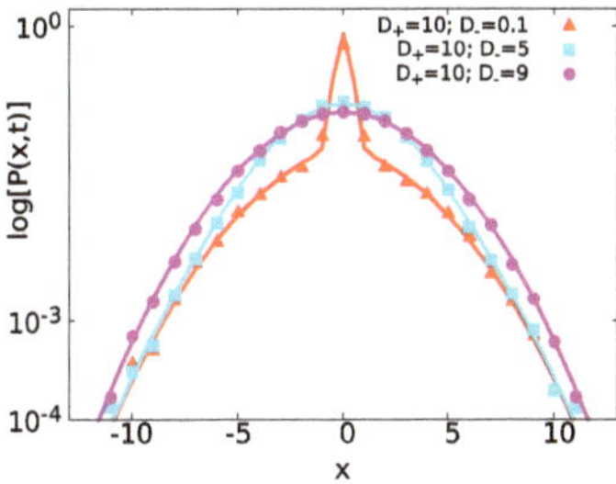

Figure A1. Distribution of displacements $P(x,t)$ obtained by simulations of a two state system with $D_+ > D_- > 0$ and gamma distributed waiting times $\tau \sim Gamma(3,1)$ at D_+ and $\tau \sim Gamma(6,1)$ at D_- following Equation (30). We compare with Equation (A4) (solid lines) with $t = 0.5$, $\langle\tau\rangle_+ = 3$, $\langle\tau\rangle_- = 6$, $D_+ = 10$. For $D_- = 0.1$ (red triangles), $D_- = 5$ (cyan squares) and $D_- = 9$ (magenta circles). Exponential like decaying is present at small values for x, when $D_+ = 10 >> D_- = 0.1$ (red solid line). In the cases when $D_- \longrightarrow D_-$ (cyan and magenta solid lines), $P(x,t)$ follows a full Gaussian distribution.

Appendix A.1.1. Short Time Regime

Substituting Equation (25) in Equation (A3), we get

$$\begin{aligned} P(x,t) &= \frac{\langle\tau\rangle_+}{\langle\tau\rangle_+ + \langle\tau\rangle_-}\left(1 - \frac{t}{\langle\tau\rangle_+}\right)\frac{e^{-\frac{x^2}{4D_+t}}}{\sqrt{4\pi D_+ t}} + \frac{\langle\tau\rangle_-}{\langle\tau\rangle_+ + \langle\tau\rangle_-}\left(1 - \frac{t}{\langle\tau\rangle_-}\right)\frac{e^{-\frac{x^2}{4D_-t}}}{\sqrt{4\pi D_- t}} \\ &+ \frac{2}{\sqrt{\pi}(\langle\tau\rangle_+ + \langle\tau\rangle_-)[D_- - D_+]}\Bigg\{\sqrt{D_- t}e^{-\frac{x^2}{4D_-t}} - \sqrt{D_+ t}e^{-\frac{x^2}{4D_+t}} \\ &- \frac{\sqrt{\pi}|x|}{2}\left[Erf\left(\frac{|x|}{\sqrt{4D_+t}}\right) - Erf\left(\frac{|x|}{\sqrt{4D_-t}}\right)\right]\Bigg\}. \end{aligned} \tag{A4}$$

In Figure A1, we compare Equation (A4) (in solid lines) and $P(x,t)$ obtained by simulations of a two state model for a fixed value of D_+ and different values of D_-, such that $D_+ > D_-$. In all cases, we used gamma distributed waiting times $\tau \sim Gamma(3,1)$ for the state with D_+ and $\tau \sim Gamma(6,1)$ for the state with D_- (the gamma distribution is defined by Equation (30)). As we can see when $D_+ >> D_-$, e.g., $D_+ = 10$ and $D_- 0.1$ (red triangles), the PDF of the displacements at small values of x has a non-Gaussian peak; thereafter, for large values of x, it follows a Gaussian distribution. When the values of D_- approach D_+ (cyan squares and magenta circles), $P(x,t)$ is fully described by Gaussian statistics even in the short time limit.

Appendix A.1.2. Long Time Regime

In the long time limit, the PDF of temporal occupation fraction is provided by Equation (28), then, according to Equation (A3), the PDF of the displacements is determined by

$$P(x,t) \sim \sqrt{\frac{\langle\tau\rangle_+ + \langle\tau\rangle_-}{4\pi t[D_+\langle\tau\rangle_+ + D_-\langle\tau\rangle_-]}}e^{-\frac{x^2(\langle\tau\rangle_+ + \langle\tau\rangle_-)}{4t[D_+\langle\tau\rangle_+ + D_-\langle\tau\rangle_-]}}. \tag{A5}$$

Thus, the Gaussian limit is also restored.

Appendix A.2. $P(x,t)$ for Exponentially Distributed Waiting Times with $\langle\tau\rangle_+ \neq \langle\tau\rangle_-$

In the short time regime, we can use the uniform approximation Equation (46) in Equation (A3), and the distribution for the displacements yields

$$\begin{aligned} P(x,t) &\sim \frac{\langle\tau\rangle_+}{\langle\tau\rangle_+ + \langle\tau\rangle_-}\frac{e^{-\frac{t}{\langle\tau\rangle_+} - \frac{x^2}{4D_+t}}}{\sqrt{4\pi D_+ t}} + \frac{\langle\tau\rangle_-}{\langle\tau\rangle_+ + \langle\tau\rangle_-}\frac{e^{-\frac{t}{\langle\tau\rangle_-} - \frac{x^2}{4D_-t}}}{\sqrt{4\pi D_- t}} \\ &+ \frac{1}{(\langle\tau\rangle_+ + \langle\tau\rangle_-)(D_- - D_+)\sqrt{\pi}}\Bigg\{\sqrt{4D_- t}e^{-\frac{x^2}{4D_-t}} - \sqrt{4D_+ t}e^{-\frac{x^2}{4D_+t}} \\ &+ \pi\left[Erf\left(\frac{|x|}{\sqrt{4D_-t}}\right) - Erf\left(\frac{|x|}{\sqrt{4D_+t}}\right)\right]\Bigg\}. \end{aligned} \tag{A6}$$

For the long time regime, we use $g_t(p_+)$ provided by Equation (28); then, according to Equation (A3), the PDF of the displacements follows Gaussian statistics described by Equation (A5).

Appendix B. A Complementary Deduction of $f_t^{\pm}(T_+|1)$

In this section, we obtain Equation (23) from the definition of conditional probability. The conditional probability of T_+, given that N jumps have been made, follows

$$f_t^{\pm}(T_+|N) = \frac{f_t^{\pm}(T_+, N)}{Q_t^{\pm}(N)}, \tag{A7}$$

with the distribution of jumps defined by [46]

$$Q_t^{\pm}(N) = \langle \mathbb{1}_{(t_N,t_{N+1})}(t) \rangle, \tag{A8}$$

with $\mathbb{1}_{(a,b)}(t)$ the indicator function, such that it is equal to one if $t \in (a,b)$ and zero if $t \notin (a,b)$. The average $\langle \cdot \rangle$ is over all the the values of τ_i's, with $i \in \{1,2,\ldots,N+1\}$.In addition, $f_t^{\pm}(T_+,N)$ is the joint probability of T_+ and N, which satisfies [46]

$$f_t^{\pm}(T_+,N) = \langle \delta(y-T_+)\mathbb{1}_{(t_N,t_{N+1})}(t) \rangle, \tag{A9}$$

with $t_N = \tau_1 + \ldots + \tau_N$ and the average $\langle \cdot \rangle$ defined as above.

Let us find Equations (A8) and (A9) and therefore Equation (A7) for the case $N = 1$. It is important to notice that, for the case of equilibrium initial conditions such as the one handled in Section 2.2, when $N = 1$, the corresponding average on τ_1 is given by the forward recurrence distribution Equation (10). Following Equation (A8), and taking the Laplace transform defined as $\hat{Q}_s^{\pm}(N) = \int_0^{\infty} e^{-st} Q_t^{\pm}(N) dt$, after simple manipulations, we obtain

$$\begin{aligned} \hat{Q}_s^{\pm}(1) &= \int_0^{\infty} e^{-s\tau_1} f_{eq}^{\pm}(\tau_1) d\tau_1 \int_0^{\infty} \left(\frac{1-e^{-s\tau_2}}{s}\right) \psi_{\mp}(\tau_2) d\tau_2, \\ &= \left(\frac{1-\hat{\psi}_{\pm}(s)}{\langle \tau \rangle_{\pm} s}\right)\left(\frac{1-\hat{\psi}_{\mp}(s)}{s}\right), \end{aligned} \tag{A10}$$

which is already the result shown in Equation (11).

For the joint distribution $f_t^{\pm}(T_+,1)$, following Equation (A9) and taking the double Laplace transform defined as $\hat{f}_s^{\pm}(u,N) = \int_0^{\infty} e^{-uT_+} \int_0^{\infty} e^{-st} f_t^{\pm}(T_+,N) dt dT_+$, after performing the corresponding integrals in the case we started from "+", we get

$$\begin{aligned} \hat{f}_s^{+}(u,1) &= \int_0^{\infty} e^{-(s+u)\tau_1} f_{eq}^{+}(\tau_1) d\tau_1 \int_0^{\infty} \left(\frac{1-e^{-s\tau_2}}{s}\right) \psi_{-}(\tau_2) d\tau_2, \\ &= \left(\frac{1-\hat{\psi}_{+}(s+u)}{\langle \tau \rangle_{+}(s+u)}\right)\left(\frac{1-\hat{\psi}_{-}(s)}{s}\right). \end{aligned} \tag{A11}$$

Following the same procedure for the case when the process started from "−", we obtain

$$\hat{f}_s^{-}(u,1) = \left(\frac{1-\hat{\psi}_{-}(s)}{\langle \tau \rangle_{-}(s)}\right)\left(\frac{1-\hat{\psi}_{+}(s+u)}{s+u}\right). \tag{A12}$$

Next, we show the connection of the joint PDf Equation (A9) with the uniform distribution of the occupation times Equation (24). Let $\hat{f}_s^{eq}(u,1)$ be the double Laplace transform of $f_t^{eq}(T_+,1)$, i.e., the joint PDF of T_+ and one single jump, starting from equilibrium. Clearly, the former follows

$$\hat{f}_s^{eq}(u,1) = \frac{\langle \tau \rangle_+}{\langle \tau \rangle_+ + \langle \tau \rangle_-} \hat{f}_s^{+}(u,1) + \frac{\langle \tau \rangle_-}{\langle \tau \rangle_+ + \langle \tau \rangle_-} \hat{f}_s^{-}(u,1). \tag{A13}$$

Using Equations (A11) and (A12) in Equation (A13), we get

$$\hat{f}_s^{eq}(u,1) = \frac{2}{\langle \tau \rangle_+ + \langle \tau \rangle_-} \left(\frac{1-\hat{\psi}_{+}(s+u)}{s+u}\right)\left(\frac{1-\hat{\psi}(s)}{s}\right). \tag{A14}$$

One of the key features of our paper is found when we consider both u and s to be large. This corresponds to the short time limit, when T_+ and t are of the same order. Then, we may use $\hat{\psi}_+(s+u), \hat{\psi}_-(s) \longrightarrow 0$ in Equation (A14), yielding to

$$\hat{\tilde{f}}_s^{eq}(u,1) \sim \frac{2}{[\langle\tau\rangle_+ + \langle\tau\rangle_-](s+u)s}. \tag{A15}$$

Equation (A15) is easy to invert, and we find in the short time limit

$$f_t^{eq}(T_+,1) \sim \frac{2}{\langle\tau\rangle_+ + \langle\tau\rangle_-}; \;\; for \;\; T_+ < t. \tag{A16}$$

This is the short time uniformity we have found that, in turn, as explained in Section 2.2, gives the cusp like shape in $P(x,t)$. Equation (A16) is the last term in Equation (24), corresponding to $N = 1$ (the first two terms in Equation (24) are contributions from $N = 0$).

Now, we are interested in inverting Equations (A10)–(A12) and then applying the definition of conditional probability Equation (A7). However, first, since we are dealing with the short time limit $t \longrightarrow 0$, in the Laplace space, this corresponds to the limit of $s \longrightarrow \infty$ and $u \longrightarrow \infty$. Thus, for this particular approximation, due to the definition of the Laplace transform $\hat{\psi}_\pm(s) = \int_0^\infty e^{-st}\psi_\pm(t)dt$, we have that $\lim_{s\to\infty}\hat{\psi}_\pm(s) \longrightarrow 0$ for a general $\psi_\pm(\tau)$. In this case, Equations (A10)–(A12) are approximated by

$$\hat{Q}_s^\pm(1) \sim \frac{1}{\langle\tau\rangle_\pm s^2}, \tag{A17}$$

$$\hat{\tilde{f}}_s^\pm(u,1) \sim \frac{1}{\langle\tau\rangle_\pm (s+u)s}. \tag{A18}$$

Inverting Equation (A17) with respect to s and Equation (A18) with respect to u and s with $0 < T_+ < t$, we obtain

$$Q_t^\pm(1) \sim \frac{t}{\langle\tau\rangle_\pm}, \tag{A19}$$

$$f_t^\pm(T_+,1) \sim \frac{1}{\langle\tau\rangle_\pm}. \tag{A20}$$

Now, substituting Equations (A19) and (A20) in the conditional probability Equation (A7) for $N = 1$, we obtain

$$f_t^\pm(T_+|1) \sim \frac{1}{t}, \tag{A21}$$

which is the same result shown in Equation (23) of Section 2.2. As expected, since the joint distribution $f_t^\pm(T_+,N)$ Equation (A20) does not depend on the time or any other variable, when it is used for computing the PDF of the occupation time, it gives the uniform distribution Equation (24). The same procedure for values of $N \geq 2$ gives a joint distribution $f_t^\pm(T_+,N)$ such that, in the double Laplace space, it is defined as

$$\begin{aligned}
\hat{f}_s^{+}(u,N) &= \left(\frac{1-\hat{\psi}_{+}(s+u)}{\langle\tau\rangle_{+}(s+u)}\right)\hat{\psi}_{-}^{k}(s)\hat{\psi}_{+}^{k}(s+u)\left(\frac{1-\hat{\psi}_{-}(s)}{s}\right); \\
& if \quad N=2k+1. && \text{(A22)}\\
\hat{f}_s^{+}(u,N) &= \left(\frac{1-\hat{\psi}_{+}(s+u)}{\langle\tau\rangle_{+}(s+u)}\right)\hat{\psi}_{+}^{k-1}(s+u)\hat{\psi}_{-}^{k}(s)\left(\frac{1-\hat{\psi}_{+}(s+u)}{s+u}\right); \\
& if \quad N=2k. && \text{(A23)}\\
\hat{f}_s^{-}(u,N) &= \left(\frac{1-\hat{\psi}_{-}(s)}{\langle\tau\rangle_{-}(s)}\right)\hat{\psi}_{-}^{k}(s)\hat{\psi}_{+}^{k}(s+u)\left(\frac{1-\hat{\psi}_{+}(s+u)}{s+u}\right); \\
& if \quad N=2k+1. && \text{(A24)}\\
\hat{f}_s^{-}(u,N) &= \left(\frac{1-\hat{\psi}_{-}(s)}{\langle\tau\rangle_{-}(s)}\right)\hat{\psi}_{-}^{k-1}(s)\hat{\psi}_{+}^{k}(s+u)\left(\frac{1-\hat{\psi}_{-}(s)}{s}\right); \\
& if \quad N=2k. && \text{(A25)}
\end{aligned}$$

In order to analyze the joint, we have to deal with the analytical expression of $\psi_{\pm}(\tau)$ defined by Equation (12). For instance for $N=2$, after substituting Equation (13) in Equations (A23) and (A25), we have that the dominant term is $\hat{f}_s^{+}(u,2) \sim [\Gamma(A_{-}+1)C_{A_{-}}^{-}]/[\langle\tau\rangle_{+}(s+u)^2 s^{A_{-}+1}]$, and $\hat{f}_s^{-}(u,2) \sim [\Gamma(A_{+}+1)C_{A_{+}}^{+}]/[\langle\tau\rangle_{-}s^2(s+u)^{A_{+}+1}]$, respectively. Inverting the double Laplace transform, in each case, it gives a positive power of T_{+} and therefore, with respect to t, e.g., $f_t^{\pm}(T_{+},2) \sim T_{+}^{A_{\mp}+1}$. $A_{\mp} \geq 0$ is a positive integer number, this correction term for $t \longrightarrow 0$ ($T+ \longrightarrow 0$) is negligible, and also the remaining terms in $f_t^{\pm}(T_{+},N)$ with $N>2$. We conclude that, for the case of equilibrium initial conditions, the uniformity in the short time limit, for the PDF of the occupation/fraction time, is always preserved, as long $\psi_{\pm}(\tau)$ is analytical.

Appendix B.1. Non-Equilibrium Initial Conditions

Still, by employing the joint distribution $f_t^{\pm}(T_{+},N)$, it can be shown, as follows that, for non-equilibrium initial conditions, when $\psi_{\pm}(\tau)$ in the Laplace space is approximated by $\hat{\psi}_{\pm}(s) \sim 1/s$ for short times, it can lead to a uniform distribution in the occupation time and therefore to a tent shape in $P(x,t)$.

In the case of non-equilibrium initial conditions, either starting just from D_{+} or from D_{-}, the averages over τ_1, within $Q_t^{\pm}(N)$ Equation (A8) and the joint distribution $f_t^{\pm}(T_{+},N)$ Equation (A9), are no longer given by $f_{eq}^{\pm}(\tau_1)$ Equation (10). Now, in the non-equilibrium case, these corresponding averages are performed using the waiting time PDF $\psi_{\pm}(\tau_1)$. Following the same procedure as above, for the case $N=1$, the double Laplace transform of the joint distribution $f_t^{\pm}(T_{+},1)$ yields to

$$\hat{f}_s^{\pm}(u,1) = \hat{\psi}_{\pm}(s+u)\left(\frac{1-\hat{\psi}_{\mp}(s)}{s}\right). \tag{A26}$$

For a system with non-equilibrium initial conditions, in order to recover the uniform distribution in the PDF of T_{+}, it is enough to ask that, for large s (short times), the PDF of the waiting times in the Laplace space follows

$$\hat{\psi}_{\pm}(s) \sim \frac{1}{s}. \tag{A27}$$

Substituting Equation (A27) in Equation (A26), we have that $\hat{f}_s^{\pm}(u,1) \sim 1/[(s+u)s]$. This implies in the real space, for $0<T_{+}<t$, that the joint distribution follows $f_t^{\pm}(T_{+},1) \sim 1$, and therefore, because of Equation (3), we also have a uniform distribution for T_{+}. As an example of a model in which $\hat{\psi}_{\pm}(s)$ goes as Equation (A27) and for non-equilibrium initial

conditions $\psi_{\pm}(\tau) \neq f_{eq}^{\pm}(\tau_1)$ as expected, we have the case in which the PDF of waiting times is the sum of two exponential functions, e.g., $\psi_{\pm}(\tau) = (1/2)\{[\exp(-\tau/C_{1\pm})/C_{1\pm}] + [\exp(-\tau/C_{2\pm})/C_{2\pm}]\}$, with $C_{1\pm}, C_{2\pm} > 0$. In this case for $s \longrightarrow \infty$, $\hat{\psi}_{\pm}(s) \sim [C_{1\pm} + C_{2\pm}]/[2C_{1\pm}C_{2\pm}s]$ and $f_{eq}^{\pm}(\tau_1) = [\exp(-\tau_1/C_{1\pm}) + \exp(-\tau_1/C_{2\pm})]/[C_{1\pm} + C_{2\pm}]$. For this latter case, since $\hat{\psi}_{\pm}(s)$ satisfies Equation (A27), following the same analysis as above, we find that the joint distribution $f_t^{\pm}(T_+, 1)$ is uniform.

In Appendix D, we show that, for a system with non-equilibrium conditions and exponentially distributed waiting times with equal and different mean values, the distribution of the occupation/fraction time is also uniform. This is mainly because, for exponentially distributed waiting times, the forward recurrence time distribution in equilibrium $f_{eq}^{\pm}(\tau_1)$ Equation (10) is equal to $\psi_{\pm}(\tau_1)$, as in the non-equilibrium case. Furthermore, for exponentially distributed sojourn times, Equation (A27) is also satisfied, and its PDF in the Laplace space for $s \longrightarrow \infty$ follows $\hat{\psi}_{\pm}(s) \sim 1/[\langle\tau\rangle_{\pm}s]$. By using Equation (3), this gives the uniform distribution shown in Equations (A35) and (A40).

For $\hat{\psi}_{\pm}(s)$ given in Equation (A27), the correction terms when $N \geq 2$, following the same analysis as in the case of equilibrium initial conditions, yield to elements of the form $f_t^{\pm}(T_+, N) \sim T_+^{N-1}$, which are negligible for $t \longrightarrow 0 \iff T_+ \longrightarrow 0$. Thus, they do not contribute in the PDF of the fraction occupation time.

Appendix C. $P(x,t)$ from Simulations with Uniform and Gamma Distributed Waiting Times within the Complete Range of x

Following Figure 3 in the left panel in which, for short time and displacements, the cusp of $P(x,t)$ is displayed. Now, from simulations (with the same parameters as above) of a two state model, with uniform (red triangles) and gamma (blue squares) distributed waiting times. In Figure A2, we show $P(x,t)$ in semi-log scale but for the whole span of x. In each case, we compare the normalized histogram of the simulation data with the short time analytical formula Equation (26), finding a perfect agreement. As we can see, the cusp is located at the origin, and, for large displacements, Gaussianity is recovered.

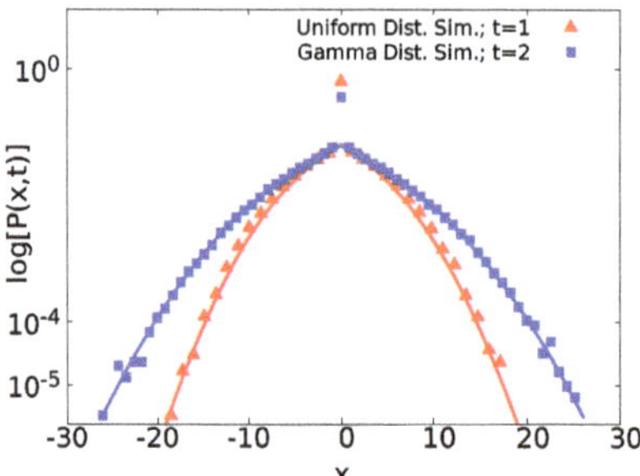

Figure A2. Distribution of displacements $P(x,t)$ in semi-log scale, obtained from simulations, of a two state system with uniform and gamma distributed waiting times within the short time limit and displaying the whole span of x. $P(x,t)$ for uniformly distributed waiting times is shown in red triangles. In addition, the case of gamma distributed waiting times is shown in blue squares. We employed the same set of parameters as those used in Figure 3 in the left panel. Both cases fit with Equation (26) (red and blue solid lines).

Appendix D. PDF of Occupation Times for Exponentially Distributed Waiting Times and Non-Equilibrium Initial Conditions

We consider the case of a system with exponentially distributed waiting times, with $\langle\tau\rangle_+ = \langle\tau\rangle_-$ in Equation (31). Here, we address the situation with non-equilibrium initial conditions. Particularly, the initial conditions are such that the probability of starting at the state with D_+ is 1 and the probability of starting from the state with D_- is 0. The PDF of T_+ then satisfies

$$f_t(T_+) = f_t^+(T_+) = \sum_{N=0}^{\infty} f_t^+(T_+, N). \tag{A28}$$

With $f_t^+(T_+, N)$ given by Equation (A9) explicitly for this case, we have [46]

$$\begin{aligned} f_t^+(T_+, 2k+1) &= \int \dots \int \delta\Big(T_+ - \sum_{i=1(odd)}^{2k+1} \tau_i\Big) \mathbb{1}_{(t_{2k+1}, t_{2k+2})}(t) \psi(\tau_1)\psi(\tau_2) \\ &\dots \quad \psi(\tau_{2k+2}) d\tau_1 d\tau_2 \dots d\tau_{2k+2} \quad if \quad N = 2k+1, \\ f_t^+(T_+, 2k) &= \int \dots \int \delta\Big(T_+ - \sum_{i=1(odd)}^{2k-1} \tau_i - \tau^*\Big) \mathbb{1}_{(t_{2k}, t_{2k+1})}(t) \psi(\tau_1)\psi(\tau_2) \\ &\dots \quad \psi(\tau_{2k+1}) d\tau_1 d\tau_2 \dots d\tau_{2k+1} \quad if \quad N = 2k, \end{aligned} \tag{A29}$$

with $\mathbb{1}_{(a,b)}(t)$ the indicator function equal to 1 if $t \in (a,b)$ and 0 if $t \notin (a,b)$. We work with the double Laplace transform $\mathcal{L}\Big\{f_t^+(T_+, N)\Big\} = f_s^+(u, N)$ with $t \iff s$ and $u \iff T_+$, which is given by $\hat{f}_s^+(u, N) = \int_0^\infty e^{-uT_+} \int_0^\infty e^{-st} f_t^+(T_+, N) dt dT_+$. Thus, taking the double Laplace transform of Equation (A29), after substitution of $\psi(\tau)$, we have

$$\begin{aligned} \hat{f}_s^+(u, 2k+1) &= \hat{\psi}^{k+1}(s+u)\hat{\psi}^k(s)\Big(\frac{1-\hat{\psi}(s)}{s}\Big) \quad if \quad N = 2k+1 \\ \hat{f}_s^+(u, 2k) &= \hat{\psi}^k(s+u)\hat{\psi}^k(s)\Big(\frac{1-\hat{\psi}(s+u)}{s+u}\Big) \quad if \quad N = 2k. \end{aligned} \tag{A30}$$

Thus, using Equation (A30) for summing over all the values of N in Equation (A28), we get

$$\hat{f}_s(u) = \left(\hat{\psi}(s+u)\frac{1-\hat{\psi}(s)}{s} + \frac{1-\hat{\psi}(s+u)}{s+u}\right)\frac{1}{1-\hat{\psi}(s+u)\hat{\psi}(s)}. \tag{A31}$$

For exponentially distributed waiting times $\hat{\psi}(s) = 1/(1+\langle\tau\rangle s)$, by substituting $\hat{\psi}(s)$ in Equation (A31), we get that the double Laplace transform of Equation (A28) is given by

$$\hat{f}_s(u) = \frac{2+\langle\tau\rangle s}{2s + \langle\tau\rangle s^2 + (1+\langle\tau\rangle s)u}. \tag{A32}$$

By the same procedures used in Appendix F, the inversion of the double Laplace transform of Equation (A32) yields

$$\begin{aligned} f_t(T_+) &= \delta(t-T_+)e^{-\frac{t}{\langle\tau\rangle}} + \frac{e^{-\frac{t}{\langle\tau\rangle}}}{\langle\tau\rangle} {}_0\tilde{F}_1\left(;1;\frac{T_+(t-T_+)}{\langle\tau\rangle^2}\right) \\ &+ \frac{e^{-\frac{t}{\langle\tau\rangle}}}{\langle\tau\rangle^2} T_+ {}_0\tilde{F}_1\left(;2;\frac{T_+(t-T_+)}{\langle\tau\rangle^2}\right). \end{aligned} \tag{A33}$$

Employing the identity $I_\nu(y) = (y/2)^\nu {}_0\tilde{F}_1(;\nu+1;y^2/4)$ [61] and changing variables, we obtain the PDF of the occupation fraction, which follows

$$g_t(p_+) = \delta(1-p_+)e^{-\frac{t}{\langle\tau\rangle}} + \frac{t}{\langle\tau\rangle}\left\{ I_0\left(\frac{2t}{\langle\tau\rangle}\sqrt{p_+(1-p_+)}\right) + p_+\frac{I_1\left(\frac{2t}{\langle\tau\rangle}\sqrt{p_+(1-p_+)}\right)}{\sqrt{p_+(1-p_+)}}\right\}e^{-\frac{t}{\langle\tau\rangle}}. \quad \text{(A34)}$$

By taking the series expansion of Equation (A34) in the limit $t \longrightarrow 0$, the PDF of p_+ can be approximated by

$$g_t(p_+) \sim \delta(1-p_+)e^{-\frac{t}{\langle\tau\rangle}} + \frac{t}{\langle\tau\rangle}. \quad \text{(A35)}$$

Therefore, for $1 > p_+ > 0$, the PDF of p_+ follows a uniform distribution (see the left panel of Figure A3), as in the case of equilibrium initial conditions (Equation (35)).

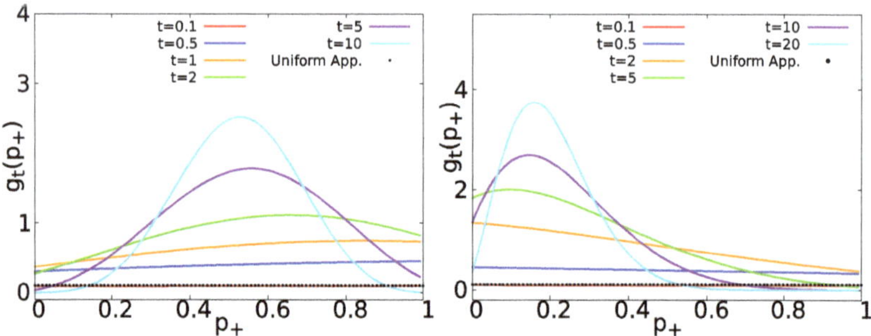

Figure A3. Left: $g_t(p_+)$ Equation (A34) for $\langle\tau\rangle = 1$ and $t \in \{0.1, 0.5, 1, 2, 5, 10\}$ and non-equilibrium initial conditions (starting from state "+"). The uniform approximation of $g_t(p_+)$ Equation (A35) for $t = 0.1$ is shown in black circles. Right: $g_t(p_+)$ Equation (A39) for $\langle\tau\rangle_+ = 1$, $\langle\tau\rangle_- = 5$ and $t \in \{0.1, 0.5, 2, 5, 10, 20\}$ and non-equilibrium initial conditions (starting from state "+"). The uniform approximation of $g_t(p_+)$ Equation (A40) for $t = 0.1$ is shown in black circles.

$P(x,t)$ is obtained by exploiting the uniform approximation of $g_t(p_+)$ in Equation (A35), i.e.,

$$P(x,t) \sim \frac{e^{-\frac{t}{\langle\tau\rangle}-\frac{x^2}{4D_+t}}}{\sqrt{4\pi D_+t}} + \frac{t}{\langle\tau\rangle}\left\{\frac{2e^{-\frac{x^2}{4D_+t}}}{\sqrt{4\pi D_+t}} - \frac{|x|}{2D_+t}\left[1 - Erf\left(\frac{|x|}{\sqrt{4D_+t}}\right)\right]\right\}. \quad \text{(A36)}$$

Equation (A36) follows the same structure as Equation (37), i.e., the case with equilibrium initial conditions.

When $\langle\tau\rangle_+ \neq \langle\tau\rangle_-$, we have to include $\psi_\pm(\tau)$ and $\hat{\psi}_\pm(s)$ in Equations (A29) and (A30). Summing the resulting expressions in Equation (A28), we obtain

$$\hat{f}_s(u) = \left(\hat{\psi}_+(s+u)\frac{1-\hat{\psi}_-(s)}{s} + \frac{1-\hat{\psi}_+(s+u)}{s+u}\right)\frac{1}{1-\hat{\psi}_+(s+u)\hat{\psi}_-(s)}. \quad \text{(A37)}$$

By employing $\hat{\psi}(s) = 1/(1+\langle\tau\rangle_\pm s)$ in Equation (A37), we obtain that the double Laplace transform of the PDF of T_+ is provided by [49]

$$\hat{f}_s(u) = \frac{\langle\tau\rangle_+ + \langle\tau\rangle_- + \langle\tau\rangle_+\langle\tau\rangle_- s}{\langle\tau\rangle_- s + \langle\tau\rangle_+(1+\langle\tau\rangle_- s)(s+u)}. \quad \text{(A38)}$$

The inverse Laplace transform of Equation (A38) is obtained by the same procedures explained above and, in Appendix F, eventually

$$g_t(p_+) = \delta(1-p_+)e^{-\frac{t}{\langle\tau\rangle_+}} + \frac{t}{\langle\tau\rangle_+}\left\{ I_0\left(2t\sqrt{\frac{p_+(1-p_+)}{\langle\tau\rangle_+\langle\tau\rangle_-}}\right) + \sqrt{\frac{\langle\tau\rangle_+}{\langle\tau\rangle_-}}p_+ \frac{I_1\left(2t\sqrt{\frac{p_+(1-p_+)}{\langle\tau\rangle_+\langle\tau\rangle_-}}\right)}{\sqrt{p_+(1-p_+)}}\right\} e^{-\frac{tp_+}{\langle\tau\rangle_+}-\frac{t(1-p_+)}{\langle\tau\rangle_-}}. \quad \text{(A39)}$$

We recover Equation (A34) when $\langle\tau\rangle_+ = \langle\tau\rangle_- = \langle\tau\rangle$. In the short time limit, Equation (A39) follows as well a uniform distribution for $1 > p_+ > 0$, see the right panel of Figure A3. In this case, the PDF of p_+ is given by

$$g_t(p_+) \sim \delta(1-p_+)e^{-\frac{t}{\langle\tau\rangle_+}} + \frac{t}{\langle\tau\rangle_+}. \quad \text{(A40)}$$

Thus, for exponentially distributed waiting times, for equilibrium and non-equilibrium conditions, in the short time regime, the PDF of the occupation fraction is always uniform. This feature is only valid for exponentially distributed waiting times, and it is not necessarily fulfilled for other distributions of waiting times.

$P(x,t)$ for the specific case of non-equilibrium initial conditions and exponentially distributed waiting times is

$$P(x,t) \sim \frac{e^{-\frac{t}{\langle\tau\rangle_+}-\frac{x^2}{4D_+t}}}{\sqrt{4\pi D_+ t}} + \frac{t}{\langle\tau\rangle_+}\left\{\frac{e^{-\frac{x^2}{4D_+t}}}{\sqrt{\pi D_+ t}} + \frac{|x|}{2D_+t}\left[1 - Erf\left(\frac{|x|}{\sqrt{4D_+t}}\right)\right]\right\}. \quad \text{(A41)}$$

Appendix E. Deduction of $g_t(p_+)$ for Waiting Times with Similar Mean Waiting Times

We can use the results found in [52,53] for inverting the double Laplace transform of S_t as provided by Equation (33). For a Fourier–Laplace transform, $\hat{\kappa}(\omega,s) = \int_{-\infty}^{\infty}\int_0^{\infty} e^{i\omega\tilde{x}}e^{-st}\kappa(\tilde{x},t)dsdx$ of the form

$$\hat{\kappa}(\omega,s) = \frac{2\tilde{\lambda}+s}{s^2+2\tilde{\lambda}s+c^2\omega^2}, \quad \text{(A42)}$$

the inversion yields the result [52,53]

$$\kappa(\tilde{x},t) = \frac{1}{2}e^{-\frac{t}{\tilde{\lambda}}}\{\delta(\tilde{x}-ct)+\delta(\tilde{x}+ct)\} + \frac{1}{2\tilde{\lambda}c}\Theta(ct-|\tilde{x}|)[I_0(z(t)) + \frac{t}{\tilde{\lambda}z(t)}I_1(z(t))], \quad \text{(A43)}$$

with $z(t) = \frac{1}{\tilde{\lambda}c}\sqrt{c^2t^2-\tilde{x}^2}$.

Now, we can compare the double Laplace transform of S_t given in Equation (33) with the results shown in Equation (A42). Concretely, we can relate the Laplace variable v with the Fourier variable ω in $\hat{\kappa}(\omega,s)$. Since the Laplace transform and the Fourier transform are exponential operators, by setting $v = ic\omega$, we can make the former equivalent to a Fourier transform, and we use the expression given by Equation (A43) for inverting $\phi_s(v)$. In our case, $c = 1$, so $S_t \Leftrightarrow \omega$ are Fourier conjugates and therefore the inversion of $\phi_s(v)$ results in

$$\phi_t(S_t) = \frac{1}{2}e^{-\frac{t}{\langle\tau\rangle}}\left\{\delta(S_t-t)+\delta(S_t+t)\right\} + \frac{\Theta(t-|S_t|)}{2\langle\tau\rangle}\left[I_0\left(\frac{\sqrt{t^2-S_t^2}}{\langle\tau\rangle}\right) + \frac{tI_1\left(\frac{\sqrt{t^2-S_t^2}}{\langle\tau\rangle}\right)}{\sqrt{t^2-S_t^2}}\right]. \quad \text{(A44)}$$

By changing variables, $S_t = 2T_+ - t = 2p_+t - t$, we obtain Equation (34) in a straightforward manner. The results in Equation (34) can also be obtained by the inversion of the double Laplace transform ($t \Leftrightarrow s$ and $T_+ \Leftrightarrow u$) of the PDF of T_+ Equation (3) (see Appendix F). In this case, the double Laplace transform of the PDF of T_+ is given by

$$\hat{f}_s(u) = \frac{4 + 2\langle\tau\rangle s + \langle\tau\rangle u}{2\langle\tau\rangle s^2 + 4s + (2 + 2\langle\tau\rangle s)u}. \tag{A45}$$

Finally, we mention that the moments of T_+ and therefore p_+ can be obtained by expanding Equation (A45) in powers of u as

$$\hat{f}_s(u) = \frac{1}{s} - \frac{1}{2s^2}u + \frac{1 + \langle\tau\rangle s}{2s^3(2 + \langle\tau\rangle s)}u^2 + O(u^3). \tag{A46}$$

The first two moments of T_+ are then

$$\langle T_+\rangle \sim \frac{t}{2}, \tag{A47}$$

$$\langle T_+^2\rangle \sim \left(\frac{\langle\tau\rangle}{4} + \frac{t}{4}\right)t + \frac{\langle\tau\rangle^2}{8}\left(e^{-\frac{2t}{\langle\tau\rangle}} - 1\right). \tag{A48}$$

For $\langle p_+\rangle = \langle T_+\rangle / t$, we obtain

$$\langle p_+\rangle \sim \frac{1}{2}, \tag{A49}$$

$$\langle p_+^2\rangle \sim \frac{1}{4} + \frac{\langle\tau\rangle}{4t} + \frac{\langle\tau\rangle^2}{8t^2}\left(e^{-\frac{2t}{\langle\tau\rangle}} - 1\right), \tag{A50}$$

$$Var(p_+) \sim \frac{\langle\tau\rangle}{4t} + \frac{\langle\tau\rangle^2}{8t^2}\left(e^{-\frac{2t}{\langle\tau\rangle}} - 1\right). \tag{A51}$$

Appendix F. Deduction of $g_t(p_+)$ for Waiting Times with $\langle\tau\rangle_+ \neq \langle\tau\rangle_-$

Here, we show the procedure for obtaining Equation (45) in Section 2.3.2. Starting from the double Laplace transform of the PDF of T_+ given by Equation (44), first by inverting with respect to $u \Longleftrightarrow T_+$, we get

$$\begin{aligned}\hat{f}_s(T_+) &= \frac{\langle\tau\rangle_-^2\delta(T_+)}{(\langle\tau\rangle_+ + \langle\tau\rangle_-)(1 + \langle\tau\rangle_- s)} \\ &+ \frac{(\langle\tau\rangle_+ + \langle\tau\rangle_- + \langle\tau\rangle_+\langle\tau\rangle_- s)^2}{\langle\tau\rangle_+(\langle\tau\rangle_+ + \langle\tau\rangle_-)(1 + \langle\tau\rangle_- s)^2} e^{-T_+ s\left(\frac{\langle\tau\rangle_+ + \langle\tau\rangle_- + \langle\tau\rangle_+\langle\tau\rangle_- s}{\langle\tau\rangle_+ + \langle\tau\rangle_+\langle\tau\rangle_- s}\right)},\end{aligned} \tag{A52}$$

the exponent in Equation (A52) can be written as $-T_+ s\left(1 + \frac{\langle\tau\rangle_-}{\langle\tau\rangle_+ + \langle\tau\rangle_+\langle\tau\rangle_- s}\right)$. The inversion of Equation (A52) with respect to $s \Leftrightarrow t$ can be expressed as

$$\hat{f}_s(T_+) = \frac{\langle\tau\rangle_- e^{-\frac{t}{\langle\tau\rangle_-}}}{\langle\tau\rangle_+ + \langle\tau\rangle_-}\delta(T_+) + \mathcal{L}^{-1}\{\hat{q}(s)\hat{h}(s)\}. \tag{A53}$$

Thus, the inversion of the second term in Equation (A53) is given by the convolution theorem, following $\mathcal{L}^{-1}\{\hat{q}(s)\hat{h}(s)\} = \int_0^t \mathcal{L}^{-1}\{\hat{q}(s)\}|_{t-t'}\mathcal{L}^{-1}\{\hat{h}(s)\}|_{t'}dt'$, with $\hat{q}(s) = \frac{(\langle\tau\rangle_+ + \langle\tau\rangle_- + \langle\tau\rangle_+\langle\tau\rangle_- s)^2}{\langle\tau\rangle_+(\langle\tau\rangle_+ + \langle\tau\rangle_-)(1 + \langle\tau\rangle_- s)^2}$ and $\hat{h}(s) = e^{-T_+ s}e^{-\frac{T_+\langle\tau\rangle_- s}{\langle\tau\rangle_+ + \langle\tau\rangle_+\langle\tau\rangle_- s}}$. The inverse Laplace transform of $\hat{q}(s)$ is given by

$$\mathcal{L}^{-1}\{\hat{q}(s)\} = \frac{2e^{-\frac{t}{\langle\tau\rangle_-}}}{\langle\tau\rangle_+ + \langle\tau\rangle_-} + \frac{te^{-\frac{t}{\langle\tau\rangle_-}}}{\langle\tau\rangle_+(\langle\tau\rangle_+ + \langle\tau\rangle_-)} + \frac{\langle\tau\rangle_+\delta(t)}{\langle\tau\rangle_+ + \langle\tau\rangle_-}. \tag{A54}$$

The inverse Laplace transform of $\hat{h}(s)$ can be obtained by rewriting the exponent in the second term of $\hat{h}(s)$ as $-\frac{T_+\langle\tau\rangle_- s}{\langle\tau\rangle_+ + \langle\tau\rangle_+\langle\tau\rangle_- s} = -\frac{T_+}{\langle\tau\rangle_+} + \frac{T_+\langle\tau\rangle_-}{\langle\tau\rangle_+\langle\tau\rangle_- + \langle\tau\rangle_-^2\langle\tau\rangle_+ s}$, then we obtain

$$
\begin{aligned}
\mathcal{L}^{-1}\{\hat{h}(s)\} &= \mathcal{L}^{-1}\Big\{e^{-\frac{T_+\langle\tau\rangle_- s}{\langle\tau\rangle_+ + \langle\tau\rangle_+\langle\tau\rangle_- s}}\Big\}\Big|_{t-T_+}\Theta(t-T_+) \\
&= e^{-\frac{T_+}{\langle\tau\rangle_+}}\mathcal{L}^{-1}\Big\{e^{\frac{T_+\langle\tau\rangle_-}{\langle\tau\rangle_+\langle\tau\rangle_- + \langle\tau\rangle_-^2\langle\tau\rangle_+ s}}\Big\}\Big|_{t-T_+}\Theta(t-T_+) \\
= e^{-\frac{T_+}{\langle\tau\rangle_+}}&\left[e^{-\frac{(t-T_+)}{\langle\tau\rangle_-}}\sqrt{\frac{T_+}{\langle\tau\rangle_+\langle\tau\rangle_-(t-T_+)}}I_1\left(2\sqrt{\frac{T_+(t-T_+)}{\langle\tau\rangle_+\langle\tau\rangle_-}}\right)\right. \\
&\left.+\frac{e^{-\frac{(t-T_+)}{\langle\tau\rangle_-}}}{\langle\tau\rangle_+\langle\tau\rangle_-^2}\delta\left(\frac{t-T_+}{\langle\tau\rangle_+\langle\tau\rangle_-^2}\right)\right]\Theta(t-T_+).
\end{aligned} \tag{A55}
$$

Substituting Equations (A54) and (A55) in Equation (A53), and, after integration, we obtain

$$
\begin{aligned}
f_t(T_+) &= \frac{\langle\tau\rangle_- e^{-\frac{t}{\langle\tau\rangle_-}}}{\langle\tau\rangle_+ + \langle\tau\rangle_-}\delta(T_+) \\
&+ \frac{\langle\tau\rangle_+ e^{-\frac{t}{\langle\tau\rangle_+}}}{\langle\tau\rangle_+ + \langle\tau\rangle_-}\delta(t-T_+) + \left\{\frac{2}{\langle\tau\rangle_+ + \langle\tau\rangle_-}{}_0\tilde{F}_1\left(;1;\frac{T_+(t-T_+)}{\langle\tau\rangle_+\langle\tau\rangle_-}\right)\right. \\
&+ \left.\left[\frac{t-T_+}{\langle\tau\rangle_+(\langle\tau\rangle_+ + \langle\tau\rangle_-)} + \frac{T_+}{\langle\tau\rangle_-(\langle\tau\rangle_+ + \langle\tau\rangle_-)}\right]{}_0\tilde{F}_1\left(;2;\frac{T_+(t-T_+)}{\langle\tau\rangle_+\langle\tau\rangle_-}\right)\right\}e^{-\frac{T_+}{\langle\tau\rangle_+}-\frac{(t-T_+)}{\langle\tau\rangle_-}}.
\end{aligned} \tag{A56}
$$

Employing the identity $I_\nu(y) = (y/2)^\nu{}_0\tilde{F}_1(;\nu+1;y^2/4)$ [61] and changing variables, we obtain the form of $g_t(p_+)$ as provided by Equation (45). This procedure can be employed for a system with the same mean waiting times, inverting the double Laplace transform Equation (A45) and obtaining $g_t(p_+)$ shown in Equation (34). In addition, also for the non-equilibrium cases treated in Appendix D, see Equations (A34) and (A39).

Finally, we show the corresponding first two moments of T_+ and p_+. As we proceeded in Appendix E, we obtain the moments of T_+ by expanding in powers of u Equation (A37), which yields

$$
\langle T_+\rangle \sim \frac{\langle\tau\rangle_+ t}{\langle\tau\rangle_+ + \langle\tau\rangle_-}, \tag{A57}
$$

$$
\langle T_+^2\rangle \sim \frac{\langle\tau\rangle_+^2 t^2}{(\langle\tau\rangle_+ + \langle\tau\rangle_-)^2} + \frac{2\langle\tau\rangle_+^2\langle\tau\rangle_-^2 t}{(\langle\tau\rangle_+ + \langle\tau\rangle_-)^3} + \frac{2\langle\tau\rangle_+^3\langle\tau\rangle_-^3}{(\langle\tau\rangle_+ + \langle\tau\rangle_-)^3}\left(e^{-\frac{(\langle\tau\rangle_+ + \langle\tau\rangle_-)t}{\langle\tau\rangle_+\langle\tau\rangle_-}} - 1\right). \tag{A58}
$$

Therefore, the moments of p_+ are

$$
\langle p_+\rangle \sim \frac{\langle\tau\rangle_+}{\langle\tau\rangle_+ + \langle\tau\rangle_-}, \tag{A59}
$$

$$
\langle p_+^2\rangle \sim \frac{\langle\tau\rangle_+^2}{(\langle\tau\rangle_+ + \langle\tau\rangle_-)^2} + \frac{2\langle\tau\rangle_+^2\langle\tau\rangle_-^2}{t(\langle\tau\rangle_+ + \langle\tau\rangle_-)^3} + \frac{2\langle\tau\rangle_+^3\langle\tau\rangle_-^3}{t^2(\langle\tau\rangle_+ + \langle\tau\rangle_-)^3}\left(e^{-\frac{(\langle\tau\rangle_+ + \langle\tau\rangle_-)t}{\langle\tau\rangle_+\langle\tau\rangle_-}} - 1\right), \tag{A60}
$$

$$
Var(p_+) \sim \frac{2\langle\tau\rangle_+^2\langle\tau\rangle_-^2}{t(\langle\tau\rangle_+ + \langle\tau\rangle_-)^3} + \frac{2\langle\tau\rangle_+^3\langle\tau\rangle_-^3}{t^2(\langle\tau\rangle_+ + \langle\tau\rangle_-)^3}\left(e^{-\frac{(\langle\tau\rangle_+ + \langle\tau\rangle_-)t}{\langle\tau\rangle_+\langle\tau\rangle_-}} - 1\right). \tag{A61}
$$

Appendix G. Deduction of the MSD in a Two State Model with $\langle\tau\rangle_+ \neq \langle\tau\rangle_-$

From Equation (A1), we can compute the second moment of $x(t)$ as

$$
\begin{aligned}
\langle x^2(t)\rangle &= \left\langle\left(\sqrt{2D_+T_+}\xi_1 + \sqrt{2D_-(t-T_+)}\xi_2\right)^2\right\rangle, \\
&= 2D_+\langle T_+\rangle + 2D_-(t-\langle T_+\rangle),
\end{aligned} \tag{A62}
$$

In the second line of Equation (A62), we have employed the linearity of $\langle\cdot\rangle$, and the properties of independent standard normal random variables, i.e., $\langle\xi_i^2\rangle = 1$ and $\langle\xi_i\xi_j\rangle = 0$ (with $i,j \in \{1,2\}$ and $i \neq j$). Thus, now we just have to find $\langle T_+\rangle$. In order to do that, we start from the definition of average occupation time

$$\begin{aligned}\langle T_+\rangle &= \int_0^\infty T_+ f_t(T_+)dT_+,\\ &= \int_0^\infty f_t(T_+)\Big(-\frac{d}{du}e^{-uT_+}\Big)\Big|_{u=0}dT_+,\\ &= -\lim_{u\to 0}\Big(\frac{d}{du}\hat{f}_t(u)\Big).\end{aligned} \tag{A63}$$

With $\hat{f}_t(u) = \int_0^\infty f_t(T_+)e^{-uT_+}$, and $T_+ \Leftrightarrow u$ Laplace conjugates. Now, for equilibrium initial conditions, the PDF of the occupation time is given by

$$f_t(T_+) = \frac{\langle\tau\rangle_+}{\langle\tau\rangle_+ + \langle\tau\rangle_-}\sum_{N=0}^\infty f_t^+(T_+,N) + \frac{\langle\tau\rangle_-}{\langle\tau\rangle_+ + \langle\tau\rangle_-}\sum_{N=0}^\infty f_t^-(T_+,N), \tag{A64}$$

with $f_t^\pm(T_+,N)$ the joint PDF of the occupation times at D_+ and the number of jumps between states during t, once started from $D_\pm$. When starting from D_+ and having $N = 2k+1$ or $N = 2k$ jumps, the occupation time in each case satisfies Equation (1). In the case when the initial state is at D_-, we have

$$\begin{aligned}T_+ &= \tau_2 + \tau_4 + \ldots + \tau^* \quad && if \;\; N = 2k+1,\\ T_+ &= \tau_2 + \tau_4 + \ldots + \tau_N \quad && if \;\; N = 2k,\end{aligned} \tag{A65}$$

with $\tau^* = t - t_N$, the backward recurrence time. The definition of the joint PDF $f_t^\pm(T_+,N)$ is already given in Equation (A9). In addition, its double Laplace transform $\hat{f}_s^\pm(u,N) = \int_0^\infty\int_0^\infty f_t(T_+,N)\exp(-uT_+ - st)\,dT_+\,dt$, is shown in Equations (A22)–(A25). When $N = 0$, we have

$$\begin{aligned}\hat{f}_s^+(u,0) &= \frac{1-\Big(\frac{1-\hat{\psi}_+(s+u)}{\langle\tau\rangle_+(s+u)}\Big)}{s+u},\\ \hat{f}_s^-(u,0) &= \frac{1-\Big(\frac{1-\hat{\psi}_-(s)}{\langle\tau\rangle_- s}\Big)}{s}.\end{aligned} \tag{A66}$$

Now, for obtaining $\hat{f}_s(u)$, we compute the double Laplace transform of Equation (A64) and then we sum Equations (A22), (A25), and (A66) for all values of N. Thereafter, we compute the derivative of $\hat{f}_s(u)$ with respect to u and its corresponding limit when $u \longrightarrow 0$. Following algebraic simplifications, we yield

$$\lim_{u\to 0}\Big(\frac{d}{du}\hat{f}_s(u)\Big) = -\frac{\langle\tau\rangle_+}{\langle\tau\rangle_+ + \langle\tau\rangle_-}\frac{1}{s^2}. \tag{A67}$$

For obtaining the average occupation time Equation (A63), we invert Equation (A67) with respect to s, having

$$\langle T_+\rangle = \left(\frac{\langle\tau\rangle_+}{\langle\tau\rangle_+ + \langle\tau\rangle_-}\right)t. \tag{A68}$$

Finally, substituting Equation (A68) in Equation (A62), we get Equation (49), which indicates that the MSD is linear with respect to t, for any value of time t.

References

1. Chaudhuri, P.; Berthier, L.; Kob, W. Universal Nature of Particle Displacements close to Glass and Jamming Transitions. *Phys. Rev. Lett.* **2007**, *99*, 060604–060608. [CrossRef] [PubMed]
2. Hapca, S.; Crawford, J.W.; Young, I.M. Anomalous diffusion of heterogeneous populations characterized by normal diffusion at the individual level. *J. R. Soc. Interface* **2009**, *6*, 111–122. [CrossRef]
3. Wang, B.; Anthony, S.M.; Bae, S.C.; Granick, S. Anomalous yet Brownian. *Proc. Natl. Acad. Sci.* **2009**, *106*, 15160–15164. [CrossRef]
4. Leptos, K.C.; Guasto, J.S.; Gollub, J.P.; Pesci, A.I.; Goldstein, R.E. Dynamics of Enhanced Tracer Diffusion in Suspensions of Swimming Eukaryotic Microorganisms. *Phys. Rev. Lett.* **2009**, *103*, 198103–198107. [CrossRef] [PubMed]
5. Wang, B.; Kuo, J.; Bae, S.C.; Granick, S. When Brownian diffusion is not Gaussian. *Nat. Mater.* **2012**, *11*, 481–485. [CrossRef]
6. Lampo, T.J.; Stylianidou, S.; Backlund, M.P.; Wiggins, P.A.; Spakowitz, A.J. Cytoplasmic RNA-Protein Particles Exhibit Non-Gaussian Subdiffusive Behavior. *Biophys J.* **2017**, *112*, 532–542. [CrossRef] [PubMed]
7. Sabri, A.; Xu, X.; Krapf, D.; Weiss, M. Elucidating the Origin of Heterogeneous Anomalous Diffusion in the Cytoplasm of Mammalian Cells. *Phys. Rev. Lett.* **2020**, *125*, 058101–058107. [CrossRef]
8. Weeks, E.R.; Crocker, J.C.; Levitt, A.C.; Schofield, A.; Weitz, D.A. Three-Dimensional Direct Imaging of Structural Relaxation Near the Colloidal Glass Transition. *Science* **2000**, *287*, 627–631. [CrossRef]
9. Kegel, W.K.; van Blaaderen, A. Direct Observation of Dynamical Heterogeneities in Colloidal Hard-Sphere Suspensions. *Science* **2000**, *287*, 290–293. [CrossRef]
10. Chakraborty, I.; Roichman, Y. Disorder-induced Fickian, yet non-Gaussian diffusion in heterogeneous media. *Phys. Rev. Res.* **2020**, *2*, 022020–022025. [CrossRef]
11. Lavaud, M.; Salez, T.; Louyer, Y.; Amarouchene, Y. Surface Force Measurements Using Brownian Particles. *arXiv* **2020**, arXiv:2012.05512.
12. Chubynsky, M.V.; Slater, G.W. Diffusing Diffusivity: A Model for Anomalous, yet Brownian, Diffusion. *Phys. Rev. Lett.* **2014**, *113*, 098302–098307. [CrossRef]
13. Miyaguchi, T.; Akimoto, T.; Yamamoto, E. Langevin equation with fluctuating diffusivity: A two-state model. *Phys. Rev. E* **2016**, *94*, 012109–012130. [CrossRef] [PubMed]
14. Chechkin, A.V.; Seno, F.; Metzler, R.; Sokolov, I.M. Brownian yet Non-Gaussian Diffusion: From Superstatistics to Subordination of Diffusing Diffusivities. *Phys. Rev. X* **2017**, *7*, 021002–021022. [CrossRef]
15. Sposini, V.; Chechkin, A.V.; Seno, F.; Pagnini, G.; Metzler, R. Random diffusivity from stochastic equations: Comparison of two models for Brownian yet non-Gaussian diffusion. *New J. Phys.* **2018**, *20*, 043044–043077. [CrossRef]
16. Lanoiselée, Y.; Grebenkov, D.S. A model of non-Gaussian diffusion in heterogeneous media. *J. Phys. A* **2018**, *51*, 145602. [CrossRef]
17. Sposini, V.; Chechkin, A.; Metzler, R. First, passage statistics for diffusing diffusivity. *J. Phys. A* **2018**, *52*, 04LT01. [CrossRef]
18. Ślęzak, J.; Burnecki, K.; Metzler, R. Random coefficient autoregressive processes describe Brownian yet non-Gaussian diffusion in heterogeneous systems. *New J. Phys.* **2019**, *21*, 073056. [CrossRef]
19. Grebenkov, D.; Sposini, V.; Metzler, R.; Oshanin, G.; Seno, F. Exact first-passage time distributions for three random diffusivity models. *J. Phys. A* **2020**. [CrossRef]
20. Wang, W.; Seno, F.; Sokolov, I.M.; Chechkin, A.V.; Metzler, R. Unexpected crossovers in correlated random-diffusivity processes. *New J. Phys.* **2020**, *22*, 083041. [CrossRef]
21. Jain, R.; Sebastian, K.L. Diffusion in a Crowded, Rearranging Environment. *J. Phys. Chem. B* **2016**, *120*, 3988–3992. [CrossRef] [PubMed]
22. Barkai, E.; Burov, S. Packets of Diffusing Particles Exhibit Universal Exponential Tails. *Phys. Rev. Lett.* **2020**, *124*, 060603. [CrossRef] [PubMed]
23. Wang, W.; Barkai, E.; Burov, S. Large Deviations for Continuous Time Random Walks. *Entropy* **2020**, *22*, 697. [CrossRef]
24. Pacheco-Pozo, A.; Sokolov, I.M. Large Deviation in Continuous Time Random Walks. *arXiv* **2020**, arXiv:2012.09656.
25. Samanta, N.; Chakrabarti, R. Tracer diffusion in a sea of polymers with binding zones: Mobile vs. frozen traps. *Soft Matter* **2016**, *12*, 8554–8563. [CrossRef]
26. Kumar, P.; Theeyancheri, L.; Chaki, S.; Chakrabarti, R. Transport of probe particles in a polymer network: Effects of probe size, network rigidity and probe–polymer interaction. *Soft Matter* **2019**, *15*, 8992–9002. [CrossRef] [PubMed]
27. Baldovin, F.; Orlandini, E.; Seno, F. Polymerization Induces Non-Gaussian Diffusion. *Front. Phys.* **2019**, *7*, 124. [CrossRef]
28. Hidalgo-Soria, M.; Barkai, E. Hitchhiker model for Laplace diffusion processes. *Phys. Rev. E* **2020**, *102*, 012109. [CrossRef]
29. Yin, Q.; Li, Y.; Marchesoni, F.; Nayak, S.; Ghosh, P. Non-Gaussian Normal Diffusion in Low Dimensional Systems. *arXiv* **2021**, arXiv:2101.06875.
30. Goswami, K.; Sebastian, K.L. Exact solution to the first-passage problem for a particle with a dichotomous diffusion coefficient. *Phys. Rev. E* **2020**, *102*, 042103. [CrossRef]
31. Bouchaud, J.; Comtet, A.; Georges, A.; Le Doussal, P. Classical diffusion of a particle in a one-dimensional random force field. *Ann. Phys.* **1990**, *201*, 285 – 341. [CrossRef]

32. Monthus, C. Anomalous diffusion, localization, aging, and subaging effects in trap models at very low temperature. *Phys. Rev. E* **2003**, *68*, 036114. [CrossRef] [PubMed]
33. Burov, S.; Barkai, E. Time Transformation for Random Walks in the Quenched Trap Model. *Phys. Rev. Lett.* **2011**, *106*, 140602. [CrossRef] [PubMed]
34. Burov, S.; Barkai, E. Weak subordination breaking for the quenched trap model. *Phys. Rev. E* **2012**, *86*, 041137. [CrossRef]
35. Luo, L.; Yi, M. Non-Gaussian diffusion in static disordered media. *Phys. Rev. E* **2018**, *97*, 042122. [CrossRef]
36. Luo, L.; Yi, M. Quenched trap model on the extreme landscape: The rise of subdiffusion and non-Gaussian diffusion. *Phys. Rev. E* **2019**, *100*, 042136. [CrossRef] [PubMed]
37. Postnikov, E.B.; Chechkin, A.; Sokolov, I.M. Brownian yet non-Gaussian diffusion in heterogeneous media: from superstatistics to homogenization. *New J. Phys.* **2020**, *22*, 063046. [CrossRef]
38. Regev, S.; Grønbech-Jensen, N.; Farago, O. Isothermal Langevin dynamics in systems with power-law spatially dependent friction. *Phys. Rev. E* **2016**, *94*, 012116. [CrossRef] [PubMed]
39. Radice, M.; Onofri, M.; Artuso, R.; Cristadoro, G. Transport properties and ageing for the averaged Lévy–Lorentz gas. *J. Phys.* **2019**, *53*, 025701. [CrossRef]
40. Barkai, E. Fractional Fokker–Planck equation, solution, and application. *Phys. Rev. E* **2001**, *63*, 046118. [CrossRef]
41. Kärger, J. NMR self-diffusion studies in heterogeneous systems. *Adv. Colloid Interface Sci.* **1985**, *23*, 129–148. [CrossRef]
42. Margolin, G.; Barkai, E. Aging correlation functions for blinking nanocrystals, and other on–off stochastic processes. *J. Chem. Phys.* **2004**, *121*, 1566–1577. [CrossRef]
43. Aharony, A.; Entin-Wohlman, O.; Chowdhury, D.; Dattagupta, S. Is Telegraph Noise A Good Model for the Environment of Mesoscopic Systems? *J. Stat. Phys.* **2019**, *175*, 704–724. [CrossRef]
44. Yamamoto, E.; Akimoto, T.; Mitsutake, A.; Metzler, R. Universal relation between instantaneous diffusivity and radius of gyration of proteins in aqueous solution. *arXiv* **2020**, arXiv:2009.06829.
45. Kanazawa, K.; Sano, T.G.; Cairoli, A.; Baule, A. Loopy Lévy flights enhance tracer diffusion in active suspensions. *Nature* **2020**, *579*, 364–367. [CrossRef]
46. Godrèche, C.; Luck, J.M. Statistics of the Occupation Time of Renewal Processes. *J. Stat. Phys.* **2001**, *104*, 489–524. [CrossRef]
47. Miyaguchi, T.; Uneyama, T.; Akimoto, T. Brownian motion with alternately fluctuating diffusivity: Stretched-exponential and power-law relaxation. *Phys. Rev. E* **2019**, *100*, 012116. [CrossRef]
48. Cox, D.R. *Renewal Theory*; Methuen Publishing: North Yorkshire, UK, 1962.
49. Bel, G.; Barkai, E. Occupation times and ergodicity breaking in biased continuous time random walks. *J. Phys. Condens. Matter* **2005**, *17*, S4287–S4304. [CrossRef]
50. Schulz, J.H.P.; Barkai, E.; Metzler, R. Aging Effects and Population Splitting in Single-Particle Trajectory Averages. *Phys. Rev. Lett.* **2013**, *110*, 020602. [CrossRef]
51. Schulz, J.H.P.; Barkai, E.; Metzler, R. Aging Renewal Theory and Application to Random Walks. *Phys. Rev. X* **2014**, *4*, 011028. [CrossRef]
52. Masoliver, J.; Weiss, G.H. Finite-velocity diffusion. *Eur. J. Phys.* **1996**, *17*, 190–196. [CrossRef]
53. Masoliver, J.; Lindenberg, K. Continuous time persistent random walk: A review and some generalizations. *Eur. Phys. J. B* **2017**, *90*, 107. [CrossRef]
54. Beck, C.; Cohen, E. Superstatistics. *Phys. A Stat. Mech. Appl.* **2003**, *322*, 267–275. [CrossRef]
55. Laplace, P.S. *Mémoirs Présentés à l' Académie des Sciences*; Académie Des Sciences: Paris, France, 1774.
56. Wilson, E.B. First, and Second Laws of Error. *J. Am. Stat. Assoc.* **1923**, *18*, 841–851. [CrossRef]
57. Burov, S.; Jeon, J.H.; Metzler, R.; Barkai, E. Single particle tracking in systems showing anomalous diffusion: The role of weak ergodicity breaking. *Phys. Chem. Chem. Phys.* **2011**, *13*, 1800–1812. [CrossRef] [PubMed]
58. Metzler, R.; Jeon, J.H.; Cherstvy, A.G.; Barkai, E. Anomalous diffusion models and their properties: Non-stationarity, non-ergodicity, and ageing at the centenary of single particle tracking. *Phys. Chem. Chem. Phys.* **2014**, *16*, 24128–24164. [CrossRef] [PubMed]
59. Grebenkov, D.S. Time-averaged mean square displacement for switching diffusion. *Phys. Rev. E* **2019**, *99*, 032133. [CrossRef]
60. Wang, W.; Cherstvy, A.G.; Chechkin, A.V.; Thapa, S.; Seno, F.; Liu, X.; Metzler, R. Fractional Brownian motion with random diffusivity: Emerging residual nonergodicity below the correlation time. *J. Phys. A* **2020**, *53*, 474001. [CrossRef]
61. From Wolfram Research. Available online: https://functions.wolfram.com/Bessel-TypeFunctions/BesselI/26/01/01/ (accessed on 24 November 2020).

Article

Local Analysis of Heterogeneous Intracellular Transport: Slow and Fast Moving Endosomes

Nickolay Korabel [1,*], Daniel Han [1,2,3], Alessandro Taloni [4], Gianni Pagnini [5,6], Sergei Fedotov [1], Viki Allan [2] and Thomas Andrew Waigh [3,*]

1 Department of Mathematics, The University of Manchester, Manchester M13 9PL, UK; daniel.han@manchester.ac.uk (D.H.); sergei.fedotov@manchester.ac.uk (S.F.)
2 School of Biological Sciences, The University of Manchester, Manchester M13 9PT, UK; viki.allan@manchester.ac.uk
3 Biological Physics, Department of Physics and Astronomy, The University of Manchester, Manchester M13 9PL, UK
4 CNR—Consiglio Nazionale delle Ricerche, Istituto dei Sistemi Complessi, Via dei Taurini 19, 00185 Roma, Italy; alessandro.taloni@isc.cnr.it
5 BCAM—Basque Center for Applied Mathematics, Mazarredo 14, 48009 Bilbao, Spain; gpagnini@bcamath.org
6 Ikerbasque—Basque Foundation for Science, Plaza Euskadi 5, 48009 Bilbao, Spain
* Correspondence: nickolay.korabel@manchester.ac.uk (N.K.); t.a.waigh@manchester.ac.uk (T.A.W.)

Citation: Korabel, N.; Han, D.; Taloni, A.; Pagnini, G.; Fedotov, S.; Allan, V.; Waigh T.A. Local Analysis of Heterogeneous Intracellular Transport: Slow and Fast Moving Endosomes. *Entropy* **2021**, *23*, 958. https://doi.org/10.3390/e23080958

Academic Editors: Janusz Szwabiński and Aleksander Weron

Received: 14 June 2021
Accepted: 23 July 2021
Published: 27 July 2021

Abstract: Trajectories of endosomes inside living eukaryotic cells are highly heterogeneous in space and time and diffuse anomalously due to a combination of viscoelasticity, caging, aggregation and active transport. Some of the trajectories display switching between persistent and anti-persistent motion, while others jiggle around in one position for the whole measurement time. By splitting the ensemble of endosome trajectories into slow moving subdiffusive and fast moving superdiffusive endosomes, we analyzed them separately. The mean squared displacements and velocity autocorrelation functions confirm the effectiveness of the splitting methods. Applying the local analysis, we show that both ensembles are characterized by a spectrum of local anomalous exponents and local generalized diffusion coefficients. Slow and fast endosomes have exponential distributions of local anomalous exponents and power law distributions of generalized diffusion coefficients. This suggests that heterogeneous fractional Brownian motion is an appropriate model for both fast and slow moving endosomes. This article is part of a Special Issue entitled: "Recent Advances In Single-Particle Tracking: Experiment and Analysis" edited by Janusz Szwabiński and Aleksander Weron.

Keywords: heterogeneous; anomalous diffusion; endosomes

1. Introduction

Intracellular transport of organelles, such as endosomes, has been described by anomalous diffusion caused by different mechanisms [1,2]. Various models have been proposed to describe it, such as fractional Brownian motion (FBM), continuous time random walks and fractional Langevin equations [3]. However, which of these models is the best is a current topic of much debate.

To decipher which mechanism is at work and determine the appropriate mathematical model to describe it, a large ensemble of trajectories is necessary. Modern experimental techniques facilitate the tracking of large ensembles of intracellular objects for considerable amounts of time. Therefore, the extraction of meaningful statistical information from trajectories is becoming an important issue. The traditional statistical analysis of trajectories includes quantification of ensemble evolution in time and space using the ensemble-averaged mean squared displacements (EMSD), time-averaged MSD (TMSD), probability density functions of displacements and correlation functions. As the accessible measurement time in experiments increases with better live-cell microscopy techniques, the accurate analysis of single trajectories has become possible [4]. New methods of trajectory analysis were developed, such as local time-averaged MSD [5], first passage probability analysis [6–8] and time-averaged diffusion coefficients [9].

Improved microscopy imaging, tracking and analysis methods revealed the intrinsic spatial and temporal heterogeneity within individual trajectories of numerous biological processes [5,10–18]. Significant progress has also been made in analysis and interpretation of superresolution single particle trajectories [19–23]. Recently, individual trajectories of quantum dots in the cytoplasm of living cultured cells were found to perform subdiffusive motion of the FBM type with switching between two distinct mobility states [24]. In contrast to homogeneous systems, heterogeneous trajectories are most prominently described by broad distributions of diffusivities and anomalous exponents, an exponential probability distribution of diffusivities and a Laplace probability distribution of displacements [25]. These observations led to the development of various heterogeneous diffusion models [26–39].

Recently, the intracellular transport of endosomes in eukaryotic cells was shown to be described by spatiotemporal heterogeneous fractional Brownian motion (hFBM) with non-constant Hurst exponents [40]. By analyzing the local motion of endosomes, we found that it is characterized by power-law probability distributions of displacements and displacement increments, exponential probability distributions of local anomalous exponents and power-law probability distributions of local generalized diffusion coefficients. In this paper, we split the ensemble of endosomes into slow and fast moving vesicles, which is the main difference between this study and that of [40]. This splitting allows us to study sub-ensembles separately in addition to studying the ATP driven active transport of endosomes. In particular, there is the central question: What is the appropriate mathematical model to describe the subdiffusive transport of slow moving endosomes? By analyzing locally the slow and fast endosomal trajectories, we find that both are characterized by exponential distributions of anomalous exponents and power-law distributions of generalized diffusion coefficients. This suggests that hFBM is an appropriate model for both slow and fast endosomes.

Endosome trajectories are composed of segments of active and passive motion, and therefore they could be further decomposed into directed runs and random motion. We segmented endosomal trajectories in this way in [10]. In this study, we separated endosome trajectories into superdiffusive trajectories and subdiffusive trajectories for their whole duration. Subdiffusive trajectories do not contain segments of directed movement and cannot be segmented further into active and passive motion. In contrast, fast superdiffusive trajectories can be further segmented. We leave the segmentation of fast trajectories into directed runs and random motion for future work.

2. Materials and Methods

2.1. Experimental Trajectories

We studied a large ensemble of two dimensional experimental trajectories, $\mathbf{r}(t) = \{x(t), y(t)\}$, of early endosomes in a stable MRC5 cell line expressing GFP-Rab5. Trajectories were obtained from tracking wide-field fluorescence microscopy videos (see [10] for experimental details). We studied 103,361 experimental trajectories of early endosomes, the same data acquired in [10]. Three live-cell microscopy videos of MRC5 cells stably expressing GFP-Rab5 could be found in the Supplementary Material (https://zenodo.org/record/5106450#.YPBsEuhKhPY, accessed on 23 July 2021). An example of experimental trajectories is shown in Figure A1. The endosomes were tracked using an automated tracking software (AITracker, based on a convolutional neural network) [41]. Currently, it is not yet feasible to determine the diameter of endosomes in these experiments, because they are diffraction limited. Thus, it was possible to track the centers of endosomes with sub-pixel accuracy, but not the sizes of the smaller endosomes (less than 200 nm). The duration of all trajectories, T, has a good fit to a power law distribution, $T^{-1.85}$ [40], which is a manifestation of the heterogeneity of the trajectories. Slow moving endosomes stay longer within the observation volume and therefore have longer trajectories than fast moving endosomes, leading to the emergence of the power-law probability distribution for the trajectories' duration.

2.2. Splitting of Ensemble into Slow and Fast Moving Endosomes

We split ensemble of trajectories into slow and fast moving endosomes using the distance traveled by endosomes:

$$R(t) = \sqrt{(x(t) - x(0))^2 + (y(t) - y(0))^2}. \tag{1}$$

Trajectories which possess active motion have periods of rapid increase or decrease of R (Figure 1A). Fast trajectories which have active motion are defined as $\max\{R(t)\} > \epsilon$ and slow trajectories which exhibit only passive motion are defined by $\max\{R(t)\} < \epsilon$. Here, $\max\{R(t)\}$ denotes the maximum values of $R(t)$ attained in the time interval $(0, t)$ and ϵ is the threshold. We choose the threshold $\epsilon = 0.25$ μm. In the Appendix, we show that changing the threshold to $\epsilon = 0.2$ μm in the splitting does not qualitatively change the results. Therefore, we define fast moving endosomes as those that, in the time interval $(0, t)$, experienced at least one period of active motion and the maximum distance travelled from the origin exceeds the threshold of $\epsilon = 0.25$ μm. Otherwise, an endosome is defined as slow moving. Small variations of the threshold value do not affect the EMSDs of slow and fast moving endosomes, which suggests that the splitting method is robust (Figure A3).

Changing the splitting threshold from $\max\{R(t)\} = 0.25$ μm to $\max\{R(t)\} = 0.2$ μm, the increase of the number of slow trajectories was 12%. Therefore, in addition to the method of splitting trajectories which uses the minimum travelled distance, we also tested a second method, which makes use of the time-dependent Hurst exponent $H(t)$ neural network (NN) estimate at the single trajectory level [10]. The procedure is as follows: (1) estimate the time-dependent anomalous exponent α^{NN} using the NN; (2) if the anomalous exponent α^{NN} is superdiffusive $\alpha^{NN}(t) > 1$ for more than 4 consecutive time points, the endosome is considered as fast moving. Otherwise, the endosome is labeled as slow moving (see Figure A2). The correct implementation of the NN procedure requires a minimum time window [10] that is larger than the duration of some of the endosomal trajectories. Hence, short trajectories were discarded in this analysis. The similarity of the distributions of generalized diffusion coefficients (Figures A3B and A4B) suggests that the chosen threshold $\max\{R(t)\} = 0.25$ μm was reasonable. Alternative methods of binary classification could be performed using the first passage probability analysis [7] or implementing the normalized radius of gyration of each trajectory [42].

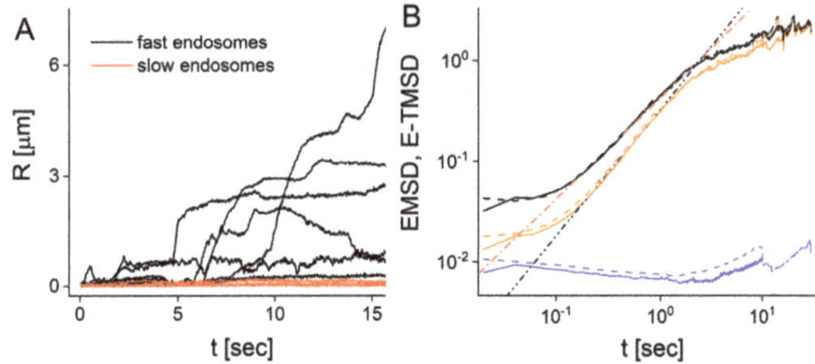

Figure 1. Endosomes are split into slow and fast moving: (**A**) Distance $R(t)$ traveled by fast (black curves) and slow (red curves) endosomes (nine sample experimental trajectories are shown). Most experimental trajectories possess active motion visible as a rapid increase or decrease of R; (**B**) EMSDs (solid curves) and E-TMSDs (dashed curves) of fast (black curves) and slow (blue curves) endosomes compared with EMSD and E-TMSD of all trajectories (orange curves). The dashed-dotted and dashed-double-dotted lines represent $t^{1.26}$ and t functions.

2.3. Ensemble and Time Averaged Mean Squared Displacements

From the two dimensional experimental trajectories $\mathbf{r}(t) = \{x(t), y(t)\}$, we calculated the ensemble-averaged mean squared displacement (EMSD) as

$$\mathrm{EMSD}(t) = \frac{\langle \mathbf{r} \rangle^2(t)}{l^2}, \tag{2}$$

where l is the length scale which we choose $l = 1$ µm,

$$\langle \mathbf{r} \rangle^2(t) = \Big\langle (x_i(t) - x_i(0))^2 + (y_i(t) - y_i(0))^2 \Big\rangle, \tag{3}$$

where the angular brackets denotes averaging over an ensemble of trajectories, $\langle A \rangle = \sum_{i=1}^{N} A_i / N$ and N is the number of trajectories in the ensemble.

By fitting the EMSD to power law functions, the anomalous exponent α and the generalized diffusion coefficient D_α can be extracted using

$$\mathrm{EMSD}(t) = 4D_\alpha \left(\frac{t}{\tau} \right)^{\alpha}, \tag{4}$$

where α and D_α are constants which characterize averaged transport properties of ensemble of endosomal trajectories. The time scale $\tau = 1$ s and the length scale $l = 1$ µm are introduced in order to make the generalized diffusion coefficient D_α dimensionless.

The time-averaged mean squared displacement (TMSD) of an individual trajectory $\{x_i, y_i\}$ of a duration T is calculated as:

$$\mathrm{TMSD}_i(t) = \frac{\overline{\delta^2(t)}}{l^2}, \tag{5}$$

where l is the length scale, for which we chose $l = 1$ µm, and

$$\overline{\delta^2(t)} = \frac{\int_0^{T-t} (x_i(t'+t) - x_i(t'))^2 + (y_i(t'+t) - y_i(t'))^2) dt'}{T - t}. \tag{6}$$

TMSDs of individual trajectories are averaged further over the ensemble of trajectories to get the ensemble-time-averaged MSD (E-TMSD):

$$\text{E-TMSD}(t) = \langle \mathrm{TMSD}_i(t) \rangle, \tag{7}$$

where the angular brackets denotes averaging over an ensemble of trajectories as before.

2.4. Local Analysis of Endosomal Trajectories

The time-local statistical analysis was implemented as follows. We considered only the portion of a single endosomal trajectory within a window of size W and centered around the time t, i.e., $(t - W/2, t + W/2)$. We calculated the TMSD within this chunk of trajectory only: this is the reason for the acronym L-TMSD, i.e., the local TMSD. The experimental detection of the endosomal motion is achieved with the frame rate $1/\Delta t$ s^{-1}, hence $t = i\Delta t$ (here, $i = 0, 1, 2, \ldots$ is the time index) and $W = N\Delta t$, with $N > 10$. The first 10 points of the L-TMSD were fitted with the power-law function

$$\text{L-TMSD} = 4D^L(t) \left(\frac{t'}{\tau} \right)^{\alpha^L(t)}, \tag{8}$$

where $t' = 10\Delta t$. $\alpha^L(t)$ and $D_{\alpha^L}(t)$ are the local anomalous exponent and generalized diffusion coefficient, respectively. We iterate this procedure by shifting the time window of a single Δt $(i \to i+1)$ until the end of the experimental endosomal trace, thus obtaining

$\alpha^L(t)$ and $D_{\alpha^L}(t)$ along the entire trajectory. Notice that α^L and D^L are not constants in time and they vary, being local properties of each endosomal trajectory.

2.5. The Time and Ensemble-Time Averaged Velocity Auto-Correlation Functions

The time averaged auto-correlation function (TVACF) along a single trajectory is defined as:

$$\mathrm{TVACF}_i(t) = \frac{\int_0^{T-t-\tau} \vec{v}(t'+t)\vec{v}(t')dt'}{T-t-\tau}, \tag{9}$$

where $\vec{v} = \frac{\vec{r}(t+\tau)-\vec{r}(t)}{\tau}$. TVACFs of individual trajectories are averaged further over the ensemble of trajectories to get the ensemble-time averaged VACF (E-TVACF):

$$\text{E-TVACF}(t) = \langle \mathrm{TVACF}_i(t) \rangle, \tag{10}$$

where the angular brackets denotes averaging over an ensemble of trajectories. The velocity autocorrelation function was suggested as a tool to distinguish between subdiffusion models [43].

3. Results

We split the ensemble of endosomes into slow and fast moving vesicles using the two methods described above (see Methods, Figures 1A and A2). For both slow and fast endosomes, the EMSDs and E-TMSDs show similar behavior, which suggests ergodicity (see Methods and Figure 1B). MSDs of slow endosomes are not increasing in time, which confirms that these trajectories have no active periods of motion. Surprisingly, we found that both EMSDs and E-TMSDs of slow endosomes are decreasing functions of time, which to our knowledge has never been observed before. We explain this behavior in terms of the coupling between the average diffusivities of slow trajectories and their duration (see Figure 4 and the discussion below). Conversely, MSDs of fast endosomes are increasing functions of time in the intermediate time scale $(0.2, 2)$ s. The anomalous exponent extracted from EMSD or E-TMSD of fast endosomes is $\alpha \simeq 1$, smaller than the anomalous exponent obtained by considering all trajectories without distinction into fast or slow, i.e., $\alpha \simeq 1.26$. Notice that two subdiffusive regimes characterize the MSD time behavior for fast and all trajectories. The first, at small time scales ($t \leq 10^{-1}$ s), can be attributed to the measurement errors [44–46]. The second, at longer time scales ($t > 10$ s), was shown to be spurious and originate from the coupling of the trajectories' duration and their diffusivities [40,47]. We suggest that, due to this coupling, the anomalous exponents deduced from the power-law fit of EMSD and E-TMSD, do not capture the essential characteristics of the endosome superdiffusive motility, nor shed light on its fundamental aspects. Therefore, to reveal the effect of the duration of trajectories on the statistical analysis, we consider only trajectories longer than a certain threshold T [40].

Figure 2A,B shows the EMSDs and E-TMSDs of slow and fast endosomes, considering only experimental trajectories with duration longer than T seconds (2 or 8 s). Unlike the slow moving endosomes, the MSDs of fast vesicles (Figure 2B) present similar qualitative behaviors by choosing $T = 2$ s, $T = 8$ s or no T at all (all the fast molecules considered as in Figure 1B, black curve). However, in the intermediate regime, the superdiffusive behavior becomes more and more apparent, $\propto t^{1.26}$, and stable. In [40] we found that this process is described by the space-time heterogeneous FBM with the Hurst exponent H that randomly switches between persistent $H > 0.5$ and anti-persistent regimes $H < 0.5$, together with the coupling between the diffusivity and duration of trajectories which account for spurious subdiffusion at longer time scales. Moreover, the EMSD curves obtained for $T = 2$ s and $T = 8$ s deviates considerably from the corresponding E-TMSD curves.

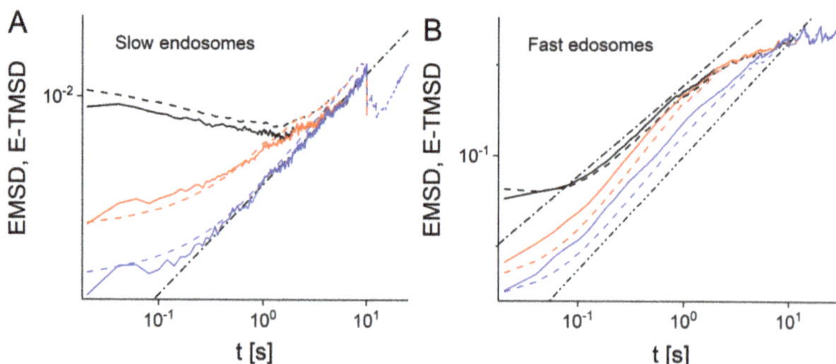

Figure 2. EMSDs and E-TMSDs (solid and dashed curves) of experimental trajectories of slow (**A**) and fast moving endosomes (**B**). Black curves correspond to $T \to \infty$ s. Red and blue curves represent EMSDs and E-TMSDs of experimental trajectories which have duration longer than 2 and 8 s, respectively. The dashed-dotted line in (**A**) represents the function $t^{0.5}$. In (**B**), the dashed-dotted line and the dashed-double-dotted line represent the linear, t, and super-linear, $t^{1.26}$, functions, respectively.

The MSDs of slow endosomes (Figure 2A) display very different, but ergodic, behavior. For $0.01 < t < 2$ s, the MSDs of all slow endosomes decreases in time. On the other side, the MSDs of the sub-ensembles of slow endosomes with $T = 2$ s and $T = 8$ s reveal subdiffusive trends with $\alpha \sim 0.5$. As in the case of fast moving endosomes, we argue that this behavior is due to the coupling between the diffusivity and duration of trajectories. Therefore, we attempt to confirm this hypothesis, by performing simulations of an ensemble of heterogeneous FBM trajectories with constant Hurst exponent $H = 0.25$ (see Figure 4).

The velocity auto-correlation functions (VACF) also confirm the effectiveness of this simple threshold splitting (Figure 3A,B). Indeed, slow and fast endosomes have very different VACFs. Ensemble-time averaged VACFs (E-TVACFs) of fast endosomes (Figure 3B) are positive as expected for superdiffusive motion. In contrast, E-TVACFs of slow endosomes have negative dips at $t = \tau$ and approach zero from negative values (Figure 3A). Such behavior is characteristic of FBM and the generalized Langevin equation but cannot be reproduced by the CTRW model [3].

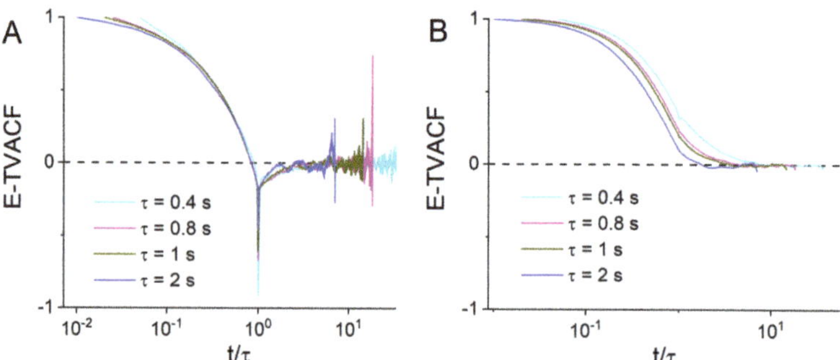

Figure 3. Time-ensemble averaged VACF (E-TVACF) of experimental trajectories of slow (**A**) and fast endosomes (**B**) calculated for different τ given in the legend.

To verify that heterogeneous FBM describes slow moving endosomes, we simulated an ensemble of hFBM trajectories. Individual hFBM trajectories were simulated with constant Hurst exponent $H = 0.25$. For standard FBM, this would correspond to subdiffusive MSDs, $\langle \mathbf{r}^2(t) \rangle \sim t^{2H} \sim t^{0.5}$. The duration of hFBM trajectories was drawn from the power-law distribution $\phi(T) \sim T^{-1.85}$, in accordance with the experimental evidence [40]. The generalized diffusion coefficients were chosen inversely proportional to the duration of trajectories, i.e., $D \sim T^{-0.6}$. As shown in Figure 4, the EMSDs of hFBM trajectories agree well with the experimental data.

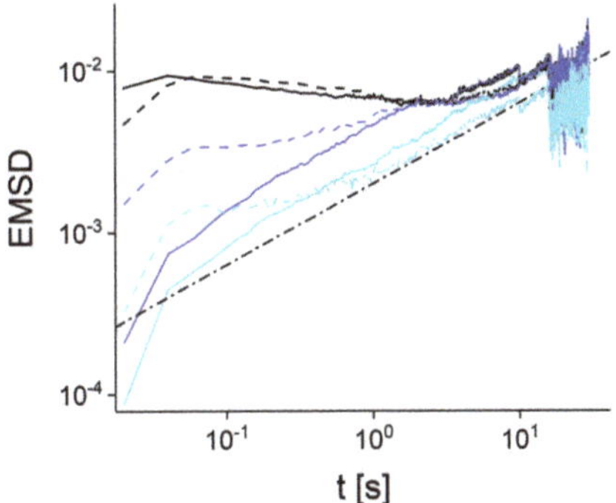

Figure 4. EMSDs calculated for simulated hFBM trajectories (solid lines) as a function of time interval. Black curves correspond to EMSDs of all trajectories of slow endosomes and hFBM, blue curves are EMSDs of trajectories longer than $T = 2$ s and cyan curves are EMSDs of trajectories longer than $T = 8$ s. The subdiffusive behavior with the anomalous exponent $\alpha = 0.5$ is shown as the dashed-dotted line. The EMSDs of slow experimental endosomal trajectories are shown for comparison (dashed lines). Notice that hFBM trajectories were simulated without external noise (measurement error), which led to discrepancy between simulated and experimental EMSDs at small time scale.

We next implemented the local analysis [40] to better characterize the slow and fast endosomal dynamics. We calculated the local TMSDs (L-TMSD) for each experimental trajectory at various times t (Methods). From the fit of L-TMSD to Equation (8), we extracted the local anomalous exponents $\alpha^L(t)$ and the local generalized diffusion coefficients $D_{\alpha^L}(t)$ for slow and fast endosomes separately. The local anomalous exponents $\alpha^L(t)$ and the local generalized diffusion coefficients $D_{\alpha^L}(t)$ appear to be positively correlated both for slow and fast endosomes (see Figure A6). The origin of these correlations is not known and will be investigated in future publications. In [40], we found that PDFs of local anomalous exponents and local generalized diffusion coefficients do not depend on the window size or the time t (stationary) and are best fitted with exponential and power law functions, respectively.

The PDFs of α^L and D_{α^L} for slow and fast endosomes are shown in Figure 5A,B. In both cases, the PDFs of α^L follow an exponential distribution, while those of D_{α^L} are best fitted with a power-law. However, the parameters characterizing the distribution shapes are very different. Furthermore the parameters for the fast endosomes' PDFs coincide with those found by considering all experimental trajectories [40]. This is in agreement with a heterogeneous FBM model of endosomal transport [40], which describes the endosome motion as FBM with non-constant Hurst exponents.

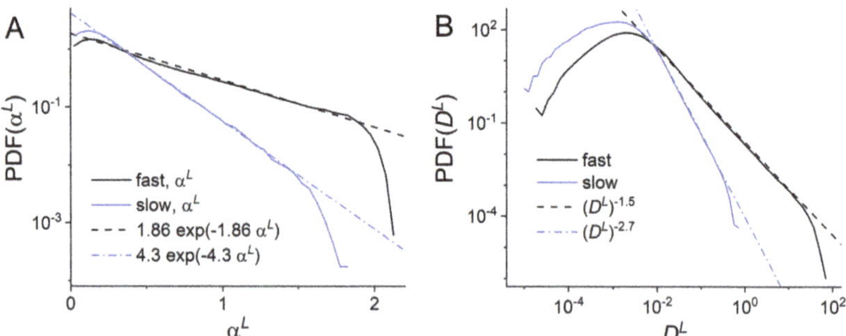

Figure 5. Distribution of local anomalous exponents α^L (**A**) and local generalized diffusion coefficients D^L (**B**) obtained from experimental trajectories of slow and fast endosomes. The dashed and dashed-dotted lines are best fits to exponential (**A**) and power-law PDFs (**B**). In (**A**), they correspond to $1.86\exp(-1.86\alpha^L)$ for PDF of α^L of fast endosomes (dashed line) and $4.3\exp(-4.3\alpha^L)$ for PDF of α^L of slow endosomes (dashed-dotted line). In (**B**) they correspond to $(D^L)^{-1.5}$ for PDF of D^L of fast endosomes (dashed line) and $(D^L)^{-2.7}$ for PDF of D^L of slow endosomes (dashed-dotted line).

Finally, we calculated propagators of experimental trajectories for slow and fast endosomes (Figure 6). Using the power-law forms of distributions of local generalized diffusion coefficients of slow $p_S(D^L) \sim (D^L)^{-1-\gamma_S}$ and fast $p_F(D^L) \sim (D^L)^{-1-\gamma_F}$ endosomes with $\gamma_S \simeq 1.7$ and $\gamma_F \simeq 0.5$ (Figure 5), we fit the propagators with the propagators of hFBM, $\mathrm{PDF}(\xi) \sim |\xi|^{-1-2\gamma}$ with $\gamma = \gamma_S$ for slow endosomes and $\gamma = \gamma_F$ for fast endosomes (see Supplementary Note and [33]). For slow endosomes (Figure 6A), we also compare the experimental PDFs with the analytical propagator for obstructed diffusion in two dimensions, $\xi^{-0.108}\exp(-|\xi|^{1.65})$ [48].

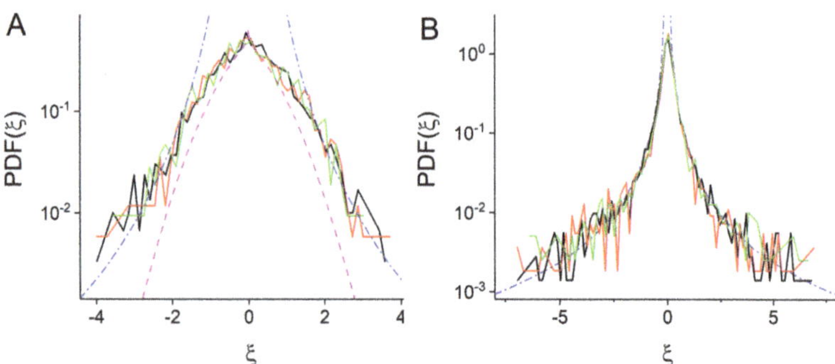

Figure 6. Distribution of scaled x-component of coordinate $\xi = x/\sigma_x$ obtained from experimental trajectories of slow (**A**) and fast (**B**) endosomes. The dashed-dotted lines correspond to power-law fit $|\xi|^{-1-2\gamma_S}$ for slow endosomes ($\gamma_S \simeq 1.7$) and $|\xi|^{-1-2\gamma_F}$ for fast endosomes ($\gamma_F \simeq 0.5$). In (**A**), we also compare PDF of slow endosomes with the analytical propagator for obstructed diffusion (dashed line) [48].

4. Discussion

In this paper, we extend our investigation of the heterogeneous intracellular transport of endosomes based on the local analysis of experimental trajectories [40]. Individual endosomes move for long distances in a heterogeneous way with short bursts of directed motility, interspersed with periods of subdiffusive motion [49,50]. The heterogeneous character of this motion is also manifested as some endosomes are less motile than others. Some endosomes look as if they are jiggling in one position for the whole period of observation. Therefore, we split the ensemble of trajectories into slow and fast moving endosomes. The distinct time behavior of mean squared displacements and velocity autocorrelation functions confirm the effectiveness of these methods. The splitting allowed us to study passive subdiffusive and active superdiffusive transport of endosomes separately.

Comparing the behavior of fast endosomes (MSDs, VACFs and propagators) to the behavior of the entire ensemble, we find that they are most consistent with FBM models [40]. Therefore, we conclude that fast endosomes follow heterogeneous FBM [40]. The ergodicity (Figure 2A) and the VACF (Figure 3A) suggest that slow endosomes are also described by the hFBM or heterogeneous generalized fractional Langevin equation motion. For slow endosomes, crowding and obstruction effects could also lead to subdiffusive behavior [2,4]. It is known that obstructed diffusion has many similarities with FBM such as stationarity of the increments and the equivalence of the time and ensemble MSDs [48,51]. The propagators provide a clear way to distinguish obstructed diffusion from FBM. Therefore, we calculated propagators of experimental slow endosomes and compared them with analytical prediction for the propagator of obstructed diffusion and prediction of heterogeneous fBM. The results shown in Figure 6 indicate that slow endosomes follow hFBM at longer time scales, while on smaller scales obstructed diffusion likely contributes to their subdiffusive behavior as well. Crowding effects remain as a possible source of anomalous diffusion of slow endosomes. Recently, in numerical simulations, lipids in crowded conditions of the membrane were shown to be multifractal and anomalous. The dynamics was no longer described by the mechanism consistent with the fractional Langevin equation or by any single known mechanism. Instead, the motion was found to be non-Gaussian and heterogeneous, yet maintains its ergodic properties [52], which is similar to what we observed for experimental trajectories of slow endosomes.

Both slow and fast endosomal trajectories are found to be highly heterogeneous in space and time. The spatial heterogeneity in the form of coupling between endosome diffusivity and duration of endosome trajectory explains the behavior of the MSDs. Longer trajectories have smaller generalized diffusion coefficients since in experiments slowly moving endosomes with smaller diffusion coefficients stay longer in the field of view, having longer durations. For slow and fast endosomes, we can conclude that EMSD and E-TMSD are not adequate to describe the large heterogeneity exhibited in space and time. Therefore, we applied a time local analysis of individual trajectories.

From the local analysis, we found that slow and fast endosomal trajectories are both characterized by exponentially distributed anomalous exponents and power-law distributed generalized diffusion coefficients. However, the parameters of these distributions are different. Although the factors that cause the power-law distributed generalized diffusion coefficients for slow and fast endosomes could be different, some common factors can exist. One of them could be the scale free properties of endosomal networks [53]. Hence, the differences in endosome diameters could generate distinct diffusive properties intrinsic to each endosome. Heterogeneous diffusion generated by the fluctuations of molecular size was found in single-molecule experiments within the cell [14,18,42]. Another common factor promoting power-law distributions of generalized diffusion coefficients could be non-specific interactions with the endoplasmic reticulum or other organelles and large intracellular structures. Recently, non-specific interactions were shown to generate heterogeneous diffusion of nanosized objects in mammalian cells [47].

Our analysis of endosomal transport would be valuable for both fundamental cell biology and nanomedicine applications such as drug and gene delivery. In these applications, nanoparticles are often used as cargo-carrying vesicles, which in turn utilize the endosomal network for their intracellular transport. For example, gold nanoparticles were shown to cluster inside endosomes and move via sub- and superdiffusion [54]. Our results would also be useful for the nanoparticle enhanced radiation therapy of cancer [55–57] where clusters of nanoparticles inside endosomes are used for dose enhancement.

In the future, we expect microscopy techniques will improve in tandem with tracking algorithms, providing datasets with larger ranges of time scales and improved resolution. Thus, further subclassification of ensembles of endosomal tracks (beyond the binary fast and slow separation) will become possible towards the ultimate goal of single molecule specificity. Increasing the dynamic range (to submillisecond time scales) will allow the stepping motion of the motor proteins (kinesin and dynein) attached to microtubules to be connected with the spectra of α and D_α for the fast moving endosomes at a fundamental level.

Supplementary Materials: The following are available online at https://zenodo.org/record/5106450#.YPBsEuhKhPY, Video S1: Videos for MRC5 cells stably expressing GFP-Rab5.

Author Contributions: N.K. analyzed the data. N.K. and A.T. performed computer simulations. N.K., A.T. and T.A.W. wrote the paper. D.H. designed and trained the neural network. N.K., D.H., A.T., G.P., S.F., V.A. and T.A.W. interpreted the data and edited and approved the final version. All authors have read and agreed to the published version of the manuscript.

Funding: N.K., S.F. and V.A. acknowledge financial support from EPSRC Grant No. EP/V008641/1. D.H. acknowledges financial support from the Wellcome Trust Grant No. 215189/Z/19/Z. G.P. is supported by the Basque Government through the BERC 2018–2021 programs and the Spanish Ministry of Economy and Competitiveness MINECO through the BCAM Severo Ochoa excellence accreditation SEV-2017-0718.

Data Availability Statement: The data that support the findings of this study are available from the corresponding author upon reasonable request.

Conflicts of Interest: The authors declare no conflict of interest.

Appendix A

Figure A1: An example of the experimental endosome trajectories measured in MRC5 cells stably expressing GFP-Rab5. See the main text for details.

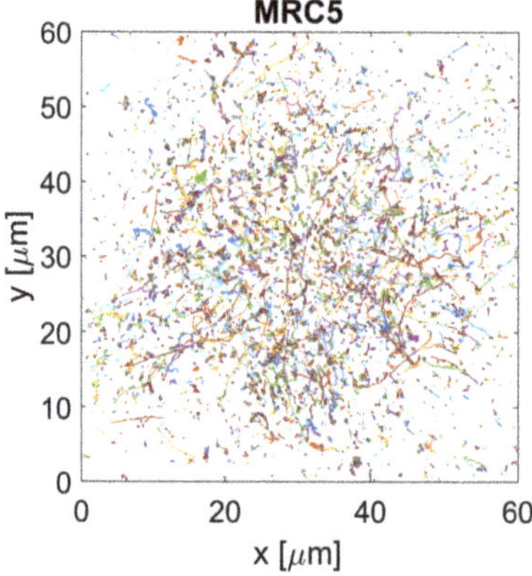

Figure A1. An example of the experimental endosomal trajectories (30,000 trajectories are shown).

Figure A2: Two splitting methods used to separate endosome trajectories into slow and fast moving endosomes. The first method uses the maximum distance traveled $R(t)$. The second method uses the time-dependent anomalous exponent $H(t)$ estimated with the neural network. An example of the two trajectories is shown, which were classified as slow and fast by both methods. See the main text for details of the methods.

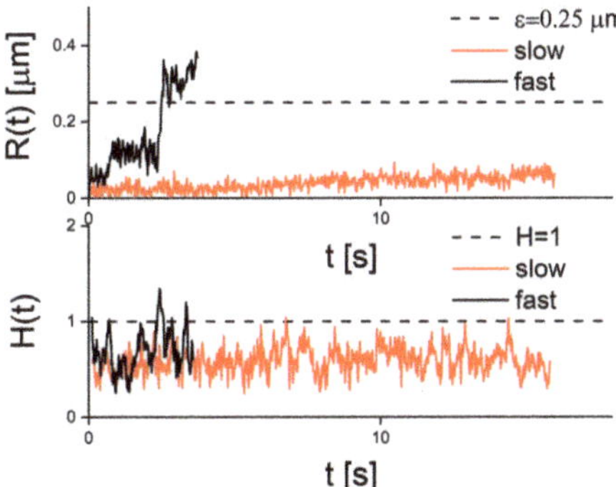

Figure A2. An example of experimental trajectories of slow and fast moving endosomes obtained using the maximum distance traveled $R(t)$ (**top**) and the time-dependent anomalous exponent $H(t)$ estimated with the neural network (**bottom**).

Figure A3: The EMSD of slow and fast moving endosomes calculated with the splitting method, which uses the maximum distance traveled $R(t)$. Two values of the threshold ϵ produce qualitatively similar results, which suggests that the splitting method is robust against small variations of the threshold.

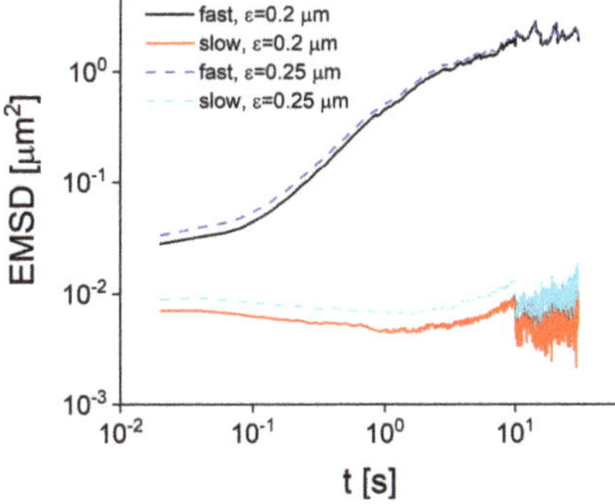

Figure A3. The EMSD of experimental trajectories of slow and fast moving endosomes calculated with the splitting method, which uses the maximum distance traveled $R(t)$ for two values of the threshold: $\epsilon = 0.2$ μm and $\epsilon = 0.25$ μm.

Figure A4: Comparison of distributions of anomalous exponents α^{NN} and generalized diffusion coefficients D^{NN} and local anomalous exponents α^{L} and D^{L} of slow moving endosomes. Anomalous exponents α^{NN} were estimated using a neural network with window size 0.26 s. The generalized diffusion coefficients D^{NN} were estimated by fitting the local TMSD of the trajectory with the power law $D^{NN}t^{\alpha^{NN}}$. The distribution of α^{NN} has a maximum of 0.6 and decays faster than the distribution of α^{L}. This may be because many short trajectories are missing in the NN analysis, since the NN could analyze trajectories with durations longer than its window size [10]. The distributions of generalized diffusion coefficients (right panel), on the other hand, are similar to each other.

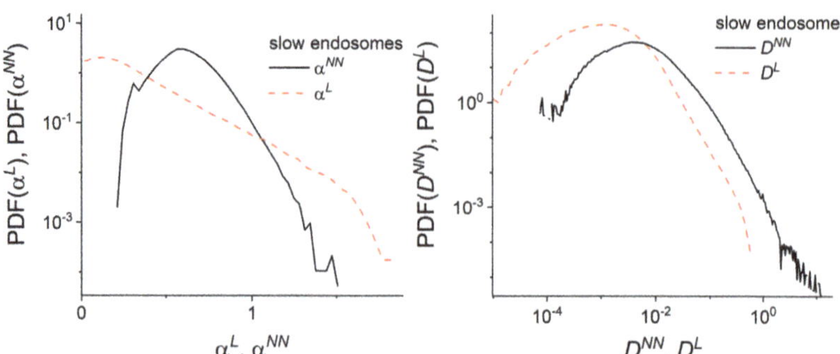

Figure A4. Slow endosomes. (**Left**) Distribution of anomalous exponents α^{NN} of slow moving endosome trajectories (the solid curve) compared with the distribution of local anomalous exponents α^L of slow moving endosome trajectories (the dashed curve); (**Right**) Distribution of generalized diffusion coefficients D^{NN} of slow moving endosome trajectories (the solid curve) compared with the distribution of local generalized diffusion coefficients D^L of slow moving endosome trajectories (the dashed curve).

Figure A5: Comparison of distributions of anomalous exponents α^{NN} and generalized diffusion coefficients D^{NN} and local anomalous exponents α^L and D^L of fast moving endosomes. Anomalous exponents $\alpha^N N$ were estimated using neural network with window size 0.26 s. The generalized diffusion coefficients D^{NN} were estimated by fitting the local TMSD of trajectory with the power law $D^{NN}t^{\alpha^{NN}}$. The distributions of anomalous exponents (Left) are similar to each other, while the distributions of generalized diffusion coefficients (Right) are almost indistinguishable.

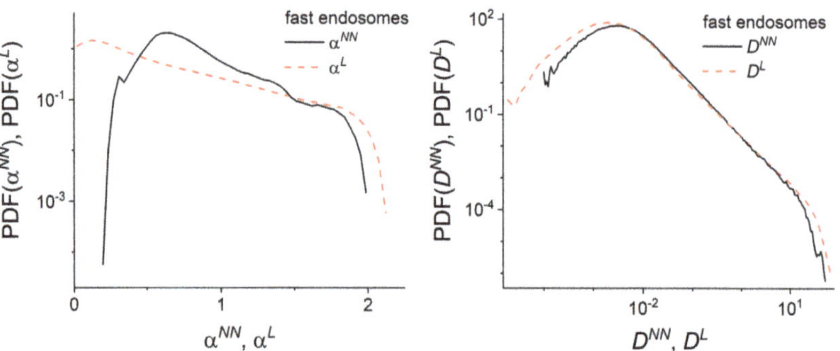

Figure A5. Fast endosomes. (**Left**) Distribution of anomalous exponents α^{NN} of fast moving endosome trajectories (the solid curve) compared with the distribution of local anomalous exponents α^L of slow moving endosome trajectories (the dashed curve); (**Right**) Distribution of generalized diffusion coefficients D^{NN} of fast moving endosome trajectories (the solid curve) compared with the distribution of local generalized diffusion coefficients D^L of fast moving endosome trajectories (the dashed curve).

Figure A6: Local anomalous exponents α^L and local generalized diffusion coefficients D^L are positively correlated for both slow and fast moving endosomes.

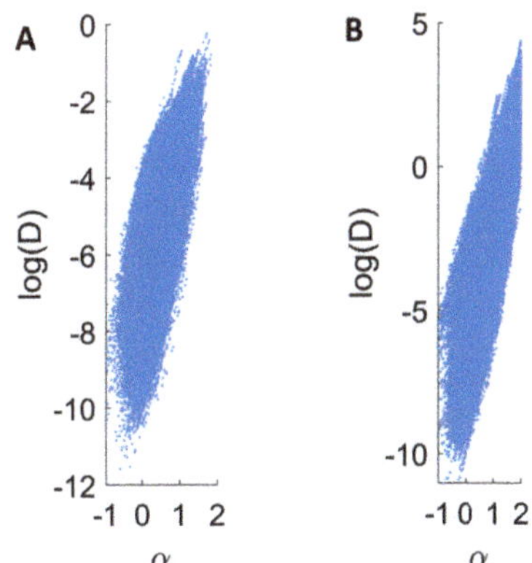

Figure A6. Correlation between local anomalous exponents α^L and local generalized diffusion coefficients D^L for slow (**A**) and fast (**B**) moving endosomes.

References

1. Klages, R.; Radons, G.; Sokolov, I.M. (Eds.) *Anomalous Transport: Foundations and Applications*; Wiley: Weinheim, Germany, 2004.
2. Hofling, F.; Franosch, T. Anomalous transport in the crowded world of biological cells. *Rep. Prog. Phys.* **2013**, *76*, 046602. [CrossRef]
3. Metzler, R.; Jeon, J.-H.; Cherstvy, A.G.; Barkai, E. Anomalous diffusion models and their properties: Non-stationarity, non-ergodicity, and ageing at the centenary of single particle tracking. *Phys. Chem. Chem. Phys.* **2014**, *16*, 24128. [CrossRef]
4. Barkai, E.; Garini, Y.; Metzler, R. Strange kinetics of single molecules in living cells. *Phys. Today* **2012**, *65*, 29–35. [CrossRef]
5. Arcizet, D.; Meier, B.; Sackmann, E.; Radler, J.O.; Heinrich, D. Temporal Analysis of Active and Passive Transport in Living Cells. *Phys. Rev. Lett.* **2008**, *101*, 248103. [CrossRef] [PubMed]
6. Waigh, T.A. *The Physics of Living Processes: A Mesoscopic Approach*; John Wiley & Sons Ltd.: Chichester, UK, 2014.
7. Rogers, S.S.; Flores-Rodriguez, N.; Allan, V.J.; Woodman, P.G.; Waigh, T.A. The first passage probability of intracellular particle trafficking. *Phys. Chem. Chem. Phys.* **2010**, *12*, 3753–3761. [CrossRef]
8. Kenwright, D.A.; Andrew, W. Harrison, A.W.; Waigh, T.A.; Woodman, P.G.; Allan, V.J. First-passage-probability analysis of active transport in live cells. *Phys. Rev. E* **2012**, *86*, 031910. [CrossRef]
9. Akimoto, T.; Yamamoto, E. Detection of transition times from single-particle-tracking trajectories. *Phys. Rev. E* **2017**, *96*, 052138. [CrossRef] [PubMed]
10. Han, D.; Korabel, N.; Chen, R.; Johnston, M.; Gavrilova, A.; Allan, V.J.; Fedotov, S.; Waigh, T.A. Deciphering anomalous heterogeneous intracellular transport with neural networks. *eLife* **2020**, *9*, e52224. [CrossRef] [PubMed]
11. Saxton, M.J. Single-particle tracking: The distribution of diffusion coefficients. *Biophys. J.* **1997**, *72*, 1744–1753. [CrossRef]
12. Wang, B.; Anthony, S.M.; Bae, S.C.; Granick, S. Anomalous yet Brownian. *Proc. Natl. Acad. Sci. USA* **2009**, *106*, 15160. [CrossRef]
13. Wang, B.; Kuo, J.; Bae, S.C.; Granick, S. When Brownian diffusion is not Gaussian. *Nat. Mater.* **2012**, *11*, 481–485. [CrossRef] [PubMed]
14. Lampo, T.J.; Stylianidou, S.; Backlund, M.P.; Wiggins, P.A.; Spakowitz, A.J. Cytoplasmic RNA-Protein Particles Exhibit Non-Gaussian Subdiffusive Behavior. *Biophys. J.* **2017**, *112*, 532–542. [CrossRef]
15. Sabri, A.; Xu, X.; Krapf, D.; Weiss, M. Elucidating the Origin of Heterogeneous Anomalous Diffusion in the cytoplasm of mammalian cells. *Phys. Rev. Lett.* **2020**, *125*, 058101. [CrossRef]
16. Ba, Q.; Raghavan, G.; Kiselyov, K.; Yang, G. Whole-cell scale dynamic organization of lysosomes revealed by spatial statistical analysis. *Cell Rep.* **2018**, *23*, 3591–3606. [CrossRef] [PubMed]
17. Manzo, C.; Torreno-Pina, J.A.; Massignan, P.; Lapeyre, G.J., Jr.; Lewenstein, M.; Garcia-Parajo, M.F. Weak Ergodicity Breaking of Receptor Motion in Living Cells Stemming from Random Diffusivity. *Phys. Rev. X* **2015**, *5*, 011021. [CrossRef]
18. Sadoon, A.A.; Wang, Y. Anomalous, non-Gaussian, viscoelastic, and age-dependent dynamics of histonelike nucleoid-structuring proteins in live Escherichia coli. *Phys. Rev. E* **2018**, *98*, 042411. [CrossRef]
19. Calderon, C.P. Motion blur filtering: A statistical approach for extracting confinement forces and diffusivity from a single blurred trajectory. *Phys. Rev. E* **2016**, *93*, 053303. [CrossRef] [PubMed]
20. Godoy, B.I.; Vickers, N.A.; Andersson, S.B. An Estimation Algorithm for General Linear Single Particle Tracking Models with Time-Varying Parameters. *Molecules* **2021**, *26*, 886. [CrossRef]
21. Hozé, N.; Holcman, D. Statistical Methods for Large Ensembles of Super-Resolution Stochastic Single Particle Trajectories in Cell Biology. *Annu. Rev. Stat. Appl.* **2017**, *4*, 189-223. [CrossRef]

22. Weron, A. Mathematical Models for Dynamics of Molecular Processes in Living Biological Cells. A Single Particle Tracking Approach. *Ann. Math. Sil* **2018** *32*, 5–41. [CrossRef]
23. Ernst, D.; Köhler, J.; Weiss, M. Probing the type of anomalous diffusion with single-particle tracking. *Phys. Chem. Chem. Phys.* **2014**, *16*, 7686-7691. [CrossRef] [PubMed]
24. Janczura, J.; Balcerek, M.; Burnecki, K.; Sabri, A.; Weiss, M.; Krapf, D. Identifying heterogeneous diffusion states in the cytoplasm by a hidden Markov mode. *New J. Phys.* **2021**, *23*, 053018. [CrossRef]
25. Metzler, R. Superstatistics and Non-Gaussian. *Eur. Phys. J. Spec. Top.* **2020**, *229*, 711–728. [CrossRef]
26. Beck, C.; Cohen, E.D.B. Superstatistics. *Phys. A Stat. Mech. Appl.* **2003**, *322*, 267. [CrossRef]
27. Molina-García, D.; Pham, T.M.; Paradisi, P.; Manzo, C.; Pagnini, G. Fractional kinetics emerging from ergodicity breaking in random media. *Phys. Rev. E* **2016**, *94*, 052147. [CrossRef] [PubMed]
28. Maćkała, A.; Magdziarz, M. Statistical analysis of superstatistical fractional Brownian motion and applications. *Phys. Rev. E* **2019**, *99*, 012143. [CrossRef] [PubMed]
29. Cherstvy, A.G.; Metzler, R. Nonergodicity, fluctuations, and criticality in heterogeneous diffusion processes. *Phys. Rev. E* **2014**, *90*, 012134. [CrossRef]
30. Spakowitz, A.J. Transient Anomalous Diffusion in a Heterogeneous Environment. *Front. Phys.* **2019**, *7*, 119. [CrossRef]
31. Chubynsky, M.V.; Slater, G.W. Diffusing diffusivity: A model for anomalous yet Brownian diffusion. *Phys. Rev. Lett.* **2014**, *113*, 098302. [CrossRef] [PubMed]
32. Sposini, V.; Chechkin, A.V.; Seno, F.; Pagnini, G.; Metzler, R. Random diffusivity from stochastic equations: Comparison of two models for Brownian yet non-Gaussian diffusion. *New J. Phys.* **2018**, *20*, 043044. [CrossRef]
33. Chechkin, A.V.; Seno, F.; Metzler, R.; Sokolov, I.M. Brownian yet Non-Gaussian Diffusion: From Superstatistics to Subordination of Diffusing Diffusivities. *Phys. Rev. X* **2017**, *7*, 021002. [CrossRef]
34. Wang, W.; Cherstvy, A.G.; Chechkin, A.V.; Thapa, S.; Seno, F.; Liu, X.; Metzler, R. Fractional Brownian motion with random diffusivity: Emerging residual nonergodicity below the correlation time. *J. Phys. A Math. Theor.* **2020**, *53*, 474001. [CrossRef]
35. Itto, Y.; Beck, C. Superstatistical modelling of protein diffusion dynamics in bacteria. *J. R. Soc. Interface* **2021**, *18*, 20200927. [CrossRef]
36. Korabel, N.; Barkai, E. Paradoxes of subdiffusive infiltration in disordered systems. *Phys. Rev. Lett.* **2010**, *104*, 170603. [CrossRef] [PubMed]
37. Fedotov, S.; Korabel, N. Self-organized anomalous aggregation of particles performing nonlinear and non-Markovian random walks. *Phys. Rev. E* **2015**, *92*, 062127. [CrossRef] [PubMed]
38. Fedotov, S.; Han, D. Asymptotic behavior of the solution of the space dependent variable order fractional diffusion equation: Ultraslow anomalous aggregation. *Phys. Rev. Lett.* **2019**, **123**, 050602. [CrossRef]
39. Sandev, T.; Chechkin, A.V.; Korabel, N.; Kantz, H.; Sokolov, I.M.; Metzler, R. Distributed-order diffusion equations and multifractality: Models and solutions. *Phys. Rev. E* **2015**, *92*, 042117. [CrossRef]
40. Korabel, N.; Han, D.; Taloni, A.; Pagnini, G.; Fedotov, S.; Allan, V.; Waigh, T.A. Unravelling Heterogeneous Transport of Endosomes. *arXiv* **2021**, arXiv:2107.07760.
41. Newby, J.M.; Schaefer, A.M.; Lee, P.T.; Forest, M.G.; Lai, S.K. Convolutional neural networks automate detection for tracking of submicron-scale particles in 2D and 3D. *Proc. Natl. Acad. Sci. USA* **2018**, *115*, 9026–9031. [CrossRef]
42. He, W.; Song, H.; Su, Y.; Geng, L.; Ackerson, B.J.; Peng, H.B.; Tong, P. Dynamic heterogeneity and non-Gaussian statistics for acetylcholine receptors on live cell membrane. *Nat. Commun.* **2016**, *7*, 11701. [CrossRef] [PubMed]
43. Weber, S.C.; Spakowitz, A.J.; Theriot, J.A. Bacterial Chromosomal Loci Move Subdiffusively through a Viscoelastic Cytoplasm. *Phys. Rev. Lett.* **2010**, *104*, 238102. [CrossRef] [PubMed]
44. Weber, S.C.; Thompson, M.A.; Moerner, W.E.; Spakowitz, A.J.; Theriot, J.A. Analytical Tools To Distinguish the Effects of Localization Error, Confinement, and Medium Elasticity on the Velocity Autocorrelation Function. *Biophys. J.* **2012**, *102*, 2443–2450. [CrossRef] [PubMed]
45. Waigh, T.A. Advances in the microrheology of complex fluids. *Rep. Prog. Phys.* **2016**, *79*, 74601–74663. [CrossRef]
46. Savin, T.; Doyle, P.S. Static and Dynamic Errors in Particle Tracking Microrheology. *Biophys. J.* **2005**, *88*, 623–638. [CrossRef]
47. Etoc, F. Non-specific interactions govern cytosolic diffusion of nanosized objects in mammalian cells. *Nat. Mater.* **2018**, *17*, 740–746. [CrossRef] [PubMed]
48. Weigel, A.V.; Ragi, S.; Reid, M.L.; Chong, E.K.P.; Tamkun, M.M.; Krapf, D. Obstructed diffusion propagator analysis for single-particle tracking. *Phys. Rev. E* **2012**, *85*, 041924. [CrossRef]
49. Flores-Rodriguez, N.; Rogers, S.S.; Kenwright, D.A.; Waigh, T.A.; Woodman, P.G.; Allan, V.J. Roles of dynein and dynactin in early endosome dynamics revealed using automated tracking and global analysis. *PLoS ONE* **2011**, *6*, e24479. [CrossRef]
50. Zajac, A.L.; Goldman, Y.E.; Holzbaur, E.L.; Ostap, E.M. Local cytoskeletal and organelle interactions impact molecular motor-driven early endosomal trafficking. *Curr. Biol.* **2013**, *23*, 1173–1180. [CrossRef]
51. Krapf, D. Mechanisms underlying anomalous diffusion in the plasma membrane. *Curr. Top. Membr.* **2015**, *75*, 167–207.
52. Jeon, J.H.; Javanainen, M.; Martinez-Seara, H.; Metzler, R.; Vattulainen, I. Protein Crowding in Lipid Bilayers Gives Rise to Non-Gaussian Anomalous Lateral Diffusion of Phospholipids and Proteins. *Phys. Rev. X* **2016**, *6*, 021006. [CrossRef]

53. Foret, L.; Dawson, J.E.; Villaseñor, R.; Collinet, C.; Deutsch, A.; Brusch, L.; Zerial, M.; Kalaidzidis, Y.; Jülicher, F. A general theoretical framework to infer endosomal network dynamics from quantitative image analysis. *Curr. Biol.* **2012**, *22*, 1381–1390. [CrossRef] [PubMed]
54. Liu, M.; Li, Q.; Liang, L. Real-time visualization of clustering and intracellular transport of gold nanoparticles by correlative imaging. *Nat Commun.* **2017**, *8*, 15646. [CrossRef] [PubMed]
55. Lin, Y.; McMahon, S.J.; Paganetti, H.; Schuemann, J. Biological modeling of gold nanoparticle enhanced radiotherapy for proton therapy. *Phys. Med. Biol.* **2015**, *60*, 4149. [CrossRef] [PubMed]
56. Villagomez-Bernabe, B.; Currell, F.J. Physical Radiation Enhancement Effects Around Clinically Relevant Clusters of Nanoagents in Biological Systems. *Sci. Rep.* **2019**, *9*, 8156. [CrossRef]
57. Sotiropoulos, M.; Henthorn, N.T.; Warmenhoven, J.W.; Mackay, R.I.; Kirkby, K.J.; Merchant, M.J. Modelling direct DNA damage for gold nanoparticle enhanced proton therapy. *Nanoscale* **2017**, *9*, 18413. [CrossRef]

MDPI

Article

Detecting Transient Trapping from a Single Trajectory: A Structural Approach

Yann Lanoiselée [1,2,*], Jak Grimes [1,2], Zsombor Koszegi [1,2] and Davide Calebiro [1,2]

1 Institute of Metabolism and Systems Research, University of Birmingham, Birmingham B15 2TT, UK; JXG937@student.bham.ac.uk (J.G.); Z.Koszegi@bham.ac.uk (Z.K.); d.calebiro@bham.ac.uk (D.C.)

2 Centre of Membrane Proteins and Receptors (COMPARE), Universities of Nottingham and Birmingham, Birmingham B15 2TT, UK

* Correspondence: y.lanoiselee@bham.ac.uk

Abstract: In this article, we introduce a new method to detect transient trapping events within a single particle trajectory, thus allowing the explicit accounting of changes in the particle's dynamics over time. Our method is based on new measures of a smoothed recurrence matrix. The newly introduced set of measures takes into account both the spatial and temporal structure of the trajectory. Therefore, it is adapted to study short-lived trapping domains that are not visited by multiple trajectories. Contrary to most existing methods, it does not rely on using a window, sliding along the trajectory, but rather investigates the trajectory as a whole. This method provides useful information to study intracellular and plasma membrane compartmentalisation. Additionally, this method is applied to single particle trajectory data of β_2-adrenergic receptors, revealing that receptor stimulation results in increased trapping of receptors in defined domains, without changing the diffusion of free receptors.

Keywords: single particle trajectory; stochastic processes; trapping; confinement

Citation: Lanoiselée, Y.; Grimes, J.; Koszegi, Z.; Calebiro, D. Detecting Transient Trapping from a Single Trajectory: A Structural Approach. *Entropy* **2021**, *23*, 1044. https://doi.org/10.3390/e23081044

Academic Editor: Janusz Szwabiński and Aleksander Weron

Received: 15 June 2021
Accepted: 4 August 2021
Published: 13 August 2021

1. Introduction

Single particle methods, which track fluorescent molecules over time, allow for the quantification of biological events with unprecedented spatial and temporal resolution. In cell biology, the complex organisation of the plasma membrane significantly impacts the lateral diffusion of membrane proteins, leading to non-stationary motion patterns. A proper interpretation of these complex trajectories requires that we take into account the changes in a molecule's underlying motion mechanism. For example, transient trapping of G-protein-coupled receptors and G-proteins is closely related to a restricted collision-coupling model [1,2]. In this model, the association rates of molecules on the plasma membrane are enhanced by the presence of confining nano-domains, where receptors and G-proteins are more likely to encounter one another. However, using analysis tools that assume the same molecular motion over time leads to incorrect interpretations of the underlying biology. An intermittent process alternating between free Brownian motion and trapping (as observed in [3]) can wrongly be interpreted as a case of anomalous diffusion with an anomalous exponent $\alpha < 1$.

In the present article, we introduce a method to detect transient trapping events within a single trajectory. An advantage of analysing transient trapping events is the possibility of quantifying the binding kinetics of a molecule through different cellular nano-domains. Additionally, this approach does not require multiple visits of independent molecules to the same nano-domain to assess trapping and does not assume trapping nano-domains to be long-lived. Our strategy is to isolate different trapped portions of trajectories by considering the spatial self-localization of consecutive points within a single trajectory. We introduce local measures computed for each trajectory point, $n \in [1, N]$, containing information on neighbouring trajectory coordinates as a way to elucidate the structure of the trajectory. For each trajectory position, the number of neighbours considered for the local measure is determined by the number of consecutive trajectory coordinates within the range of the test lengthscale.

The detection of trapping is challenging and has been the subject of investigation by several authors. A possible strategy, based on an ensemble of trajectories, consists of evaluating trapping domains from the evaluation of local confining force [4–8]. On the side of single trajectory analysis, techniques were based on the maximum square displacement [9–11], although they are generally too sensitive to noise and local fluctuations of trajectory dynamics. Following this, a number of methods were developed, including: image analysis techniques [12], a model specific maximum likelihood estimator [13], random forest models [14], back propagation neural network approaches [15], moment-scaling spectrum analysis [16], and standardized maximum distance [17]. Another approach proposes to detect confinement size based on first-passage times [18]. Closer to our approach, Sikora et al. [19,20] have developed a method for transient confinement identification based on recurrence statistics, and Verdier et al. [21] used a graphical representation of trajectories to identify the diffusion mode of a whole trajectory. Most of the above-mentioned techniques [9–11,14–16,19,20] rely on time window approaches. Alternatively, our method is based on a recurrence matrix and investigates a trajectory as a whole, whilst still determining sub-trajectory dynamics.

Recurrence matrices are used in various areas of science. They can be used to reconstruct protein structure [22] and are even used to detect structural changes in reaction-diffusion systems [23]. In general, they are used for quantifying non-linear time-series derived from dynamical systems, such as detecting protein conformation changes in molecular dynamics [24] or for quantifying physiological measurements [25]. In the context of dynamical systems, it has been shown that one can reconstruct the chaotic attractor associated with a time-series [26]. Additionally, the influence of observational noise on recurrence plots has been previously investigated [27], in addition to recurrence plots being used for testing time-series stationarity [28]. Although the concept of a recurrence matrix is not new, we construct it in a modified way that greatly limits the effect of outliers and localisation error. Our central hypothesis is that a trapping event within a trajectory is translated into a recurrence matrix as a square block structure along the diagonal of the recurrence matrix. We introduce 3 new local measures that are particularly relevant in detecting block structures along the diagonal, which are characteristic signatures of molecular trapping. From both our construction and these newly introduced measures, we derive a quantity that is invariant when the molecule is trapped and close to zero everywhere else.

In Section 3, we performed extensive simulations and tests to assess the reliability of our method and its robustness to noise for both 2D and 3D trajectories, comparing our method to the 'Divide and Conquer Moment Scaling Spectrum' (DC-MSS) [16]. Finally, in Section 4, we apply our method to single particle tracking data to trajectories of a prototypical G-protein-coupled receptor (β_2 adrenergic receptor), analysing the effects of different pharmacological treatments on receptor trapping.

2. Methods

We consider either 2 or 3-dimensional trajectories composed of N successive coordinates $\{\mathbf{x_1}, \ldots, \mathbf{x_N}\}$, where bold face emphasises the multi-dimensionality of each data point. To make our analysis independent of the trajectory scale, trajectory increments (one-step displacements) are rescaled on each coordinate by their empirical standard deviation. Therefore, the results obtained for Brownian motion are independent of its diffusion coefficient. A recurrence matrix is then calculated from the distance between each pair of points within the trajectory. For each trajectory (see Figure 1a), we construct a positive matrix with Gaussian weights (see Figure 1b):

$$M_{i,j} = \exp\left(-\frac{1}{2}\left(\frac{|\mathbf{x_i} - \mathbf{x_j}|}{\lambda}\right)^2\right), \tag{1}$$

where $|\mathbf{x_i} - \mathbf{x_j}|$ denotes distance between two points i and j, and λ is the test lengthscale. Each element $M_{i,j}$ is in the range $[0,1]$, taking values close to 0 when the distance $|r_i - r_j|$ is larger than λ. The weights are chosen to be Gaussian in such a way that each element, $M_{i,j}$, remains close to 1 for $|\mathbf{x_i} - \mathbf{x_j}|/\ \lambda < 1$ and decays fast when $|\mathbf{x_i} - \mathbf{x_j}|/\ \lambda > 1$. Therefore, the presence of a trapped portion of a trajectory of size $\approx \lambda$ translates in the matrix M as a square block of near 1 entries whose diagonal is aligned to the matrix diagonal. Due to the random nature of molecule displacement, $M_{i,j}$ entries are noisy, making it difficult to determine the transition between different phases of motion. To overcome this, a local smoothing of the matrix is performed. This operation can be done with computational efficiency by convolving the matrix M by a normalized and constant square matrix $(2\mu + 1) \times (2\mu + 1)$, where μ is the smoothing parameter, through a fast Fourier transform (FFT). An advantage of this is to greatly limit the effect of outliers, such as one-step large jumps in position due to tracking errors within a particle's trajectory. Whereas locally averaging trajectory coordinates would greatly disturb the shape of the trajectory and enhance the effect of outliers, zeroes in the Laplacian matrix induced by an outlier are removed by local averaging if an outlier lies inside a trapping block.

The smoothed recurrence matrix is then thresholded to obtain a binary matrix B by setting to one all the values larger than a critical value p_c (see Figure 1c). Here, we choose the critical value to be $p_c = \exp(-1)$, so two points of a trajectory are considered colocalizing if they are within a distance $\lambda\sqrt{2}$ from each other. The consequence of these manipulations turns the problem of finding trapped regions in the trajectories into finding square block structures along the diagonal of the binary matrix B.

From matrix B, one has to identify the individual block structures. This could be achieved by employing a clustering algorithm, such as k-means or k-medoids algorithms; however, these require known numbers of clusters. Even though empirical methods exist to estimate the number of clusters, such as the 'elbow' or 'silhouette' methods, they do not perform well when clusters are of a greatly differing number of entities [29]. Although spectral clustering [29] does not suffer from these limitations on cluster sizes, detection of cluster numbers relies on spectral gap detection, which fails when blocks overlap. Thus, it would not be suited for situations where a molecule jumps from one trap to another (Hop-diffusion, described in [10]). We therefore introduce a new methodology that is specific to the detection of block structures and solves all of the aforementioned issues.

We wish to detect if any trajectory step $n \in [1, \ldots, N]$ is a part of a block or not (i.e., trapped or not). For this purpose we define three measures that can be constructed from each point along the matrix's diagonal $B_{n,n}$ (see Figure 1d for visual illustration). (i) $t_{|}(n)$: The block time, which is the approximate trapping duration seen from the n-th trajectory coordinate. It is computed as the number of matrix elements being both equal to 1 and connected to $B_{n,n}$ along the vertical line $B_{n \pm k,n}$. (ii) $t_{\perp}(n)$: The neighbouring time, which is related to the size of the window $2t_{\perp}(n) + 1$ centred on time point n for which all points colocalize. The neighbouring time is computed as the number of connected matrix elements being equal to 1 along the line perpendicular to the matrix diagonal and going through $B_{n,n}$. (iii) $t_{\parallel}(n)$: The persistence time, which is the segment formed by connected matrix elements being equal to 1 that are parallel to the matrix diagonal and starting from the extremity of the segment used to compute $t_{\perp}$. This determines how many m frames in the future the lower bound $t_{\perp}(n) \leq t_{\perp}(n+m)$ holds.

Let us consider an ideal case where the whole trajectory is trapped such that the recurrence matrix is an $N \times N$ square with matrix elements being equal to 1 everywhere. From these three measures, one can deduce an invariant quantity that is valid for any point along the matrix diagonal (see proof in Appendix A):

$$\nu(n) = \frac{t_{|}(n)}{t_{\parallel}(n) + t_{\perp}(n) - 1} = 1. \tag{2}$$

Figure 1e illustrates the computed block time as a function of time (red) and how neighbouring (cyan) and persistence (purple) time compensate each other to verify the equality. This equality is a necessary condition for the point $\mathbf{x_n}$ to belong to a square block. Specific events or features related to trapping interruption will cause violation of this equality.

For an ideal free portion, the matrix is diagonal (let us call it $1-$diagonal). In the case of a trapping event followed by free diffusion at $n+1$, there is a sharp transition from $\nu(n-1) = 1$ to $\nu(n+1) = 1/N$. Let us consider a special case, where two successive trapping events are spatially separated such that their corresponding blocks, of size s_1 and s_2, respectively, share only a single point (the transition point n) that lies on the matrix diagonal. Given that the trajectory is longer than the two trapping events, $N > s_1 + s_2$, there is a sharp transition at n because the increase of $t_{\|}$ from $s_1 - 2$ to N is not compensated by the increase of $t_{|}$ from s_1 to $s_1 + s_2 - 1$ (see Figure 1f). In the case where two trapping events occur successively at even closer locations, their corresponding blocks will overlap. Even though the equality would be broken at the transition point, departure from $\nu = 1$ may not be very sharp because the transition point is no longer on the matrix diagonal, and accordingly, the persistence time is bounded by blocks sizes $t_{\|} < s_1 + s_2$. Adding n_d diagonal lines along each side of the matrix diagonal, such that an ideal free motion for which B is a $1-$diagonal matrix would become a $(2n_d + 1)-$diagonal matrix, helps to enhance the variation in ν at the transition point by changing the bound to $t_{\|} < N - 2n_d$, where n_d is the number of diagonals. Adding sufficient numbers of lines makes the persistence time $t_{\|}(n)$ almost as long as the trajectory duration itself, so that when the invariant is violated, ν becomes very close to 0.

The number of diagonal lines that should be added depends on the lengthscale λ and the smoothing parameter μ. In general, adding more diagonal lines allows one to distinguish between traps that are very close to each other. In turn, a large number of diagonal lines reduces the precision of change-point detection for isolated traps. In order to decide the number of diagonal lines used, we performed simulations. Block time has been calculated from $M = 10^3$ simulated trajectories of $N = 2 \times 10^3$ steps drawn from two reference types of motion that mimic free diffusion.

In the first case, we simulated Brownian motion (Bm) as the classical model of a freely moving molecule in a homogeneous medium. In the second case, we simulated subdiffusive, fractional Brownian motion (fBm) [30] with anomalous exponent $\alpha = 0.7$ (Hölder exponent $H = 0.35$) as a prototypical diffusion in a crowded environment at percolation threshold [31–34]. In both cases, trajectories were simulated in both 2 and 3 dimensions. As a compromise between sensitivity and precision, the tenth percentile values of block time are used for the rest of the paper for the numbers of diagonal lines to be filled, independent of the dimension of the problem and of the user's choice of reference model (see Appendix D for a comparison of the effect on detection results of the number of added lines).

It is possible that our invariant is broken because of lacunarities inside blocks due to the random nature of molecules' displacements, which can easily be avoided by filling lacunarities inside block components along the diagonal (function imfill in MATLAB). Figure 1g presents the three measures for the trajectory in Figure 1a. The graph shows that inside a block, the pattern is very similar to the one presented in Figure 1e for the ideal case and shows the large change in persistence time at a block transition.

We claim that the n-th point of the trajectory is in a block when $\nu(n)$ is larger than a critical value ν_c. In practice, blocks are never perfect squares, so we choose $\nu_c = 3/4$ as a criterion such that blocks can be deformed as illustrated in Figure 1h. However, even in the case of purely free motion (e.g., Brownian motion), some blocks would still be detected because it takes a random finite time to escape a region of size λ. To ensure that a detected block is due to trapping and not due to chance, we chose a p-value approach. For each test lengthscale and each type of test motion (2D and 3D Brownian motion and fractional Brownian motion), we simulated 10^3 trajectories and computed the matrices B_{ij} before adding diagonal lines based on our previous simulations. Those simulated trajectories were very long (10^4 steps each) in order to ensure the capture of very large potential blocks as test lengthscale λ increases. Block size was computed as the number of consecutive points for which our criterion $\nu(n) > \nu_c$ is verified. From these simulations, we estimated in each case the empirical cumulative probability density of block size. From a given p-value p_{val}, the hypothesis that a detected block is a real trapping event (compared to the reference simulated motion: Bm or fBm) is then rejected if the cumulative probability density associated with the tested block size is smaller than $1 - p_{val}$.

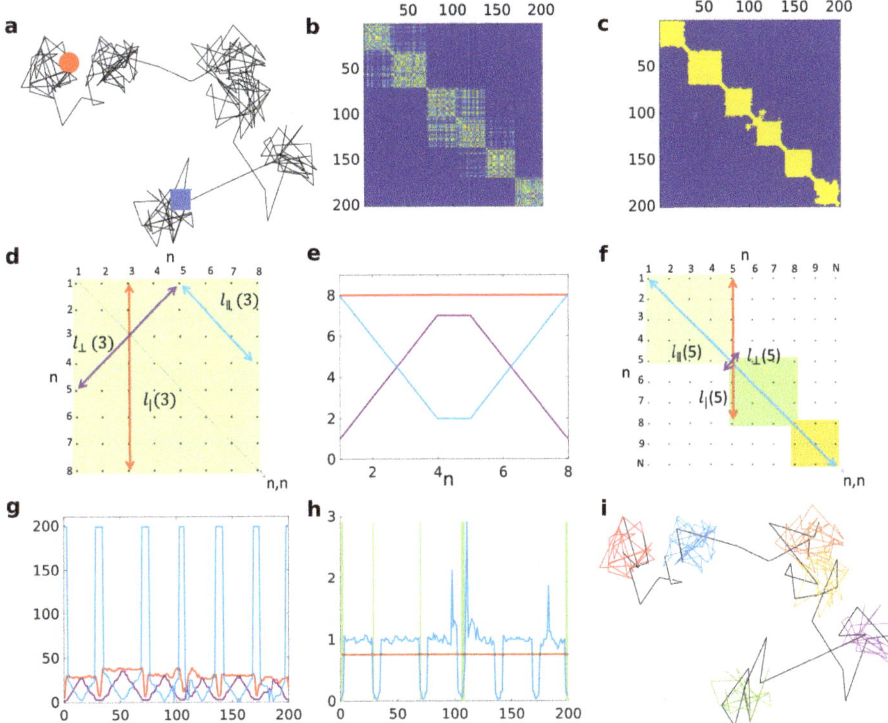

Figure 1. (**a**). Simulated 2D trajectory alternating between free Brownian motion and reflected diffusion in a disk of radius $R = 1$. Diffusion coefficient is $D = 1/2$ in both cases and duration of states in both cases is a Poisson distributed duration with mean $T_f = 5$ and $T_f = 30$ for free and reflected motions, respectively. Red circle denotes beginning and blue square the end of the trajectory. (**b**). Matrix M computed from trajectory in (**a**) with a test lengthscale $\lambda = 1$. (**c**). Binary matrix B after thresholding M in (**b**), filling the lacunarities and adding diagonal lines. (**d**). Illustration of the the block time $t_|$ (red), the neighbouring time $t_\perp(n)$ (purple), and the persistence time $t_{||}(n)$ (cyan) computed at the step $n = 3$ of an ideal B matrix illustrating a fully trapped trajectory of 8 steps. (**e**). Illustration of the inequality along the diagonal B_{nn} for a perfect block $t_| = t_{||}(n) + t_\perp(n) - 1$. (**f**). Illustration demonstrating that at transition between two blocks, the persistence time $t_{||}(n)$ becomes as long as the trajectory itself. (**g**). Computation of $t_|$ (red), $t_\perp(n)$ (purple), and $t_{||}(n)$ (cyan) based on (**c**). (**h**). Block invariant $\nu(n)$ (blue) computed over time based on (**g**) against the threshold value $\nu_c = 0.75$; green rectangles underline misclassified trajectory portions. (**i**). Classified trajectory, where black represents free portions and different colours represent distinct detected trapped portions.

3. Simulations

In this section, we present performance tests for our algorithm in 2D and 3D and compare it to an alternative algorithm, DC-MSS [16], where possible (in 2D).

3.1. Fixed Parameters

For the analysis, the smoothing parameter μ, the number of lines to be filled along the matrix diagonal, and the p-value needed to be set. We chose $\mu = 2$ in such a way that high-frequency variability in the matrix M_{ij} would be dampened without significantly affecting the precision of change-point detection. Then, based on simulations on the effect of different additional diagonal lines (see Appendix D), we added a number of diagonal lines corresponding to the tenth percentile of block-times. Finally, reasoning that the tail of a Brownian motion's first-passage-time distribution from the centre to the border of a disk spans over multiple timescales, choosing a p-value very close to 1 would exclude many transient trapping events. Accordingly, we fixed the p-value $p_{val} = 0.05$ as a compromise between the sensitivity and reliability of the method and then varied simulation parameters to assess the potential of our approach.

3.2. Simulation

To test our methodology, for each data point presented below, we simulated 10^3 of either 2D or 3D trajectories of 10^3 steps each. Molecules alternate between a free diffusive state and a trapped state in which the molecule remains within a region of set size. We chose the free state to be Brownian motion with one-step diffusion lengthscale $\sigma = 1$, corresponding to a diffusion coefficient $D = 1/2$. The trapped state was chosen to be reflected Brownian motion inside a disk (2D) or sphere (3D) of radius R with the same diffusion coefficient. Similarly, we produced another dataset (2D and 3D), where instead of Brownian motion, we modelled free portions with fractional Brownian motion with Hölder exponent $H = 0.35$, corresponding to an anomalous exponent $\alpha = 0.7$. In all cases, the random duration of each state was chosen to be Poisson distributed with mean τ_{Bm} and τ_{trap} for the free and trapped states, respectively. White noise with standard deviation σ_{err} was added to trajectory coordinates to model the effect of the localisation error, starting from low noise $\sigma_{err} = \sigma/10$ to mild noise $\sigma_{err} = \sigma/2$ and finally strong noise with an equivalent standard deviation of trajectory one-step displacements $\sigma_{err} = \sigma$.

3.3. Results

Figure 2a shows the results where both the time spent in free duration and in trapping were varied while the trapping radius was always $R = 1$, and the test lengthscale was $\lambda = 1$. Different levels of noise $\sigma_{err} = \sigma/10$, $\sigma/2$, σ were added, respectively, in Figure 2a–c. In these cases, the minimal duration for detecting a trapping event is $\tau_{p0.05} = 9$ frames (see table in Appendix E). In these three cases, when there is no confinement at all ($\tau_{conf}/\tau_{p0.05} = 0$), the recognition score is close to 1, meaning that the algorithm is robust to false negatives and is able to confirm the absence of trapping. In most cases, more than 90% of trajectories are correctly assigned to their state. The method performs poorly when the trapping duration is close to or shorter than $\tau_{p0.05}$ or when the time spent between two trapping events is smaller than the time it takes to explore a distance larger than the trap size. Both mild and strong noise does lower the recognition score, but only marginally. Figure 2d–f test cases when radius $R = 3$ and the test lengthscale is $\lambda = 3$. In this case, the conclusions are the same, but one has to keep in mind that the durations are much longer because the minimum duration for detecting trapping with $\lambda = 3$ is $\tau_{p0.05} = 42$ (see table in Appendix E).

The above presented cases are idealised because, except when searching for a particular trap size, one does not precisely know the size of traps a priori. A reasonable range can instead be determined by observation of the experimental data. Taking advantage of the robustness to false negatives offered by our p-value approach, we propose combining the recognition for each lengthscale into a single one. We combine results by taking the union of detected trapped frames, considering lengthscales in the range $\lambda \in [1, \lambda_{max}]$ by increments of 0.5. We simulated trajectories alternating between free motion and trapping of distributed sizes. Possible trap radii are uniformly distributed in the range $[1, R_{max}]$, where $R_{max} = 1, 2, 3$ in Figure 2g–i, respectively. The duration in each trapped state is set to be $\tau_{conf} = 6R^2 + 50$, so the trapping time takes into account the radius of the trapping area plus an offset of 50 frames. Trapping was simulated as reflected Brownian motion with an integration step $dt = 1/2$ unless the diffusion length during a step was larger than a third of the radius $\sqrt{2Ddt} > R/3$, in which case positions were approximated as being uniformly distributed inside the trap.

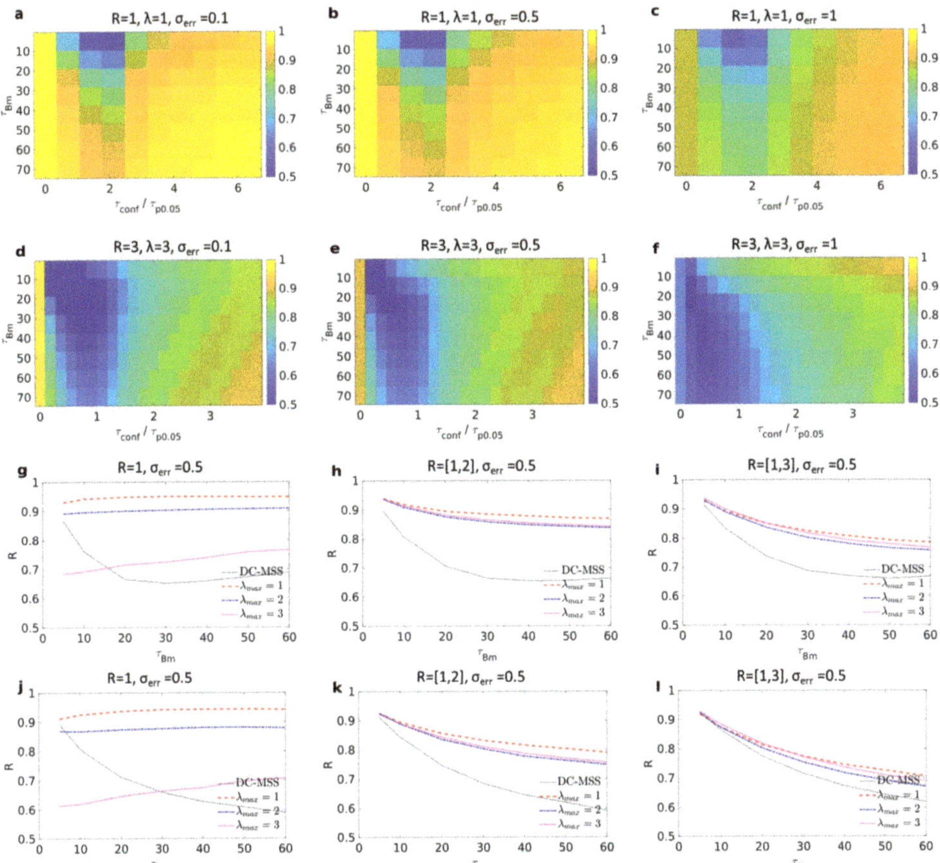

Figure 2. Each panel presents the recognition score $\in [0, 1]$ for 2D trajectories alternating between free and trapped motions. **(a–c)** Trapping radius is $R = 1$, and test lengthscale is $\lambda = 1$. Shown are tested combinations of free Brownian motion of mean duration $\tau_{Bm} = [5, 10, 20, \ldots, 70]$ and mean trapping duration $\tau_{conf} \in [0, 60]$; coordinates are perturbed with white noise of level $\sigma_{err} = \sigma \times [0.1, 0.5, 1]$ **(a–c)**. **(d–f)** $R = 3$ and $\lambda = 3$. Each rectangle represents a combination of free Brownian motion of mean duration $\tau_{Bm} = [5, 10, 20, \ldots, 70]$ and mean trapping duration $\tau_{conf} \in [0, 210]$; coordinates are perturbed with white noise of level $\sigma_{err} = \sigma \times [0.1, 0.5, 1]$. **(g–i)** Free motion is Brownian motion; noise level $\sigma_{err} = 0.5\sigma$ was added to trajectories. Trapping radius is in the range $R \in [1, R_{max}]$, where $R_{max} = 1, 2, 3$ in **(g–i)**. In each case, test lengthscales from $1/2$ to λ_{max} by increments of $1/2$ are combined where $\lambda_{max} = 1, 2, 3$ (dashed red, dotted-dashed blue, and dotted magenta). Black line shows the recognition score obtained from DC-MSS algorithm. **(j–l)** Same as for **(g–i)** except that the free motion is replaced by subdiffusive fractional Brownian motion with Hölder exponent $H = 0.35$.

For each of these three cases, noise level $\sigma_{err} = 0.5\sigma$ was added to trajectories, and we then tested our combination scheme with three possible $\lambda_{max} = 1, 2, 3$. For comparison, we applied the DC-MSS algorithm [16] to our simulated data with the default parameters. DC-MSS separates the data into four categories: immobile, confined, free, and superdiffusive. To make it comparable to our scheme, we considered the two first categories as being 'trapped' and the two latter as being 'free'. In Figure 2g the performance of DC-MSS is better than ours when we overestimate the maximum test lengthscale $\lambda_{max} = 3$, which overestimates three times the maximum trap size $R_{max} = 1$. In turn, choosing $\lambda_{max} = 2$ already significantly improves our classification, and $\lambda_{max} = 1$ gives close to perfect recognition. Then, in cases $R_{max} = 2$ (Figure 2h) and $R_{max} = 3$ (Figure 2i), DC-MSS had a consistently lower score for any choice of parameters λ_{max}. It can be surprising that in Figure 2h,i, $\lambda_{max} = 1$ outperforms the other λ_{max} in all cases, while the size of the traps can be larger than this. We explain this by the fact that the test lengthscale does not specify the trap size to be discovered and instead describes distances between points to be considered 'in the vicinity'. When a molecule spends enough time inside a trap of radius $R = 3$, then even with $\lambda_{max} = 1$, any trajectory point will colocalize with many other points in such a way that the recurrence matrix $M_{i,j}$ will be in 'quasi-block' form (a block with many holes). In this case, the combination of the smoothing step and the lacunarities-filling will complete the block and allow for accurate detection. In turn, larger lengthscales λ_{max} will tend to include, along with a trap, some free points in the vicinity of the confinement area, thus lowering the recognition score.

We also considered the case in which trajectories alternate between subdiffusive fractional Brownian motion and trapping. In the case of a single trap size, results were similar to those obtained in Figure 2a–f (data not shown). In the case of multiple traps' radii (see Figure 2j–l) similar results were obtained, meaning that our approach can distinguish subdiffusion due to molecular crowding from actual trapping in a nano-domain. In comparison, the DC-MSS algorithm tends to misclassify free portions as being trapped, thus giving lesser scores. In Appendix C, additional simulations performed in 3D gave similar results for both diffusive Brownian motion and subdiffusive fractional Brownian motions as 'free states' (see Figure A1).

Lastly, we verified that trajectory duration has only negligible effects as long as trajectory duration is longer than the minimum duration for trapping detection (not shown).

4. Application to Experimental Data

Based on our methodology, with $\lambda = [0.5, 1, 1.5, 2]$, smoothing parameter $\mu = 2$ and $p_{val} = 0.05$, and subdiffusive fBm as our reference for free motion, we investigated the effect of different drugs on the diffusion and trapping of β_2 adrenergic receptor (β_2AR) on the plasma membrane. We recorded fluorescently labelled β_2AR molecules with total internal reflection microscopy, as they diffuse in the plasma membrane of living cells (2D recording) (see Appendix B for experimental methods). We first characterized receptors under basal conditions (36 cells), without pharmacological stimulus. Next, we treated the cells with a gold-standard agonist (isoproterenol) that activates receptors (47 cells). Additionally, we probed receptors with a neutral antagonist (propranolol), which prevents ligand-dependent receptor activation (29 cells). Figure 3a–c show, respectively, all of the trajectories longer than 50 frames (for improved visibility) from a single cell for each described treatment. Portions of trajectories are coloured according to their identified state (trapped in red and free in blue).

It clearly appears that, although trapping is present in all cases, the prevalence of trapping is increased upon agonist stimulation. This is quantitatively supported in Figure 3d, where it is shown that under basal conditions, 39.2% of receptors at each frame were trapped on the plasma membrane. Upon agonist stimulation, this percentage increased to 52.2%, while it remained similar (45.5%) after neutral antagonist treatment. To test the relevance of the observed change, we used a non-parametric Kruskal–Wallis test with Tukey–Kramer correction for multiple comparisons. We found the change between basal and agonist stimulation to be significant ($p = 2 \times 10^{-4}$), clearly demonstrating an effect of agonist stimulation on receptor diffusion dynamics. Contrarily, the change between basal and neutral antagonist treatment was not significant ($p = 0.74$, while the difference between agonist and neutral antagonist was significant ($p = 9 \times 10^{-3}$), suggesting that the drug employed directly influences the receptor trapping, an increase in which correlates with activation of the receptors.

We then sought to further explore the differences observed between these cases. For each trapped trajectory portion, we computed the trapped radius as the distance from the estimated centre of the trap (evaluated as the median of x and y coordinates for a trapped portion) and the point further away than 95% of points within the trapped portion. In Figure 3e, we binned all of the trapped radii into an empirical probability density function (pdf) which was revealed to be similar for the three conditions, suggesting that the trapping domains are of the same nature in all cases. In all cases, the pdf of trapped radii could be fitted approximately with a Gamma distribution, highlighting the exponential decay of the tail of the distribution. This was further reinforced by the computation of the empirical pdf of trapped portions' durations (see Figure 3f), from which we again obtained a similar empirical pdf for all three conditions. The tails of the trapping duration pdf were fitted to a stretched exponential distribution, thus encompassing the wide (yet finite) range of trapping durations.

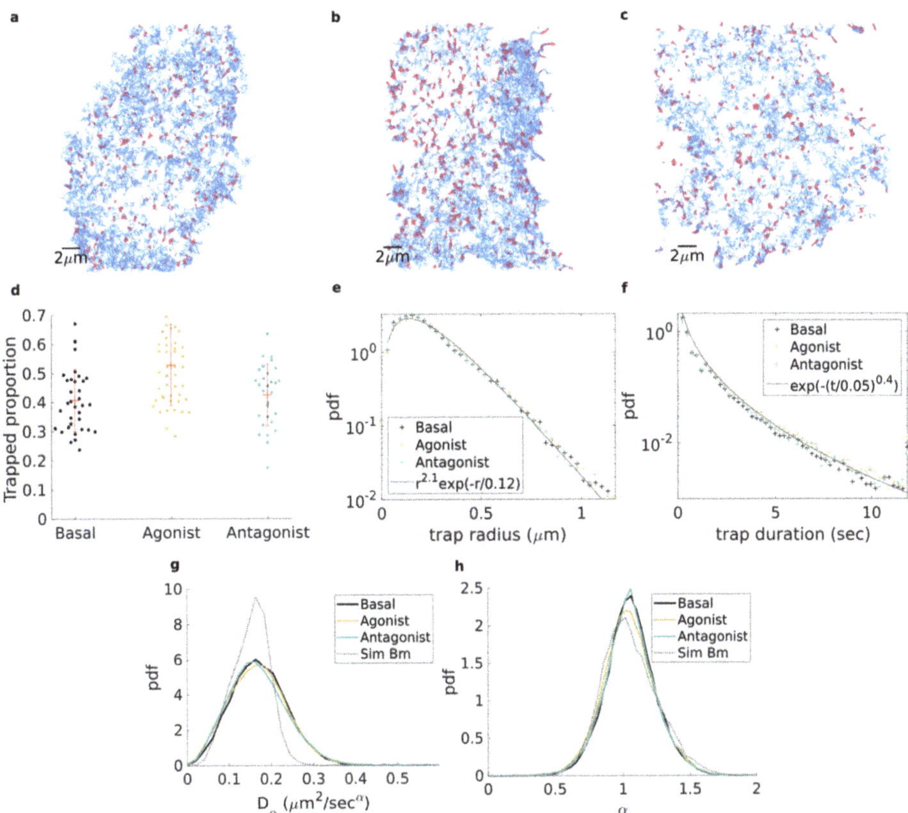

Figure 3. (**a**–**c**). All receptor trajectories longer than 50 frames from a single cell in each group; trajectory portions are coloured according to whether they are detected as free (blue) or trapped (red). Cells are, respectively: (**a**) in basal state, (**b**) stimulated with agonist, and (**c**) treatment with neutral antagonist. (**d**). Proportion of trapped molecules per frame; each point corresponds to a cell for basal (black), agonist stimulated (yellow), and neutral antagonist treated (green). (**e**,**f**). Empirical probability density function for basal (black), agonist stimulated (yellow), and neutral antagonist treated (green) of (**e**) trap radius and (**f**) trapping duration. Grey lines denote fitting with Gamma distribution (**e**) and stretched exponential (**f**). (**g**,**h**). Empirical probability density estimated for free trajectory portions longer than 50 frames of (**g**) the anomalous diffusion exponent α and (**h**) the corresponding generalized diffusion coefficient D_α.

Finally, we enquired into the dynamics of free trajectory portions. To do so, following [35], we computed the time-averaged mean square displacement (TAMSD) of each portion on each coordinate as

$$\delta^2(n,N) = \frac{1}{N-n}\sum_{k=1}^{N-n}(x_{k+n}-x_k)^2, \tag{3}$$

and summed the result for both coordinates before performing a non-linear fitting, over the lag-time range $n \in [1,5]$, with the formula for ensemble-averaged TAMSD for a $2D$ ergodic anomalous diffusion process (e.g., fractional Brownian motion), with localisation error σ_{err}

$$\langle\delta^2(n,N)\rangle = 4D_\alpha n^\alpha + 4\sigma_{err}^2, \tag{4}$$

where α is the anomalous exponent, and D_α is the generalized diffusion coefficient. From this, we obtained the empirical pdf for both anomalous exponent and generalized diffusion coefficients for each condition and observed once again that it was remarkably consistent among the tested conditions. The exponents for free portions of trajectories (see Figure 3g) were distributed slightly over $\alpha = 1$ (average exponent $\langle\alpha\rangle = 1.04, 1.05, 1.04$), corresponding to simple Brownian motion. The generalized diffusion coefficients were very similar in all tested conditions (see Figure 3h) with an average $\langle D_\alpha\rangle = 0.173, 0.169, 0.168\ \mu\text{m}^2\ \text{s}^{-1}$ for basal, agonist, and antagonist, respectively. For comparison, we computed the pdf of exponent and D_α from simulated Brownian motion (see Figure 3g,h), using the same parameters for trajectory duration and mean diffusion coefficient as the free portions found in the case of the tested agonist. The distributions obtained from simulations match the experimental for exponent (average exponent from simulation is $\langle\alpha_{sim}\rangle = 1.05$). However the experimental distributions of diffusion coefficients are wider that the simulated one. We conclude that the distributed nature of the estimated exponent is mainly due to the intrinsic randomness of the TAMSD applied to random trajectories [36,37] while the spread of D_α highlights the heterogeneous nature of cell membrane.

Altogether, these results shed light on the effects of different drug treatments on receptor dynamics. We observe that receptors do not slow down after agonist stimulation. In fact, the change we observe is that receptors are more likely to be trapped, with the nature of the trapping domains remaining the same. For the case of the antagonist, we do not find a significant difference compared to the basal condition, which correlates with the proposed model where neutral antagonists impart no intrinsic activity on the receptor in the absence of an accompanying agonist. We conclude that on timescales longer than our exposure time frame (30 ms), receptors alternate between free lateral diffusion that could be modelled by Brownian motion with fluctuating diffusion coefficient [38–49] and transient trapping in nano-domains of distributed size.

5. Conclusions

In conclusion, we present an algorithm (Code availability: MATLAB code can be downloaded from https://github.com/YannLanoiselee/Transient_trapping_analysis, accessed on 9 August 2021) that can accurately detect transient trapping events from a single trajectory either in two or three dimensions. Our approach is based on recognizing block structures along the diagonal of a thresholded, smoothed recurrence matrix. To this end, we introduced three local measures to be computed along the diagonal of the matrix from which we deduced an invariant quantity inside blocks (trapped portions).

Then, based on a set of user-inputted test lengthscales and on simulations of Brownian and fractional Brownian motions in 2D and 3D as reference processes, we could assess the minimal size of blocks that could be interpreted as the molecule actually being trapped and not a block due to chance, depending on a p-value. We tested our method on a set of simulated data and verified the good performance in 2D and 3D when the free type of motion is either Brownian motion of sub-diffusive fractional or Brownian motion with anomalous exponent $\alpha = 0.7$. We checked the robustness of our results against increasing magnitudes of localisation error. We also compared our 2D results with the classification obtained from the DC-MSS algorithm [16] and showed that our method is more accurate in the task of detecting trapping in all tested cases.

Finally, we applied our analysis to single-particle trajectories of β_2 Adrenergic G-protein-coupled receptors recorded through total internal reflection microscopy. Three conditions were tested: the basal state, stimulated with an agonist, and treatment with a neutral antagonist. In all cases, we found that molecules explore traps with similar distributions of size and duration. Instead, it was only the frequency with which molecules were trapped that was different. TAMSD analysis of the free portions of trajectories led to the conclusion that molecules were mostly undergoing Brownian motion, with a variety of parameters indicative of cell membrane heterogeneity. The demonstration of this technique on real biological data and delineation of pharmacological principles using it (agonist = activation, antagonist = net 0 effect) suggest that our methodology to detect trapping events can be used to study the complexity of both intracellular (3D) and membrane proteins (2D) in live cells.

Author Contributions: Conceptualization, Y.L.; methodology, Y.L.; software, Y.L.; validation, Y.L., J.G. and Z.K.; formal analysis, Y.L.; investigation, Y.L.; resources, J.G. and Z.K.; data curation, J.G. and Z.K.; writing—original draft preparation, Y.L.; writing—review and editing, Y.L., J.G., Z.K. and D.C.; supervision, D.C.; project administration, D.C.; funding acquisition, D.C. All authors have read and agreed to the published version of the manuscript.

Funding: This work was partially supported by a Wellcome Trust Senior Research Fellowship (212313/Z/18/Z to D.C.).

Institutional Review Board Statement: Not applicable.

Informed Consent Statement: Not applicable.

Data Availability Statement: The data presented in this study are openly available in Lanoiselée, Yann; Grimes, Jak; Koszegi, Zsombor; Calebiro, Davide (2021): Trajectory of individual of beta-2 adrenergic receptors at the plasma membrane of Chinese hamster ovary cells (CHO-K1) obtained from TIRF microscope. figshare. Dataset. (https://doi.org/10.6084/m9.figshare.15157410, accessed on 9 August 2021).

Conflicts of Interest: The authors declare no conflict of interest.

Appendix A. Proof of the Square Block Invariant

To prove the equality in Equation (2), we proceed by two inductions. For a square block of fixed side length c, we start by noting the symmetry with respect to the line perpendicular to the matrix diagonal going through point $c/2$ (that lies between two points for odd c). Then, we define $n \in [1, c/2]$; our relationship is verified for $n = 1$, and we suppose it true for n. Then, observing that $t_{|}(n+1) = t_{|}(n)$, $t_{||}(n+1) = t_{||}(n) - 2$, and $t_{\perp}(n+1) = t_{\perp}(n) + 2$, we deduce $\nu(n+1) = \nu(n) = 1$. For the second induction, we start by noting that our relationship is valid for $c = 1$, and we suppose it is true for $c = k$. Then, for $c = k+1$, we find that $t_{\perp}(n)$ remains unchanged, while both $t_{|}(n)$ and $t_{||}(n)$ increase by one, thus again verifying our equality. The relationship is thus valid for arbitrary block sizes and at any point along the diagonal within the block.

Appendix B. Experimental Methods

Appendix B.1. Materials

Cell culture reagents, Lipofectamine 2000, and TetraSpeck fluorescent beads were purchased from Thermo Fisher Scientific. Isoproterenol and Propranolol were from Tocris Bioscience. The fluorescent SNAP-Surface 549 was from New England Biolabs. Ultra-clean glass coverslips were obtained as previously described [50]. For single-molecule experiments, Chinese hamster ovary K1 (CHO-K1) cells (ATCC) were cultured in phenol red-free Dulbecco's modified Eagle's medium (DMEM)/F12, supplemented with 10% FBS, penicillin, and streptomycin at 37 °C, 5% CO_2. Cells were seeded onto ultraclean 25 mm round glass coverslips at a density of 3×10^5 cells per well. On the following day, cells were transfected using Lipofectamine 2000 with N-terminally SNAP-tagged human β_2AR (SNAP-β_2AR) [50] and N-terminally GFP-tagged clathrin light chain (GFP-CCP) (kindly provided by Emanuele Cocucci and Tom Kirchhausen), following the manufacturer's protocol. Cells were labelled with 1 µM SNAP-Surface 549 in complete culture medium for 20 min at 37 °C and imaged by single-molecule microscopy ≈ 4 h after transfection to obtain low physiological protein expression levels [2,50]. Cells were washed with complete culture medium and imaged in Hank's balanced salt solution (HBSS) supplemented with 10 mM HEPES. The labelling efficiency was ≈ 90% ([50]) with non-specific labelling < 1%. β_2ARs were stimulated with either 10 µM Isoproterenol or treated with 10 µM Propranolol.

Appendix B.2. Single-Molecule Microscopy

Single-molecule microscopy experiments were performed using total internal reflection fluorescence (TIRF) microscopy on a custom system, based on an Eclipse Ti2 microscope (Nikon, Japan) equipped with a 100× oil-immersion objective (NA 1.49, Nikon); 405, 488, 561, and 637 nm diode lasers; an iLas TIRF illuminator; quadruple band excitation and dichroic filters; a quadruple beam splitter; 1.5× tube lens (Cairn Research); four EMCCD cameras (iXon Ultra 897, Andor); and hardware focus stabilization. The sample and objective were maintained at 37 °C throughout the experiments. Multicolour single-molecule image sequences were acquired simultaneously at full frame in frame transfer mode, corresponding to one image every 30 ms. Automated single-particle detection and tracking were performed with the u-track software [51], and the obtained trajectories were further analysed using custom algorithms in MATLAB environment as previously described [2].

Appendix C. Simulations for the 3D Case

In this section, we present the simulation results obtained in three dimensions.

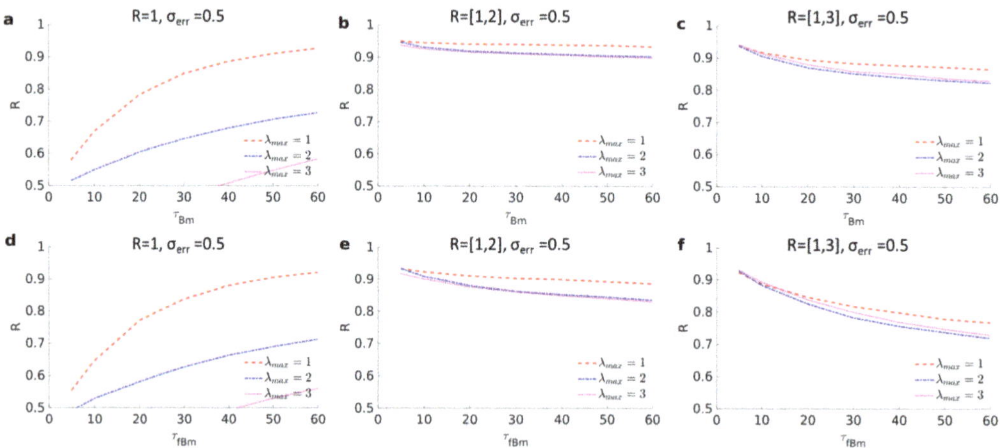

Figure A1. Each panel presents the recognition score ∈ [0, 1] for 3D trajectories alternating between free and trapped motions. (**a–c**) Free motion is Brownian motion with added noise level $\sigma_{err} = 0.5\sigma$. Trapping radius is in the range $R \in [1, R_{max}]$, where $R_{max} = 1, 2, 3$ in (**a–c**). In each case, test lengthscales from 1/2 to λ_{max} by increments of 1/2 are combined, where $\lambda_{max} = 1, 2, 3$ (dashed red, dotted-dashed blue, and dotted magenta). (**d–f**) Same as for (**a–c**) except that the free motion is replaced by subdiffusive fractional Brownian motion with Hölder exponent $H = 0.35$.

Appendix D. Effect of the Number of Diagonal Filled

In Figure A2, we analysed 2D and 3D simulated trajectories alternating between either Bm or fBm and reflected Brownian motion. Then, we computed the recognition score for three possible maximum lengthscales $\lambda_{max} = 1, 2, 3$. Then, for each of these λ_{max}, we computed results depending on the number of lines that were added along to the matrix diagonal of the binary matrix B. We considered no diagonal added at all (d_0) or the tenth percentile d_{10} or median d_{50} of the block time obtained from simulations of Bm or fBm in either 2D or 3D. In all considered cases, the best maximum lengthscale was $\lambda_{max} = 1$, and the best recognition score was obtained with d_{10}.The case d_0 was similar to d_{10} for $R_max = 1$ but failed for larger trapping radii. On the other hand, d_{50} could not capture change-points for a short duration of the free states as well as d_{10}.

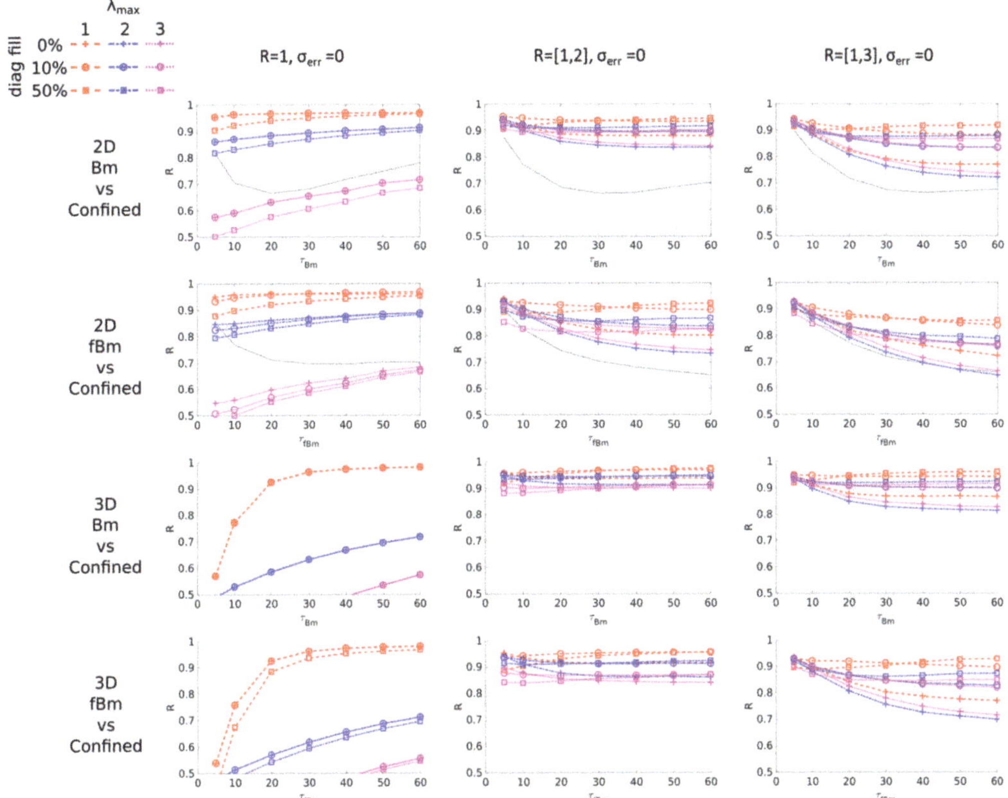

Figure A2. Each panel presents the recognition score $\in [0, 1]$ for $2-3$D trajectories alternating between free and trapped motions. Each column corresponds to a trapping radius range $R \in [1, R_{max}]$, where $R_{max} = 1, 2, 3$, respectively. Rows corresponds to different dimensionality and types of free motion 2D and Bm, 2D and fBm, 3D and Bm, and 3D and fBm, respectively. For the two first rows, black lines correspond to predictions from DC-MSS. Red, blue, and magenta lines correspond to $\lambda_{max} = 1, 2, 3$, while crosses, circles, and squares indicate that lines along the diagonal of the binary matrix B have been filled according to the zeroth, tenth, and median percentiles of block times computed from simulations for either Bm or fBm according to the situation.

Appendix E. p Value Tables

In this section, we present the minimal trapped duration including the number of filled diagonal lines (for d_{10}) corresponding to p-values $[0.1, 0.05, 0.01]$ for different test lengthscales λ for fixed smoothing parameter $\mu = 2$ and $\nu_c = 0.75$ (see Table A1). Values corresponding to each test lengthscale λ are obtained from 10^3 simulated trajectories of 10^4 steps. Simulations have been performed on 2-dimensional Brownian motion with diffusion coefficient $D = 1/2$ (although the result is independent of D because of rescaling) and on 2-dimensional fractional Brownian motion (each coordinate generated independently) with diffusion coefficient $D = 1/2$ and a Hölder exponent $H = 0.35$ corresponding to an anomalous diffusion exponent $\alpha = 0.7$, similar to what is found in diffusion in a crowded molecular environment.

In the case of 3D diffusion (see Table A2), similar simulations have been performed with one extra dimension. A table of the minimal trapped duration including the number of filled diagonal lines corresponding to p-values $[0.1, 0.05, 0.01]$ can be seen below. They are generally shorter because a trajectory has one more degree of freedom to escape. For Brownian motion, the mean square-displacement is increased 50%.

Table A1. Minimal size for a block to be considered a trapped portion for when reference motion is 2D Brownian motion (left) and 2D fractional Brownian motion with $H = 0.35$ (right).

	2D Bm			**2D fBm $H = 0.35$**		
λ	$p_{val} = 0.1$	$p_{val} = 0.05$	$p_{val} = 0.01$	$p_{val} = 0.1$	$p_{val} = 0.05$	$p_{val} = 0.01$
0.5	4	5	6	5	5	7
1	8	10	14	10	13	18
1.5	17	20	28	28	35	52
2	26	32	46	54	68	103
2.5	38	46	68	88	112	176
3	45	57	87	131	169	272

Table A2. Minimal size for a block to be considered a trapped portion for when refence motion is 3D Brownian motion (left) and 3D fractional Brownian motion with $H = 0.35$ (right).

	3D Bm			**3D fBm $H = 0.35$**		
λ	$p_{val} = 0.1$	$p_{val} = 0.05$	$p_{val} = 0.01$	$p_{val} = 0.1$	$p_{val} = 0.05$	$p_{val} = 0.01$
0.5	4	4	5	4	4	5
1	5	6	8	6	7	8
1.5	10	11	15	13	15	21
2	14	17	23	24	28	40
2.5	21	25	34	39	48	69
3	28	33	46	60	74	109

References

1. Gross, W.; Lohse, M.J. Mechanism of activation of A2 adenosine receptors. II. A restricted collision-coupling model of receptor-effector interaction. *Mol. Pharmacol.* **1991**, *39*, 524–530.
2. Sungkaworn, T.; Jobin, M.L.; Burnecki, K.; Weron, A.; Lohse, M.J.; Calebiro, D. Single-molecule imaging reveals receptor-G protein interactions at cell surface hot spots. *Nature* **2017**, *550*, 543–547. [CrossRef]
3. Dietrich, C.; Yang, B.; Fujiwara, T.; Kusumi, A.; Jacobson, K. Relationship of Lipid Rafts to Transient Confinement Zones Detected by Single Particle Tracking. *Biophys. J.* **2002**, *82*, 274–284. [CrossRef]
4. Hoze, N.; Nair, D.; Hosy, E.; Sieben, C.; Manley, S.; Herrmann, A.; Sibarita, J.B.; Choquet, D.; Holcman, D. Heterogeneity of AMPA receptor trafficking and molecular interactions revealed by superresolution analysis of live cell imaging. *Proc. Natl. Acad. Sci. USA* **2012**, *109*, 17052–17057. [CrossRef]
5. El Beheiry, M.; Dahan, M.; Masson, J.B. InferenceMAP: Mapping of single-molecule dynamics with Bayesian inference. *Nat. Methods* **2015**, *12*, 594–595. [CrossRef]
6. Floderer, C.; Masson, J.B.; Boilley, E.; Georgeault, S.; Merida, P.; El Beheiry, M.; Dahan, M.; Roingeard, P.; Sibarita, J.B.; Favard, C.; et al. Single molecule localisation microscopy reveals how HIV-1 Gag proteins sense membrane virus assembly sites in living host CD4 T cells. *Sci. Rep.* **2018**, *8*, 16283. [CrossRef]
7. Briane, V.; Salomon, A.; Myriam, V.; Kervrann, C. A computational approach for detecting micro-domains and confinement domains in cells: A simulation study. *Phys. Biol.* **2020**, *17*, 025002. [CrossRef]
8. Serov, A.S.; Laurent, F.; Floderer, C.; Perronet, K.; Favard, C.; Muriaux, D.; Westbrook, N.; Vestergaard, C.L.; Masson, J.B. Statistical Tests for Force Inference in Heterogeneous Environments. *Sci. Rep.* **2020**, *10*, 3783. [CrossRef]

9. Simson, R.; Sheets, E.D.; Jacobson, K. Detection of temporary lateral confinement of membrane proteins using single-particle tracking analysis. *Biophys. J.* **1995**, *69*, 989–993. [CrossRef]
10. Kusumi, A.; Nakada, C.; Ritchie, K.; Murase, K.; Suzuki, K.; Murakoshi, H.; Kasai, R.S.; Kondo, J.; Fujiwara, T. Paradigm shift of the plasma membrane concept from the two-dimensional continuum fluid to the partitioned fluid: High-speed single-molecule tracking of membrane molecules. *Annu. Rev. Biophys. Biomol. Struct.* **2005**, *34*, 351–378 [CrossRef]
11. Meilhac, N.; Le Guyader, L.; Salomé, L.; Destainville, N. Detection of confinement and jumps in single-molecule membrane trajectories. *Phys. Rev. E* **2006**, *73*, 011915 [CrossRef] [PubMed]
12. Weihs, D.; Gilad, D.; Seon, M.; Cohen, I. Image-based algorithm for analysis of transient trapping in single-particle trajectories. *Microfluid. Nanofluid.* **2012**, *12*, 337–344. [CrossRef]
13. Koo, P.K.; Mochrie, J.K. Systems-level approach to uncovering diffusive states and their transitions from single-particle trajectories. *Phys. Rev. E* **2016**, *94*, 052412. [CrossRef]
14. Wagner, T.; Kroll, A.; Haramagatti, C.R.; Lipinski, H.G.; Wiemann, M. Classification and Segmentation of Nanoparticle Diffusion Trajectories in Cellular Micro Environments. *PLoS ONE* **2017**, *12*, e0170165. [CrossRef]
15. Dosset, P.; Rassam, P.; Fernandez, L.; Espenel, C.; Rubinstein, E.; Margeat, E.; Milhiet, P.E. Automatic detection of diffusion modes within biological membranes using back-propagation neural network. *BMC Bioinform.* **2016**, *17*, 197. [CrossRef]
16. Vega, A.R.; Freeman, S.A.; Grinstein, S.; Jaqaman, K. Multistep Track Segmentation and Motion Classification for Transient Mobility Analysis. *Biophys. J.* **2018**, *114*, 1018–1025. [CrossRef]
17. Briane, V.; Kervrann, V.; Vimond, M. Statistical analysis of particle trajectories in living cells. *Phys. Rev. E* **2018**, *97*, 062121. [CrossRef]
18. Rajani, V.; Carrero, G.; Golan, D.E.; de Vries, G.; Cairo, C.W. Analysis of Molecular Diffusion by First-Passage Time Variance Identifies the Size of Confinement Zones. *Biophys. J.* **2011**, *100*, 1463–1472. [CrossRef]
19. Sikora, G.; Wyłomańska, A.; Gajda, J.; Solé, L.; Akin, E.J.; Tamkun, M.M.; Krapf, D. Elucidating distinct ion channel populations on the surface of hippocampal neurons via single-particle tracking recurrence analysis. *Phys. Rev. E* **2017**, *96*, 062404 [CrossRef]
20. Sikora, G.; Wyłomańska, A.; Krapf, D. Recurrence statistics for anomalous diffusion regime change detection. *Comput. Stat. Data Anal.* **2018**, *128*, 380–394 [CrossRef]
21. Verdier, H.; Duval, M.; Laurent, F.; Cassé, A.; Vestergaard, C.L.; Masson, J.B. Learning physical properties of anomalous random walks using graph neural networks. *J. Phys. A* **2021**, *54*, 234001. [CrossRef]
22. Kloczkowski, A.; Jernigan, R.L.; Wu, Z.; Song, G.; Yang, L.; Kolinski, A.; Pokarowski, P. Distance matrix-based approach to protein structure prediction. *J. Struct. Funct. Genom.* **2009**, *10*, 67–81. [CrossRef]
23. Mocenni, C.; Facchini, A.; Vicino, A. Identifying the dynamics of complex spatio-temporal systems by spatial recurrence properties. *Proc. Nat. Acad. Sci. USA* **2010**, *107*, 8097–8102. [CrossRef]
24. Karain, W.I. Detecting transitions in protein dynamics using a recurrence quantification analysis. *BMC Bioinform.* **2017**, *18*, 525. [CrossRef]
25. Webber, C.L., Jr.; Zbilut, J.P. Dynamical assessment of physiological systems and states using recurrence plot strategies. *J. Appl. Physiol.* **1994**, *76*, 965–973. [CrossRef]
26. Thiel, M.; Romano, M.C.; Kurths, J. How much information is contained in a recurrence plot? *Phys. Lett. A* **2004**, *330*, 343–349. [CrossRef]
27. Thiel, M.; Romano, M.C.; Kurths, J.; Meucci, R.; Allaria, E.; Arecchi, F.T. Influence of observational noise on the recurrence quantification analysis. *Phys. D Nonlinear Phenom.* **2002**, *171*, 138–152. [CrossRef]
28. Manuca, R.; Savit, R. Stationarity and nonstationarity in time series analysis. *Phys. D Nonlinear Phenom.* **1996**, *99*, 134–161. [CrossRef]
29. Von Luxburg, U. A tutorial on spectral clustering. *Stat. Comput.* **2007**, *17*, 395–416. [CrossRef]
30. Mandelbrot, B.B.; Van Ness, J.W. Fractional Brownian motions, fractional noises and applications. *SIAM Rev.* **1968**, *10*, 422–437. [CrossRef]
31. Weiss, M.; Elsner, M.; Kartberg, F.; Nilsson, T. Anomalous subdiffusion is a measure for cytoplasmic crowding in living cells. *Biophys. J.* **2004**, *87*, 3518–3524. [CrossRef]
32. Sokolov, I.M. Models of anomalous diffusion in crowded environments. *Soft Matter* **2012**, *8*, 9043–9052. [CrossRef]
33. Ernst, D.; Hellmann, M.; Khler, J.; Weiss, M. Fractional Brownian motion in crowded fluids. *Soft Matter* **2012**, *8*, 4886. [CrossRef]
34. Weiss, M. Single-particle tracking data reveal anticorrelated fractional Brownian motion in crowded fluids. *Phys. Rev. E* **2013**, *88*, 010101. [CrossRef]
35. Lanoiselée, Y.; Sikora, G.; Grzesiek, A.; Grebenkov, D.S.; Wyłomańska, A. Optimal parameters for anomalous-diffusion-exponent estimation from noisy data. *Phys. Rev. E* **2018**, *98*, 062139. [CrossRef]
36. Grebenkov, D.S. Probability distribution of the time-averaged mean-square displacement of a Gaussian process. *Phys. Rev. E* **2011**, *84*, 031124. [CrossRef]
37. Sikora, G.; Teuerle, M.; Wyłomańska, A.; Grebenkov, D.S. Statistical properties of the anomalous scaling exponent estimator based on time-averaged mean-square displacement. *Phys. Rev. E* **2017**, *96*, 022132. [CrossRef]
38. Chubynsky M.V.; Slater G.W. Diffusing Diffusivity: A Model for Anomalous, yet Brownian, Diffusion. *Phys. Rev. Lett.* **2014**, *113*, 098302. [CrossRef]
39. Jain R.; Sebastian K.L. Diffusion in a Crowded, Rearranging Environment. *J. Phys. Chem. B* **2016**, *120*, 3988–3992. [CrossRef]

40. Tyagi N.; Cherayil B.J.; Non-Gaussian Brownian Diffusion in Dynamically Disordered Thermal Environments. *J. Phys. Chem. B* **2017**, *121*, 7204–7209. [CrossRef]
41. Jain R.; Sebastian K.L.; Diffusing diffusivity: A new derivation and comparison with simulations. *J. Chem. Sci.* **2017**, *126*, 929–937. [CrossRef]
42. Chechkin A.V.; Seno F.; Metzler R.; Sokolov I.M. Brownian yet Non-Gaussian Diffusion: From Superstatistics to Subordination of Diffusing Diffusivities. *Phys. Rev. X* **2017**, *7*, 021002. [CrossRef]
43. Lanoiselée Y.; Grebenkov D.S. A model of non-Gaussian diffusion in heterogeneous media. *J. Phys. A Math. Theor.* **2018**, *51*, 145602. [CrossRef]
44. Lanoiselée Y.; Moutal N.; Grebenkov D.S. Diffusion-limited reactions in dynamic heterogeneous media. *Nat. Commun.* **2018**, *9*, 4398. [CrossRef]
45. Sposini V.; Chechkin A.V.; Seno F.; Pagnini G.; Metzler R. Random diffusivity from stochastic equations: comparison of two models for Brownian yet non-Gaussian diffusion. *New J. Phys.* **2018**, *20*, 043044. [CrossRef]
46. Jain R.; Sebastian K.L. Diffusing diffusivity: Fractional Brownian oscillator model for subdiffusion and its solution. *Phys. Rev. E* **2018**, *98*, 052138. [CrossRef]
47. Lanoiselée Y.; Grebenkov D.S. Non-Gaussian diffusion of mixed origins. *J. Phys. A Math. Theor.* **2019**, *52*, 304001. [CrossRef]
48. Barkai E.; Burov S. Packets of Diffusing Particles Exhibit Universal Exponential Tails. *Phys. Rev. Lett.* **2020**, *124*, 060603. [CrossRef] [PubMed]
49. Ślęzak J.; Burov S. From diffusion in compartmentalized media to non-Gaussian random walks. *Sci. Rep.* **2021**, *11*, 5101. [CrossRef]
50. Calebiro, D.; Rieken, F.; Wagner, J.; Sungkaworn, T.; Zabel, U.; Borzi, A.; Cocucci, E.; Zurn, A.; Lohse, M.J. Single-molecule analysis of fluorescently labeled G-protein-coupled receptors reveals complexes with distinct dynamics and organization. *Proc. Nat. Acad. Sci. USA* **2013**, *110*, 743–748. [CrossRef] [PubMed]
51. Jaqaman, K.; Loerke, D.; Mettlen, M.; Kuwata, H.; Grinstein, S.; Schmid, S.L.; Danuser, G. Robust single-particle tracking in live-cell time-lapse sequences. *Nat. Methods* **2008**, *5*, 695–702. [CrossRef] [PubMed]

Article

Look at Tempered Subdiffusion in a Conjugate Map: Desire for the Confinement

Aleksander Stanislavsky * and Aleksander Weron

Faculty of Pure and Applied Mathematics, Hugo Steinhaus Center, Wrocław University of Science and Technology, Wyb. Wyspiańskiego 27, 50-370 Wroclaw, Poland; aleksander.weron@pwr.edu.pl
* Correspondence: astex@ukr.net

Received: 27 October 2020; Accepted: 16 November 2020; Published: 18 November 2020

Abstract: The Laplace distribution of random processes was observed in numerous situations that include glasses, colloidal suspensions, live cells, and firm growth. Its origin is not so trivial as in the case of Gaussian distribution, supported by the central limit theorem. Sums of Laplace distributed random variables are not Laplace distributed. We discovered a new mechanism leading to the Laplace distribution of observable values. This mechanism changes the contribution ratio between a jump and a continuous parts of random processes. Our concept uses properties of Bernstein functions and subordinators connected with them.

Keywords: anomalous diffusion; statistical analysis; single-particle tracking; trajectory classification

1. Introduction

Based on a myriad of examples in the physical sciences, 1963 Nobel Prize winner in physics H.P. Wigner emphasized the exceptional role of mathematics in understanding the physical structure of the world around us [1]. Indeed, mathematics is a kind of mental tool created for this purpose, and the world is organized in a logical pattern very similar to mathematics [2]. Thus, mathematics turns out to be the language of science and technology. In many experiments the single-molecule motion manifests anomalous diffusion, absolutely not like the classical Brownian diffusion having the mean-squared displacement (MSD) linear in time [3]. To describe the data, a number of theoretical models was developed. The most popular of them are: continuous-time random walk (CTRW) and Fractional Fokker–Planck equation (FFPE) [4–7], fractional Klein–Kramers equation [8], obstructed diffusion (OD) [9,10], random walk on random walk (RWRW) [11,12], fractional Brownian motion (FBM) [13–16], fractional Lévy α-stable motion (FLSM) [17–19], fractional Langevin equation (FLE) [20,21] and autoregressive fractionally integrated moving average (ARFIMA), see [22] and references therein. The ARFIMA model [23–25] is a discrete time analogue of the overdamped fractional Langevin equation [26] responsible for the non-Gaussian law (Lévy α-stable) and a long memory. Moreover, the ARFIMA process is a universal and simple discrete time model for fractional dynamics of empirical data. Recall also that the celebrated FBM and FLSM is nothing but the limiting case of ARFIMA. Since the ARFIMA models were successful in analyzing data from other fields (econometrics, see 2003 Nobel Prize in Economic Sciences for C.W.J. Granger and R. Engel; finance and engineering [27–29]), many statistical tools (and computer packages, e.g., ITMS [24]) are widely available for users, see [25,30].

A relation beetwen physical environment and mathematical models is crucial [1,30]:

- Trapping, crowded environment (CTRW, FFPE, subordinated BM);
- Labyrinthine environment (OD, percolation, RWRW);
- Viscoelastic system (FBM, FLSM, FLE, ARFIMA);
- System with time-dependent diffusion (scaled BM, scaled FBM, ARFIMA);
- System with transient diffusion (BM with transient subordinators).

By using the conjugated Bernstein function theory [31] for a subordinated diffusion, we uncover here a general universal behavior for the pairs of conjugated subordinators. Namely, one can connect the tempered subdiffusion with the diffusion-limited aggregation. Moreover, for the pure Brownian motion this is the Laplace distribution whereas for the Lévy flights is its generalization, the Linnik distribution. It should be also noticed that a large part of well-known anomalous diffusion processes can be represented as time-changed Brownian motion. Thus, by employing the I. Monroe result [32] we find that an anomalous diffusion process is represented as time-changed Brownian motion if and only if it is a semimartingale, [33]. Randomizing the time of the Brownian motion $B(t)$ by using the independent random process $U(t)$, we obtain a new process $X(t) = B(U(t))$. Such an operation is called subordination, first introduced by S. Bochner [34], and see also [35]. The process B, called the parent process, is directed by the new operational time clock U called subordinator. The first reasonable usage of subordination in physics dates back to [36], see also [37,38]. Later, physicists developed a considerable intuition on subordination [39,40]. For the very recent results, see [41].

In the method of single-particle tracking (SPT) a major result is that motion in cell membranes is not limited to pure diffusion. Several modes of motion have been detected including such as immobile, confined, tethered, directed, normal diffusion, and anomalous diffusion [42]. After an ensemble average, the time dependence of the MSD for pure modes of motion is well recognized from others [43,44]. One of important phenomena related to the classification of modes of motion is that practically all experimental results show apparent transitions among modes of motion [45]. If a transition is real, it causes this nonclassical behavior to be different. Their studies are of great interest [46–48]. In cell membranes, anomalous diffusion is most likely the result of both obstacles to diffusion and traps with a distribution of binding energies or escape times [49]. Confined motion may result from corrals formed by cytoskeletal proteins near the membrane, from tethering to immobile species, or from restrictions to motion imposed by lipid domains [50]. The confined diffusion of plasma membrane proteins or lipids can be regarded as a special case of subdiffusion [51]. Analytical treatments have been provided for certain shapes of the confinement zones and the characteristic mobilities [52]. The motion of single biomolecules inside a living cell often exhibits subdiffusion in the confined and crowded environment [53]. For the interpretation of experimental results and quantitative predictions of the diffusion behavior, theoretical models can be extremely helpful. One of them is presented in this paper. It describes the transition from subdiffusion to a confined state. It is interesting that the model has a one-to-one connection with the well-known tempered subdiffusion which demonstrates the transition from subdiffusion to normal diffusion in condensed matter physics [54] and geophysics [55], respectively. Such characteristic crossover from subdiffusion to normal diffusion has been observed also in lipid bilayer systems [56–58]. The tempered stable process in different guises has been intensively researched recently [59–67]. The aim of this work is to get the stochastic representation of anomalous diffusion tending to the confinement. We compare its properties with similar ones for the tempered subdiffusion. Finally, our results are applied for relaxation processes with non-exponential decay as well as for the analysis of the experimental data with confined random trajectories of G proteins and receptors in living cells.

2. Conjugate Laplace Exponents and Stochastic Representation of Anomalous Diffusion

The subdiffusive dynamics is fruitfully modeled as a diffusive motion $X(\tau)$ subordinated by a wide class of random processes subject to infinitely divisible distributions. If the stochastic process $X(\tau)$ has the probability density function (PDF) $h(x,\tau)$, it is a solution of the ordinary Fokker–Planck (FP) equation

$$\partial h(x,\tau)/\partial\tau = \hat{L}(x)\,h(x,\tau)\,. \tag{1}$$

where $\hat{L}(x)$ is the time-independent FP operator (for example, $-\frac{\partial}{\partial x}F(x) + D\frac{\partial^2}{\partial x^2}$ with a force F). Generally, the operator $\hat{L}$ can be both multidimensional and fractional in space, but with no loss of generality we will consider the one-dimensional case. Infinitely divisible distributions, following the Lévy–Khintchine formula [68], are characterized by the exponentially weighted function

$$\langle e^{-u\,T_\Psi(\tau)}\rangle = e^{-\tau\Psi(u)} = \int_0^\infty e^{-ut}\, g_\Psi(t,\tau)\, dt\,, \tag{2}$$

where $\Psi(u)$ is called the Laplace exponent, and $g_\Psi(t,\tau)$ is the PDF of this process. Note, the Laplace exponents may be only Bernstein functions. This is a very extensive class of functions [31]. In the theory of Bernstein functions a special role is played by so-called conjugate pairs [69]. If one of them is $\Psi(s)$, then another will take the form $\Phi(s) = s/\Psi(s)$. The parent process $X(\tau)$ may be subordinated by $S_\Psi(t) = \inf\{\tau > 0 : T_\Psi(\tau) > t\}$ as well as $S_\Phi(t) = \inf\{\tau > 0 : T_\Phi(\tau) > t\}$, where T_Φ is a conjugate subordinator. Both cases lead to the FP equation in a general form

$$p(x,t) = q(x) + \int_0^t d\tau\, M(t-\tau)\, \hat{L}(x)\, p(x,\tau)\,, \tag{3}$$

where $M(t)$ is the memory function [54,70]. The kernel $M(t)$ has a simple expression after the Laplace transform. Denote the inverse Laplace transform L_t^{-1}. This gives

$$M_\Psi(t) = \frac{1}{2\pi i}\int_{c-i\infty}^{c+i\infty} e^{st}\,\frac{ds}{\Psi(s)} = L_t^{-1}\frac{1}{\Psi(s)}\,, \tag{4}$$

$$M_\Phi(t) = \frac{1}{2\pi i}\int_{c-i\infty}^{c+i\infty} e^{st}\,\Psi(s)\,\frac{ds}{s} = L_t^{-1}\frac{\Psi(s)}{s}\,, \tag{5}$$

where c is large enough that $1/\Psi(s)$ (for the first case) and $\Psi(s)/s$ (for the second) are defined for $\Re s \geq c$, and $i^2 = -1$. The PDF of the operational time $S_\Phi(t)$ is simply written as a Laplace image

$$\tilde{f}(\tau,s) = \frac{1}{\Psi(s)}\, e^{-\tau s/\Psi(s)}\,. \tag{6}$$

The solution (propagator) of Equation (3) takes the form of a subordination integral

$$p(x,t) = \int_0^\infty h(x,\tau)\, f(\tau,t)\, d\tau\,. \tag{7}$$

Using the Brownian motion as a parent process

$$h_B(x,\tau) = \frac{1}{\sqrt{2\pi D\tau}}\exp\left(-\frac{x^2}{2D\tau}\right)\,, \tag{8}$$

the Laplace image $\tilde{p}(x,s)$ is written as the tabulated integral [71], expressed in terms of the modified Bessel function of the third kind. As its index is equal to $1/2$, we get the following propagator

$$\tilde{p}(x,s) = \frac{1}{\sqrt{2D}}\,\frac{1}{\sqrt{s\Psi(s)}}\exp\left(-2\frac{|x|\sqrt{s}}{\sqrt{2D\Psi(s)}}\right)\,. \tag{9}$$

Similar calculations can be fulfilled for $S_\Psi(t)$, which we will not present here. If the moments of a parent process $X(\tau)$ are known exactly, as in the case of the Brownian motion, the moments of the process $X[S_\Phi(t)]$ can be found analytically. Using the MSD of Brownian motion in the form $\langle B^2(\tau)\rangle = D\tau$, where D is a diffusive constant, the MSD of $Y(t) = B[S_\Psi(t)]$ and $Y(t) = B[S_\Phi(t)]$ reads

$$\left\langle B^2[S_\Psi(t)]\right\rangle = D\, L_t^{-1}\frac{1}{s\Psi(s)} = D\int_0^t M_\Psi(y)\, dy\,, \tag{10}$$

$$\left\langle B^2[S_\Phi(t)]\right\rangle = D\, L_t^{-1}\frac{\Psi(s)}{s^2} = D\int_0^t M_\Phi(y)\, dy\,. \tag{11}$$

It is clear that the MSD is depended on the function $\Psi(s)$, but in different ways it manifests in a conjugate pair. Similar analysis of two different forms of the Fokker–Planck equation where the memory kernels are conjugate pairs has been done in [72,73]. When the memory kernel is an exponentially truncated power-law, the MSD can approach to saturation. In the next section, we will look at specific examples.

3. Tempered α-Stable Process and Its Conjugate Partner

An important exemplar of infinitely divisible subordinators is tempered α-stable processes, having all moments of operational time [74]. In this case the diffusive motion demonstrates an intermediate behavior between subdiffusion and normal diffusion [54,55]. Then the Laplace exponent is $\Psi_{\text{temp}}(s) = (s+\delta)^\alpha - \delta^\alpha$, where δ is a positive constant and $0 < \alpha < 1$. If δ equals to zero, the tempered α-stable process becomes ordinary α-stable. Let the Brownian motion be a parent process, and the inverse tempered α-stable process is directing. The MSD of the subordinated diffusion is

$$\left\langle x^2(t)\right\rangle = D\int_0^t e^{-\delta y}\, y^{\alpha-1}\, E_{\alpha,\alpha}(\delta^\alpha y^\alpha)\, dy\,, \tag{12}$$

where $E_{\alpha,\beta}(x) = \sum_{k=0}^{\infty} x^k/\Gamma(\alpha k+\beta)$ is the two-parameter Mittag–Lefeffler function [75]. If $t \ll 1$ (or $\delta \to 0$), this value strives for $Dt^\alpha/\Gamma(\alpha+1)$, whereas for $t \gg 1$ (or $\alpha \to 1$) it is linear in time $D\delta^{1-\alpha}t/\alpha$ as expected for normal diffusion shown in Figure 1a. From the asymptotic values for $\langle x^2(t)\rangle$ it is easy enough to obtain the crossover time $t_\times = \left[\alpha\,\delta^{\alpha-1}/\Gamma(\alpha+1)\right]^{1/(1-\alpha)}$ between the two diffusive modes, also shown in Figure 1a. This diffusion behaves anomalous at short time and almost normal at long times.

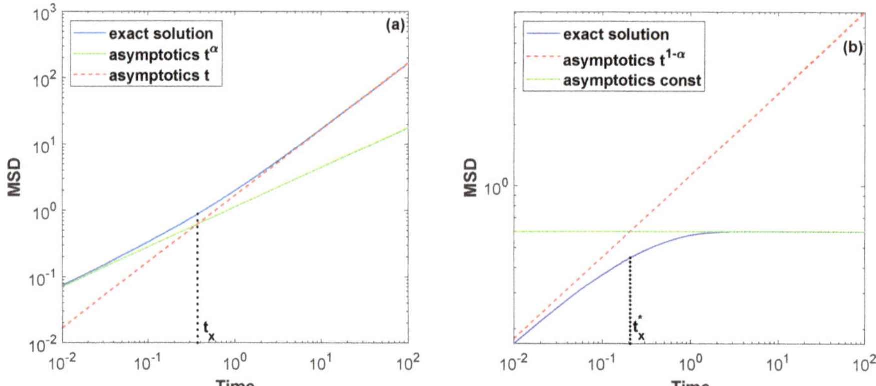

Figure 1. (Color online) Mean squared displacement of tempered subdiffusion (**a**) and its conjugate partner (**b**) with $\alpha = 0.6$ and $\delta = 1$ (for $D = 1$). The dashed red and dash-dot green lines show asymptotic behavior of the values. If the panel (**a**) indicates a transition of the subdifussion into normal diffusion at long times, whereas the panel (**b**) shows the emergence of diffusion-limited aggregation.

Next, we study the diffusion motion with the conjugate Laplace exponent $s/\Psi_{\text{temp}}(s)$. Its MSD is not difficult to find. It is expressed in terms of the three-parameter Mittag–Leffler function [75], having the following Taylor series

$$E^{\rho}_{\alpha,\beta}(x) = \sum_{k=0}^{\infty}\frac{(\rho,k)\,x^k}{\Gamma(\alpha k+\beta)k!}\,,\quad \alpha,\beta>0\,, \tag{13}$$

where $(\rho,k) = \rho(\rho+1)(\rho+2)\dots(\rho+k-1)$ is the Appell's symbol with $(\rho,0) = 1, \rho \neq 0$. The MSD has also an analytical form

$$\begin{aligned}\left\langle x^2(t)\right\rangle &= D\int_0^t e^{-\delta y}\, y^{-\alpha}\, E_{1,1-\alpha}(\delta y)\, dy - D\delta^\alpha t \\ &= De^{-\delta t}\, t^{1-\alpha}\, E^2_{1,2-\alpha}(\delta t) - D\delta^\alpha t\,, \end{aligned} \tag{14}$$

that gives the short- and long-time behavior

$$\left\langle x^2(t)\right\rangle = \begin{cases} Dt^{1-\alpha}/\Gamma(2-\alpha) & \text{if } t\to 0\,, \\ D\alpha\delta^{\alpha-1} & \text{if } t\to\infty\,. \end{cases} \tag{15}$$

The interrelation in the conjugate pair between each other is quite non-trivial. If for the Laplace exponent $\Psi_{\text{temp}}(s)$ the pure subduffusion evolves to normal diffusion in time, then for the conjugate case $s/\Psi_{\text{temp}}(s)$ the subdiffusion transforms into diffusion-limited aggregation. Figure 1b just presents the evolution. It has a simple explanation. As the normal diffusion is characterized by the Laplace exponent $\Psi(s) = s$, its conjugate partner has $\Phi(s) = s/\Psi(s) = 1$. This clearly implies the confinement. Using asymptotic behavior of the MSD, we determine the crossover time $t^\star_\times = \left[\Gamma(2-\alpha)\,\alpha\,\delta^{\alpha-1}\right]^{1/(1-\alpha)}$ between the diffusive regimes. Consequently, the duality relation between infinitely divisible subordinators allows one to generate a new impact scenario of traps, in which diffusion behaves less anomalous at short time and extremely anomalous at long times.

Using numerical methods, the propagator under the conjugate Laplace exponent $s/\Psi_{\text{temp}}(s)$ is shown in Figure 2. The propagator has a cusp which is saved for $t\to\infty$. Recall that the tempered subdiffusion loses this feature at long times. If the axis y is logarithmic, as in Figure 2, the propagator of tempered subdiffusion goes to a parabola (see the panel a), whereas in the confined case it takes a triangular shape (panel b). This is not surprising because for $t\to\infty$ the propagator of diffusion motion with the conjugate Laplace exponent $s/\Psi_{\text{temp}}(s)$ can be found analytically, and its form corresponds to the well-known Laplace (or double exponential) distribution [76,77], namely

$$\lim_{t\to\infty} p(x,t) = \frac{1}{\sqrt{2D\alpha\delta^{\alpha-1}}}\exp\left(-\frac{2|x|}{\sqrt{2D\alpha\delta^{\alpha-1}}}\right) \tag{16}$$

with a location parameter $\mu = 0$ (in general, it may be nonzero) and a scale parameter $\theta = \sqrt{\alpha\delta^{\alpha-1}D/2} > 0$. Although the PDF of the Laplace distribution is reminiscent of the normal distribution, they are different: the normal distribution is expressed in terms of the squared difference from the mean whereas the Laplace distribution is expressed in terms of the absolute difference from the mean. Therefore, the Laplace distribution has fatter (more precisely, moderate) tails than the normal distribution (with thin tails always) [68]. To get the Laplace distribution as the average value of elementary Gaussians, the necessary ("superstatistical") distribution of the diffusivities is exponential [78–80]. It should be noticed that the Brownian yet non-Gaussian diffusion is not the same considering in this paper. The stationary Laplace distribution of particles' motion also takes place in compartmentalized media [81]. The inverse cumulative distribution function of the Laplace distribution is equal to $x_c = -\theta\ln(2-2q)$. The value x_c is such that any observation from this distribution with the scale parameter θ falls in the range $[0\ x_c]$ with probability $0 < q < 1$ [77]. This allows one to estimate borders of the confinement region, taking into account the values D, α and δ. The Laplace distribution as a confined case is characteristic for the Brownian motion as a parent process. If the parent process becomes infinitely divisible, the confined distribution will be other and presented in the next section.

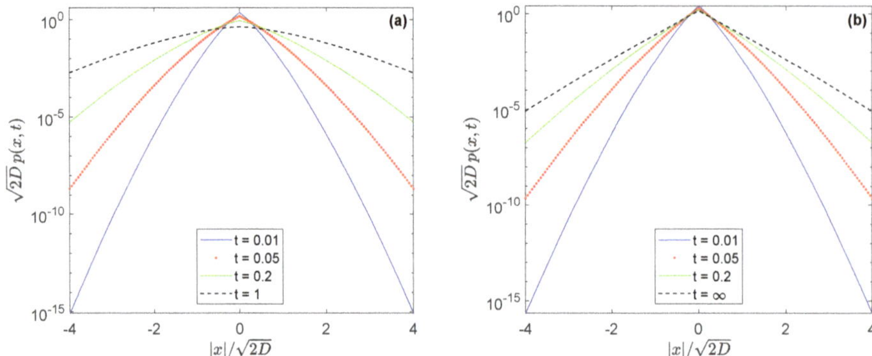

Figure 2. (Color online) Propagator $p(x,t)$ for the tempered subdiffusion (**a**) and its conjugate partner (**b**), tending to the confinement, with a constant potential, $\alpha = 0.5$ and $\delta = 1$, drawn for consecutive dimensionless instances of time. Starting with the Dirac delta-function and passing to the subdiffusive PDF, for $t \to \infty$ the value $p(x,t)$ becomes the normal distribution, shown by black dotted line on the panel (**a**), and the Laplace distribution (black dotted line) on the panel (**b**).

Note that for $\alpha = 1/2$ the MSD of the tempered subdiffusion and its conjugate partner coincide with each other at short times. The point is that the Laplace exponent $s^{1/2}$ is the only one convertible into itself by the duality relation between conjugate pairs of Laplace exponents [82]. If $\alpha > 1/2$, then for the same values α the MSD of tempered subdiffusion less anomalous than the MSD of its conjugate partner at short times. For $\alpha < 1/2$ the opposite happens. Usually the duality relation accelerates the subdiffusion more anomalous (in the sense of $\alpha < 1/2$) and slows down too fast subdiffusion (with $\alpha > 1/2$). This is especially evident for multi-scale anomalous diffusion [82].

4. Confined Distributions for Infinitely Divisible Motion

Now we apply our approach for infinitely divisible motion as a parent process, whereas the subordinator has the Laplace exponent $s/[(s+\delta)^\alpha - \delta^\alpha]$ leading to a confined distribution for $t \to \infty$. Without loss of generality the one-dimensional case will be represented. Consider any infinitely divisible motion by using the characteristic function in the form

$$\hat{h}(k,t) = \int_{-\infty}^{\infty} e^{ikx}\, h(x,t)\, dx = e^{-D^* t\, \Xi(|k|)/2}, \tag{17}$$

where D^* is a generalized diffusive constant. In the case of β-stable Lévy motion the characteristic exponent $\Xi(|k|)$ is equal to $|k|^\beta$ with $\beta \in (0,2)$. There are also other well-known examples of the characteristic exponent: (i) $(|k|^2 + m^{\beta/2})^{2/\beta} - m$, $\beta \in (0,2)$; (ii) $\log(1+|k|^\beta)$, $\beta \in (0,2]$; (iii) $b|k|^2 + |k|^\beta$; (iv) $\log((1+|k|^2) + \sqrt{(1+|k|^2)^2 - 1})$ and so on [69].

The next development is to consider the subordination of such parent processes. For this purpose we use the same subordinator led to the Laplace distribution above from Brownian motion. Based on the simple forms of $\hat{h}(k,\tau)$ and $\tilde{f}(\tau,s)$, the solution (propagator) of a subordinated infinitely divisible motion is convenient to write as the Laplace–Fourier transform, taking the form

$$\tilde{\hat{p}}(k,s) = \frac{1}{s}\frac{s/((s+\delta)^\alpha - \delta^\alpha)}{[D^*\Xi(|k|)/2 + s/((s+\delta)^\alpha - \delta^\alpha)]}. \tag{18}$$

As $\lim_{s\to 0} s/((s+\delta)^\alpha - \delta^\alpha) = \delta^{1-\alpha}/\alpha$, using the final value theorem ($\lim_{t->\infty} \hat{p}(k,t) = \lim_{s->0} \tilde{\hat{p}}(k,s)$), the confined characteristic function is written as

$$\hat{p}(k,\infty) = \frac{1}{1 + D^* \alpha \delta^{\alpha-1}\, \Xi(|k|)/2}\,. \tag{19}$$

For the ordinary Brownian motion it is not difficult to check this result by the inverse Fourier transform clearly, as this is the tabulated integral [71]. Other forms of the characteristic exponent $\Xi(|k|)$ do not lead to so simple analytical expressions, but then they can be evaluated numerically. Note that $1/[1 + A\Xi(|k|)]$ with $A = D^*\alpha\delta^{\alpha-1}/2 > 0$ is an even function, and thus its Fourier transform is equivalent to the cosine transform.

Taking the β-stable Lévy motion with the characteristic exponent $\Xi(|k|) = |k|^\beta$ under $\beta \in (0,2)$, the confined characteristic function $\hat{p}_\beta(k,\infty)$ manifests the characteristic function of the symmetric Linnik distribution [83] (or the β-Laplace distribution, following Pillai [84]), namely

$$\begin{aligned} p_\beta(x,\infty) &= \frac{1}{\pi}\int_0^\infty \frac{\cos(k|x|)}{1 + Ak^\beta}\, dk \\ &= \frac{1}{\pi}\int_0^\infty \frac{\sin(z^{1/\beta}\,|x|)}{(1 + Az)^2}\, dz\,. \end{aligned} \tag{20}$$

The last expression was obtained from the integration by parts and has a better convergence in numerical integration. Examples are shown in Figure 3. The symmetric Linnik distribution attracted considerable attention from researchers [85–88]. Generally, the PDF is unimodal [89], geometrically stable [90] and can be expressed in terms of Meijer's G-function [91]. Moreover, the peak of the density is finite for $1 < \beta \le 2$ (see Figure 3a), it becomes infinite for $0 < \beta \le 1$ (shown in Figure 3b) [92]. Based on the tabular integral of [93], its value yields

$$p_\beta(0,\infty) = \frac{1}{\pi}\int_0^\infty \frac{dk}{1 + Ak^\beta} = \frac{1}{\beta A^{1/\beta}\sin(\pi/\beta)}\,. \tag{21}$$

A series expansion for small x is written as

$$p_\beta(x,\infty) = \frac{1}{\beta}\sum_{n=0}^{\infty} \frac{(-1)^n}{(2n)!}\, \frac{x^{2n}}{A^{(1+2n)/\beta}}\, \frac{1}{\sin[\pi(1+2n)/\beta]}\,. \tag{22}$$

If $x = 0$, only the $n = 0$ term is saved, and one obtains the previous expression. According to [94], the asymptotic expansion for large x reads

$$p_\beta(x,\infty) = \frac{1}{\pi}\sum_{n=1}^{\infty} (-1)^{n+1}\, \frac{\Gamma(1+n\beta)\, A^n}{|x|^{1+\beta n}}\, \sin(\pi\beta n/2)\,. \tag{23}$$

Consequently, the leading term of this expansion becomes

$$p_\beta(x,\infty) \sim \frac{\Gamma(1+\beta)\, \sin(\pi\beta/2)}{\pi}\, \frac{A}{|x|^{1+\beta}}\,. \tag{24}$$

There are some specific examples representable by tabular integrals [93] that will be considered elsewhere.

Since $\Xi(|k|)$ is a Bernstein function (or otherwise the function having a complete monotone derivative), the characteristic function $1/[1 + A\Xi(|k|)]$ is typical for a geometrically infinitely divisible PDF [95]. In any case the PDF form is symmetric and unimodal. In dependence of $\Xi(|k|)$ it has a finite or infinite maximum. This is because the integral $\int_0^\infty dk/[1 + A\Xi(k)]$ has a single improper point, namely $k \to \infty$, where the integral is convergent or divergent.

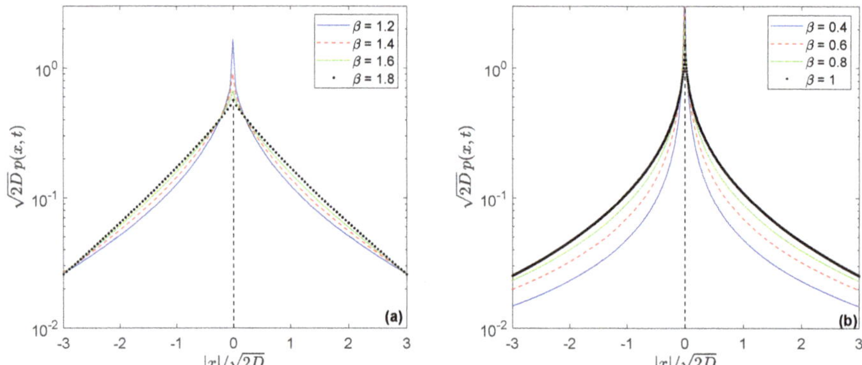

Figure 3. (Color online) Propagators $p(x,t)$ from the parent processes, having the β-stable Lévy distribution: (**a**) $1 < \beta < 2$; (**b**) $0 < \beta \leq 1$; under the subordinator, conjugate to a tempered random process in the sense of Bernstein functions, for $t \to \infty$. The value $A = D^*\alpha\delta^{\alpha-1}/2$ is taken equal to 1.

We can formulate the following **Confinement Principle:** Any subordinated infinitely divisible motion, in which the subordinator is characterized by the Laplace exponent conjugate to a tempered α-stable process, has a confined probability distribution. By the infinitely divisible motion we mean a wide class of infinitely divisible processes, including Brownian motion (as a marginal case), Lévy stable motion (Lévy flight) and many other processes with jumps. It is important that each case of characteristic exponents in such an infinitely divisible motion determines its confined probability distribution. For the pure Brownian motion this is the Laplace distribution whereas for the Lévy flights its generalization is the Linnik distribution. This procedure covers a class of geometrically infinitely divisible distributions as a confined case of the infinitely divisible motion subordinated by a special subordinator responsible for the confinement.

5. Conditionally Non-Exponential Decay of Relaxation

Our comparative analysis of tempered and confined diffusion may be pretty simple extended to relaxation processes with non-exponential decay. As is well known [96,97], the manifestations of many-body effects in anomalous dynamics of relaxing systems, independent of the physical and chemical structures of their interacting entities, are successfully described by stochastic tools. Then the relaxation function of non-exponential relaxation is written as

$$\phi_\Psi(t) = \int_0^\infty e^{-b\tau} f_\Psi(\tau,t)\,d\tau\,, \tag{25}$$

where b is a constant, and $f_\Psi(\tau,t)$ is the PDF of an inverse subordinator $S_\Psi(t) = \inf\{\tau > 0 : T_\Psi(\tau) > t\}$. The Laplace image $\phi_\Psi(t)$ takes the simple form

$$\tilde{f}_\Psi(\tau,s) = \frac{\Psi(s)}{s}\,e^{-\tau\Psi(s)}\,. \tag{26}$$

Then the Laplace transform of $\phi_\Psi(t)$ in time yields

$$\tilde{\phi}_\Psi(s) = \frac{1}{s}\frac{\Psi(s)}{\Psi(s)+b}\,. \tag{27}$$

Similar conversions that we omit can be done for $s/\Psi(s)$.

Based on the Laplace exponent of tempered diffusion $\Psi_{\text{temp}}(s) = (s+\delta)^\alpha - \delta^\alpha$ and its conjugate partner $s/\Psi_{\text{temp}}(s)$, it is not difficult to obtain their relaxation functions numerically. They are presented in Figure 4.

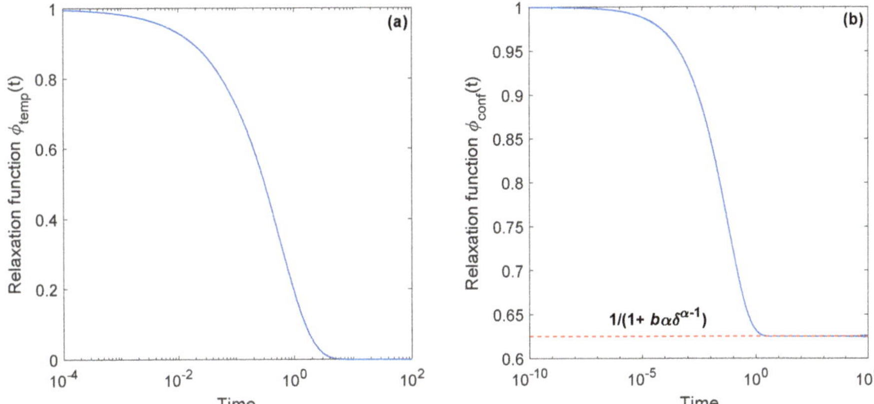

Figure 4. (Color online) Relaxation functions, caused by the inverse tempered subordinator (**a**) and its conjugate partner (**b**) respectively, with $\alpha = 0.6$, $\delta = 1$ and $b = 1$. The first represents the tempered relaxation, and the second is confined. The dashed red line shows a conditionally non-exponential decay due to the confinement effect ($\lim_{t\to\infty} \phi_{\text{conf}}(t) = \text{const} \neq 0$).

Using the above relationship, we have found asymptotic behavior of the functions. They read

$$\begin{aligned} \lim_{t\to 0}(1-\phi_{\text{temp}}(t)) &= bt^\alpha/\Gamma(\alpha+1), \\ \lim_{t\to 0}(1-\phi_{\text{conf}}(t)) &= bt^{1-\alpha}/\Gamma(2-\alpha) \end{aligned} \tag{28}$$

at short times and accordingly

$$\begin{aligned} \lim_{t\to\infty}\phi_{\text{temp}}(t) &= \lim_{t\to\infty} e^{-bt\delta^{1-\alpha}/\alpha} = 0, \\ \lim_{t\to\infty}\phi_{\text{conf}}(t) &= \frac{1}{1+\alpha\delta^{\alpha-1}b} = \text{const} \end{aligned} \tag{29}$$

at long times. Note that both types of relaxation start with 1 as a power function in time. However, tempered relaxation tends to zero exponentially, whereas the confined relaxation does not reach zero at all. From the physical point of view this latter model can be interpreted in the following way. Dipoles ordered by the external field do not fall into disorder with probability 1 after removing the field as t tends to infinity. Therefore, we believe that this model demonstrates a conditionally non-exponential decay. In this context it should be mentioned that the conditionally exponential decay model is a key for the concept of clusters and their dynamics to an imperfectly ordered state, used for the explanation of relaxation in dielectric materials [98–100]. During the relaxation process the strongly coupled local (intracluster) motions are expected to be generated first and then followed by the weakly coupled (intercluster) motions which produce the partial long-range structure. Each of these motions, those leading to the local structure order and those leading to the cluster ordering in general, has its own perceptible contribution to the observed features such as the relaxation function in time domain and the susceptibility in frequency domain.

6. G-Proteins vs. a2AR Receptors from the Analysis of the SPT Data

As an example for detecting the Laplace confinement in experimental data, we present our analysis of random trajectories obtained from a recent SPT study on G protein-coupled receptors, namely the motion and interaction of individual receptors and G proteins on the surface of living cells [44]. Two types of particles and only the basal case (without drug stimulation) data were studied: G-protein coupled receptors (further we will call them simply receptors) and the G proteins with which the receptors interact. The sample data consisted of 20,000 trajectories of 30 sets for G proteins and more 35,000 trajectories of 30 sets for receptors which randomly walk along x and y coordinates.

The first aim of the data study is to classify the dynamics for both types of particles. It is based on the standardized maximal distance T_n of random processes from its starting point [101]. This approach is quite justified. Really, if the motion is driven by the fractional Brownian motion, then the best among the available methods is the one based on the *p-VAR* test, especially for longer trajectories. But, if the particle dynamics can be described by the Ornstein–Uhlenbeck or diffusive Brownian motion process, then the method yielding the smallest errors is based on the *MAX* test [102–105]. Following the procedure, we use the statistical test:

- H_0—an observed trajectory $X = \{X_1, X_2, \ldots, X_n\}$ comes from Brownian motion,
- H_1—the trajectory looks like a confined or directed diffusion.

Then T_n is estimated with respect to the quantiles of order $\alpha/2$ and $1-\alpha/2$ (for example, α = 5%) for different trajectory lengths n. The decision rule is as follows: $T_n < q_n(\alpha/2)$ means a confined motion, whereas $T_n > q_n(1-\alpha/2)$ is superdiffusion (or directed motion). If $q_n(\alpha/2) < T_n < q_n(1-\alpha/2)$, then $X = \{X_1, X_2, \ldots, X_n\}$ is Brownian motion. Consequently, this permits us to classify the trajectories available for processing. The results are shown in Figure 5. As seen from this figure, the most of trajectories is Brownian motion: 69% for G proteins and 78% for receptors. The contribution of superdiffusion is the smallest, 2%. The rest corresponds to confined motion. This part is especially interesting to us. Next, we are going to estimate the statistics of such trajectories. It is assumed that the confined random walks can occur in two cases. The first of them is classical, the Ornstein–Uhlenbeck model. It gives the normal distribution for $t \to \infty$. The second case leads to the Laplace confinement for $t \to \infty$, considered above. Possible transitions of the particles' diffusion type within single trajectories are noted and investigated. For example, in [104] it has been proposed a statistical procedure for detecting transitions of the MSD exponent value within a single trajectory.

Discriminating the statistics of G-protein and receptor confined trajectories between the normal and Laplace distribution functions, we apply the logarithm of the ratio of their maximized likelihoods [106]. The approach leads to the calculations of means, medians, sample variances and averages of the absolute difference between data values and the median. This statistical test gives the ratio $Q > 0$ for the normal distribution, otherwise the Laplace distribution is preferred. After applying the second statistical test, its results together with the first test results are also presented in Figure 5. This shows that for G proteins the confined trajectories obey equally the normal and Laplace distributions, whereas for receptors the normal distribution is approximately twice as common as the Laplace distribution. But it should be pointed out also that the share of confined trajectories with normal statistics remains unchanged for both G proteins and receptors. Judging by the contribution of Brownian motion in all the sets of trajectories, the difference between the percentage ratio of confined trajectories with the normal and Laplace distributions for G proteins and receptors can indicate greater mobility of receptors over G proteins. It should be mentioned that Laplace distributions were detected in the complex diffusive behavior of RNA-protein particles [107].

The occurrence of the Laplace distribution for confined trajectories in the experimental data used by us seems to be natural. First, the most part of the trajectories is Brownian motion. What could be a parent process for subordination in this environment? Brownian motion is preferred. Why? Since we observe a following **Competition Principle** *between parent processes: Brownian motion, Lévy motion or other infinitely divisible process even for any fixed subordinator conjugated one to tempered α-stable responsible* *for confinement*. If Brownian motion is parent, the confined distribution from our subordination approach can have only the Laplace form. In the above data sets any feature, for example, typical for Lévy motion, is not detected. If this was true, it would be a chance for the play of generalized Laplace distributions as a confined distribution. Another case is the Ornstein–Uhlenbeck process leading to the normal statistics in confined trajectories, it has the same (Brownian) roots too. Therefore, the presence of normal and Laplace distributions together into confined trajectories is quite logical and justified physically.

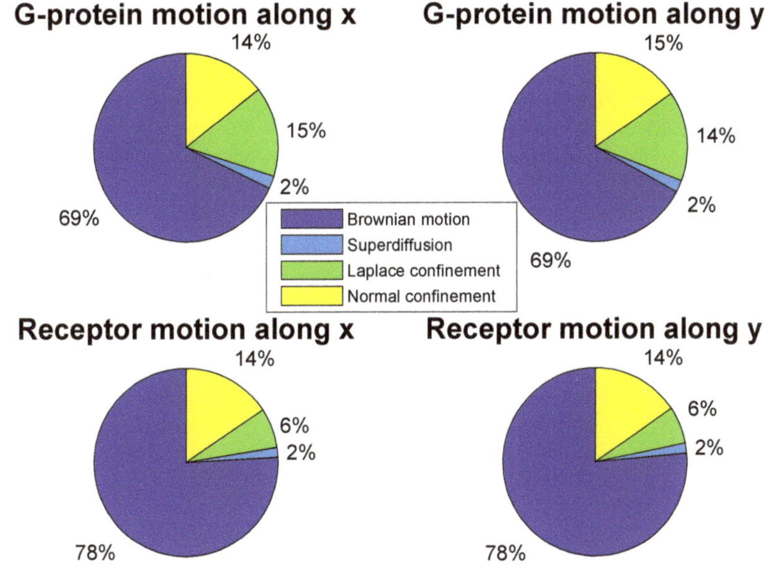

Figure 5. (Color online) Analysis of the experimental data as applied to G-protein and receptor random-walk trajectories along the coordinates x and y with the cutoff length of trajectories more and equal to 50.

7. Discussion

We have revealed that the conjugate property of Bernstein functions connects the tempered stable subdiffusion with the diffusion-limited aggregation by an one-to-one mapping (in fact, a bijection). If the pure subdiffusion is characterized by multiple trapping events with infinite mean sojourn time, and the power function exponent of MSD is constant in time, then a truncated power-law distribution of trapping times leads to tempered subdiffusion, in which diffusion is anomalous at short times and normal (contribution of traps seems to disappear) at long times [45]. The interpretation of anomalous diffusion tending to the confinement is that the trap impact has the opposite tendency, long waiting times in traps dominate more and more so that it becomes impossible to leave such traps. This model, just like the tempered one, is applicable for the analysis of SPT. Its effects are present in confined random motions of G proteins and receptors in living cells. We have established that the confined distribution form depends on the PDF of the parent process under subordination. If the parent process is Brownian motion, the confined distribution has only the Laplace form. If the Lévy motion is directed, the confined distribution takes the Linnik case. If the support of the parent process is changed from $(-\infty,\infty)$ to $(0,\infty)$, as a confined limit, the Mittag–Leffler distribution arises. All this manifests that the presented method has ample opportunities for the study of confined random walks in complex systems.

Concerning to relaxation phenomena, complete disorder (e. g. in the form of charge neutralization in dielectrics) does not occur in the relaxing system with confinement features. This concept can be used for developing new cluster models of non-exponential relaxation. It will be considered in more detail elsewhere. Our new methodology is generally valid in a wide class of problems of transport in random media that include live cells, relaxation in heterogeneous substances, and jump-diffusion.

Author Contributions: A.S. analyzed tempered subdiffusion in a conjugate map based on Brownian motion and performed the analysis of corresponding relaxation decay and experimental data. A.W. invented the confined distributions for infinitely divisible motion. A.S. and A.W. wrote the text. A.S. prepared Figures 1, 2 and 4, and Figures 3 and 5 were prepared by the authors together. Both authors reviewed the manuscript. All authors have read and agreed to the published version of the manuscript.

Funding: A.S. is grateful to the Hugo Steinhaus Center for pleasant hospitality during his visit in Wrocław University of Science and Technology as well kindly acknowledges a support of NAWA PPN/ULM/2019/1/00087/DEC/1. A.W. would like to thank for support of Beethoven Grant No. DFG-NCN 2016/23/G/ST1/04083.

Acknowledgments: The authors would like to thank T. Sungkaworn and D. Calebiro for providing the experimental data analyzed in Section 6.

Conflicts of Interest: The authors declare no conflict of interest.

References

1. Wigner, E.P. The unreasonable effectiveness of mathematics in the natural sciences. *Commun. Pure Appl. Math.* **1960**, *13*, 1–14. [CrossRef]
2. Hamming, R.W. The unreasonable effectiveness of mathematics. *Am. Math. Mon.* **1980**, *87*, 81–90. [CrossRef]
3. Metzler, R.; Jeon, J.H.; Cherstvy, A.G.; Barkai, E. Anomalous diffusion models and their properties: Non-stationarity, non-ergodicity, and ageing at the centenary of single particle tracking. *Phys. Chem. Chem. Phys.* **2014**, *16*, 24128. [CrossRef]
4. Metzler, R.; Klafter, J. The random walk's guide to anomalous diffusion: A fractional dynamics approach. *Phys. Rep.* **2000**, *339*, 1–77. [CrossRef]
5. Manzo, C.; Garcia-Parajo, M.F. A review of progress insingle particle tracking: From methods to biophysical insights. *Rep. Prog. Phys.* **2015**, *78*, 124601. [CrossRef]
6. Metzler, R.; Barkai, E.; Klafter, J. Anomalous diffusion and relaxation close to thermal equilibrium: A fractional Fokker–Planck equation approach. *Phys. Rev. Lett.* **1999**, *82*, 3563–3567. [CrossRef]
7. Bel, G.; Barkai, E. Weak ergodicity breaking in the continuous-time random walk. *Phys. Rev. Lett.* **2005**, *94*, 240602. [CrossRef]
8. Magdziarz, M.; Weron, A. Numerical approach to the fractional Klein–Kramers equation. *Phys. Rev. E* **2007**, *76*, 066708. [CrossRef] [PubMed]
9. Havlin, S.; Ben-Avraham, D. Diffusion in disordered media. *Adv. Phys.* **1987**, *36*, 695–798. [CrossRef]
10. Saxton, M.J. Anomalous diffusion due to obstacles: A Monte Carlo study. *Biophys. J.* **1994**, *66*, 394–401. [CrossRef]
11. Kehr, K.W.; Kutner, R. Random walk on a random walk. *Phys. A* **1982**, *110*, 535–549. [CrossRef]
12. Schulz, J.H.P.; Barkai, E.; Metzler, R. Aging renewal theory and application to random walks. *Phys. Rev. X* **2014** , *4*, 011028. [CrossRef]
13. Guigas, G.; Kalla, C.; Weiss, M. Probing the nanoscale viscoelasticity of intracellular fluids in living cells. *Biophys. J.* **2007**, *93*, 316–323. [CrossRef] [PubMed]
14. Burnecki, K.; Kepten, E.; Janczura, J.; Bronshtein, I.; Garini, Y.; Weron, A. Universal algorithm for identification of fractional Brownian motion. A case of telomere subdiffusion. *Biophys. J.* **2012**, *103*, 1839–1847. [CrossRef]
15. Manzo, C.; Torreno-Pina, J.A.; Massignan, P.; Lapeyre, G.J., Jr.; Lewenstein, M.; Garcia Parajo, M.F. Weak ergodicity breaking of receptor motion in living cells stemming from random diffusivity. *Phys. Rev. X* **2015**, *5*, 011021. [CrossRef]
16. Bronstein, I.; Israel, Y.; Kepten, E.; Mai, S.; Shav-Tal, Y.; Barkai, E.; Garini, Y. Transient anomalous diffusion of telomeres in the nucleus of mammalian cells. *Phys. Rev. Lett.* **2009**, *103*, 018102. [CrossRef]

17. Burnecki, K.; Weron, A. Fractional Lévy stable motion can model subdiffusive dynamics. *Phys. Rev. E* **2010**, *82*, 021130. [CrossRef]
18. Krapf, D.; Lukat, N.; Marinari, E.; Metzler, R.; Oshanin, G.; Selhuber-Unkel, C.; Squarcini, A.; Stadler, L.; Weiss, M.; Xu, X. Spectral content of a single non-Brownian trajectory. *Phys. Rev. X* **2019**, *9*, 011019. [CrossRef]
19. Dybiec, B.; Gudowska-Nowak, E.; Barkai, E.; Dubkov, A.A. Lévy flights versus Lévy walks in bounded domains. *Phys. Rev. E.* **2017**, *95*, 052102. [CrossRef]
20. Kou, S.C.; Xie, X.S. Generalized Langevin equation with fractional Gaussian noise: Subdiffusion within a single protein molecule. *Phys. Rev. Lett.* **2004**, *93*, 180603. [CrossRef]
21. Sadegh, S.; Higgins, J.L.; Mannion, P.C.; Tamkun, M.M.; Krapf, D.Plasma membrane is compartmentalized by a self-similar cortical actin meshwork. *Phys. Rev. X* **2017**, *7*, 011031. [CrossRef] [PubMed]
22. Burnecki, K.; Sikora, G.; Weron, A. Fractional process as a unified model for subdiffusive dynamics in experimental data. *Phys. Rev. E* **2012**, *86*, 041912. [CrossRef] [PubMed]
23. Granger , C.W.J.; Joyeux, R. An introduction to long memory time series models and fractional differencing. *J. Time Ser. Anal.* **1980**, *1*, 15–29. [CrossRef]
24. Brockwell, P.; Davis, R. *Introduction to Time Series and Forecasting*; Springer: New York, NY, USA, 2002.
25. Beran, J. *Statistics for Long-Memory Processes*; Chapman and Hall: New York, NY, USA, 1994.
26. Magdziarz, M.; Weron, A. Fractional Langevin equation with α-stable noise. *Stud. Math.* **2007**, *181*, 47–60. [CrossRef]
27. Crato, N.; Rothman, P. Fractional integration analysis of long-run behavior for US macroeconomic time series. *Econom. Lett.* **1994**, 45, 287–291. [CrossRef]
28. Fouskitakis, G.; Fassois, S. Pseudolinear estimation of fractionally integrated ARMA (ARFIMA) models with automatic applicatins. *IEEE Trans. Signal Process.* **1999**, *47*, 3365–3380. [CrossRef]
29. Gil-Alana, L.A. A fractionally integrated model for the Spanish real GDP. *Econom. Bull.* **2004**, *3*, 1–6.
30. Burnecki, K.; Weron, A. Algorithms for testing of fractional dynamics: A practical guide to ARFIMA modelling. *J. Stat. Mech.* **2014**, *2014*, P10036. [CrossRef]
31. Schilling, R.L.; Song, R.; Vondraček, Z. *Bernstein Functions: Theory and Applications*; de Gruyter Studies: Berlin, Germany, 2010.
32. Monroe, I. Processes that can be embedded in Brownian motion. *Ann. Probab.* **1978**, *6*, 42–56. [CrossRef]
33. Weron, A.; Magdziarz, M. Anomalous diffusion and semimartingales. *Europhys. Lett.* **2009**, *86*, 60010. [CrossRef]
34. Bochner, S. Diffusion equation and stochastic processes. *Proc. Natl. Acad. Sci. USA* **1949**, *35*, 368–370. [CrossRef] [PubMed]
35. Meerschaert, M.M.; Benson, D.A.; Scheffler, H.P.; Baeumer, B. Stochastic solution of space-time fractional diffusion equations. *Phys. Rev. E* **2002**, *65*, 041103. [CrossRef] [PubMed]
36. Blumen, A.; Klafter, J.; White, B.S.; Zumofen, G. Continuous-time random walks on fractals. *Phys. Rev. Lett.* **1984**, *53*, 1301–1304. [CrossRef]
37. Fogedby, H.C. Langevin equations for continuous time Lévy flights. *Phys. Rev. E* **1994**, *50*, 1657–1660. [CrossRef] [PubMed]
38. Eule , S.; Friedrich, R. Subordinated Langevin equations for anomalous diffusion in external potentials—Biasing and decoupled external forces. *Europhys. Lett.* **2009**, *86*, 30008. [CrossRef]
39. Sokolov, I.M.; Klafter, J. From diffusion to anomalous diffusion: A century after Einstein's Brownian motion. *Chaos* **2005**, *15*, 026103. [CrossRef]
40. Stanislavsky, A.; Weron, K.; Weron, A. Anomalous diffusion with transient subordinators: A link to compound relaxation laws. *J. Chem. Phys.* **2014**, *140*, 054113. [CrossRef]
41. Barkai, E.; Burow, S. Packets of diffusing particles exhibit universal exponential tails. *Phys. Rev. Lett.* **2020**, *124*, 060603. [CrossRef]
42. Saxton, M.J. Single Particle Tracking. In *Fundamental Concepts in Biophysics. Handbook of Modern Biophysics*; Jue, T., Ed.; Humana Press: Totowa, NJ, USA, 2009; pp. 147–179.
43. Weron, A.; Burnecki, K.; Akin, E.J.; Solé, L.; Balcerek, M.; Tamkun, M.M.; Krapf, D. Ergodicity breaking on the neuronal surface emerges from random switching between diffusive states. *Sci. Rep.* **2017**, *7*, 5404. [CrossRef]
44. Sungkaworn, T.; Jobin, M-L.; Burnecki, K.; Weron, A.; Lohse, M.J.; Calebiro, D. Single-molecule imaging reveals receptor-G protein interactions at cell surface hot spots. *Nature* **2017**, *550*, 543–547. [CrossRef]

45. Saxton, M.J. A biological interpretation of transient anomalous subdiffusion. I. Qualitative model. *Biophys. J.* **2007**, *92*, 1178–1191. [CrossRef] [PubMed]
46. Magdziarz, M.; Klafter, J. Detecting origins of subdiffusion: P-variation test for confined systems. *Phys. Rev. E* **2010**, *82*, 011129. [CrossRef] [PubMed]
47. Saxton, M.J. Wanted: A positive control for anomalous subdiffusion. *Biophys. J.* **2012**, *103*, 2411–2422. [CrossRef] [PubMed]
48. Stanislavsky, A.; Weron, A. Control of the transient subdiffusion exponent at short and long times. *Phys. Rev. Res.* **2019**, *1*, 023006. [CrossRef]
49. Saxton, M.J.; Jacobson, K. Single-particle tracking: Applications to membrane dynamics. *Annu. Rev. Biophys. Biomol. Struct.* **1997**, *26*, 373–399. [CrossRef]
50. Saxton, M.J. Lateral diffusion in an archipelago. Single-particle diffusion. *Biophys. J.* **1993**, *64*, 1766–1780. [CrossRef]
51. Schütz, G.J.; Kada, G.; Pastushenko, V.P.; Schindler, H. Properties of lipid microdomains in a muscle cell membrane visualized by single molecule microscopy. *EMBO J.* **2000**, *19*, 892–901. [CrossRef]
52. Schmidt, T.; Schütz, G.J. Single-Molecule Analysis of Biomembranes. In *Handbook of Single–Molecule Biophysics*; Hinterdorfer, P., Oijen, A., Eds.; Springer: New York, NY, USA, 2009; pp. 19–42.
53. Basak, S.; Sengupta, S.; Chattopadhyay, K. Understanding biochemical processes in the presence of sub-diffusive behavior of biomolecules in solution and living cells. *Biophys. Rev.* **2019**, *11*, 851–872. [CrossRef]
54. Stanislavsky, A.; Weron, K.; Weron, A. Diffusion and relaxation controlled by tempered α-stable processes. *Phys. Rev. E* **2008**, *78*, 051106. [CrossRef]
55. Meerschaert, M.M.; Zhang, Y.; Baeumer, B. Tempered anomalous diffusion in heterogeneous systems. *Geophys. Res. Lett.* **2008**, *35*, L17403. [CrossRef]
56. Flenner, E.; Das, J.; Rheinstädter, M.C.; Kosztin, I. Subdiffusion and lateral diffusion coefficient of lipid atoms and molecules in phospholipid bilayers. *Phys. Rev. E* **2009**, *79*, 011907. [CrossRef] [PubMed]
57. Jeon, J.H.; Monne, H.M.S.; Javanainen, M.; Metzler, R. Anomalous diffusion of phospholipids and cholesterols in a lipid bilayer and its origins. *Phys. Rev. Lett.* **2012**, *109*, 188103. [CrossRef] [PubMed]
58. Molina Garcia, D.; Sandev, T.; Safdari, H.; Pagnini, G.; Chechkin, A.; Metzler, R. Crossover from anomalous to normal diffusion: Truncated power-law noise correlations and applications to dynamics in lipid bilayers. *New J. Phys.* **2018**, *20*, 103027. [CrossRef]
59. Gajda, J.; Magdziarz, M. Fractional Fokker–Planck equation with tempered α-stable waiting times: Langevin picture and computer simulation. *Phys. Rev. E* **2010**, *82*, 011117. [CrossRef]
60. Stanislavsky, A.; Weron, K. Tempered relaxation with clustering patterns. *Phys. Lett. A* **2011**, *375*, 4244–4248. [CrossRef]
61. Janczura, J.; Wyłomańska, A. Anomalous diffusion models: Different types of subordinator distribution. *Acta Phys. Pol. B* **2012**, *43*, 1001–1016. [CrossRef]
62. Stanislavsky, A.; Weron, K. Anomalous diffusion approach to dielectric spectroscopy data with independent low- and high-frequency exponents. *Chaos Solitons Fractals* **2012**, *45*, 909–913. [CrossRef]
63. Wyłomańska, A. Arithmetic Brownian motion subordinated by tempered stable and inverse tempered stable processes. *Physica A* **2012**, *391*, 5685–5696. [CrossRef]
64. Wyłomańska, A. The tempered stable process with infinitely divisible inverse subordinators. *J. Stat. Mech.* **2013**, *2013*, P10011. [CrossRef]
65. Kumar, A.; Vellaisamy, P. Inverse tempered stable subordinators. *Stat. Probab. Lett.* **2015**, *103*, 134–141. [CrossRef]
66. Sandev, T.; Chechkin, A.; Kantz, H.; Metzler, R. Diffusion and Fokker–Planck-Smoluchowski equations with generalized memory kernel. *Fract. Calc. Appl. Anal.* **2015**, *18*, 1006–1038. [CrossRef]
67. Alrawashdeh, M.S.; Kelly, J.F.; Meerschaert, M.M. Applications of inverse tempered stable subordinators. *Comput. Math. Appl.* **2017**, *73*, 892–905. [CrossRef]
68. Feller, W. *Introduction to Probability Theory and Its Application*; John Wiley & Sons Inc.: New York, NY, USA, 1967; Volume II.
69. Song, R.; Vondraček, Z. Potential Theory of Subordinate Brownian Motion. In *Potential Analysis of Stable Processes and Its Extensions*; Graczyk, P.; Stos, A., Eds.; Lecture Notes in Mathematics 1980; Springer: Berlin/Heidelberg, Germany, 2009; pp. 87–176.

70. Magdziarz, M. Langevin picture of subdiffusion with infinitely divisible waiting times. *J. Stat. Phys.* **2009**, *135*, 763–772. [CrossRef]
71. Abramowitz, M.; Stegun, I.A. *Handbook of Mathematical Functions with Formulas, Graphs, and Mathematical Tables*; Dover: New York, NY, USA, 1964.
72. Sandev, T.; Sokolov, I.M.; Metzler, R.; Chechkin, A. Beyond monofractional kinetics. *Chaos Solitons Fractals* **2017**, *102*, 210–217. [CrossRef]
73. Sandev, T.; Metzler, R.; Chechkin, A. From continuous time random walks to the generalized diffusion equation. *Fract. Calc. Appl. Anal.* **2018**, *21*, 10–28. [CrossRef]
74. Rosiński, J. Tempering stable processes. *Stoch. Proc. Appl.* **2007**, *117*, 677–707. [CrossRef]
75. Mathai, A.M.; Haubold, H.J. *Special Functions for Applied Scientists*; Springer: New York, NY, USA, 2008.
76. Johnson, N.L.; Kotz, S.; Balakrishnan, N. *Continuous Univariate Distributions*; John Wiley & Sons Inc.: New York, NY, USA, 1994.
77. Kotz, S.; Kozubowski, T.; Podgórski, K. *The Laplace Distribution and Generalizations: A Revisit with Applications to Communications, Economics, Engineering, and Finance*; Birkhauser: Boston, MA, USA, 2001.
78. Chubynsky, M.V.; Slater, G.W. Diffusing diffusivity: A model for anomalous, yet Brownian, diffusion. *Phys. Rev. Lett.* **2014**, *113*, 098302. [CrossRef]
79. Chechkin, A.V.; Seno, F.; Metzler, R.; Sokolov, I.M. Brownian yet non-Gaussian diffusion: From superstatistics to subordination of diffusing diffusivities. *Phys. Rev. X* **2017**, *7*, 021002. [CrossRef]
80. Metzler, R. Superstatistics and non-Gaussian diffusion. *Eur. Phys. J. Spec. Top.* **2020**, *229*, 711–728. [CrossRef]
81. Ślęzak, J.; Burov, S. From diffusion in compartmentalized media to non-Gaussian random walks. *arXiv* **2019**, arXiv:1909.11395.
82. Stanislavsky, A.; Weron, A. Accelerating and retarding anomalous diffusion: A Bernstein function approach. *Phys. Rev. E* **2020**, *101*, 052119. [CrossRef] [PubMed]
83. Linnik, Yu.V. Linear forms and statistical criteria, I, II. In *English Translations in Mathematical Statistics and Probability, 3*; American Mathematical Society: Providence, RI, USA, 1962; pp. 1–90.
84. Pillai, R.N. Semi α-Laplace distributions. *Commun. Stat. Theory Methods* **1985**, *14*, 991–1000. [CrossRef]
85. Kotz, S.; Ostrovskii, I.V. A. Hayfavi, Analytic and asymptotic properties of Linnik's probability densities. *J. Math. Anal. Appl.* **1995**, *193*, 353–371. [CrossRef]
86. Kozubowski, T.J. Geometric stable laws: Estimation and applications. *Math. Comput. Model.* **1999**, *29*, 241–253. [CrossRef]
87. Kozubowski, T.J. Fractional moment estimation of Linnik and Mittag–Leffler parameters. *Math. Comput. Model.* **2001**, *34*, 1023–1035. [CrossRef]
88. Kotz , S.; Ostrovskii, I.V. A mixture representation of the Linnik distribution. *Statist. Probab. Lett.* **1996**, *26*, 61–64. [CrossRef]
89. Lukacs, E. *Characteristic Functions*, 2nd ed.; Charles Griffin and Co.: London, UK, 1970.
90. Anderson, D.N.; Arnold, B.C. Linnik distributions and processes. *J. Appl. Probab.* **1993**, *30*, 330–340. [CrossRef]
91. George, S.; Pillai, R.N. Multivariate α-Laplace distributions. *J. Nat. Acad. Math.* **1987**, *5*, 13–18.
92. Devroye, L. A note on Linnik's distribution. *Stat. Prob. Let.* **1990**, , 305–306. [CrossRef]
93. Gradshteyn, I.S.; Ryzhik, I.M. *Table of Integrals, Series, and Products*; Academic Press: New York, NY, USA, 2007.
94. Wintner, A. The singularities of Cauchy's distributions. *Duke Math. J.* **1941**, *8*, 678–681. [CrossRef]
95. Pillai, R.N.; Sandhya, E. Distributions with complete monotone derivative and geometric infinite divisibility. *Adv. Appl. Prob.* **1990**, *22*, 751–754. [CrossRef]
96. Weron, K. A probabilistic mechanism hidden behind the universal power law for dielectric relaxation: General relaxation equation. *J. Phys. Condens. Matter* **1991**, *3*, 9151–9162. [CrossRef]
97. Stanislavsky, A.; Weron, K. Stochastic tools hidden behind the empirical dielectric relaxation laws. *Rep. Prog. Phys.* **2017**, *80*, 036001. [CrossRef] [PubMed]
98. Weron, K.; Jurlewicz, A. Two forms of self–similarity as a fundamental feature of the power–law dielectric response. *J. Phys. A Math. Gen.* **1993**, *26*, 395–410. [CrossRef]
99. Jurlewicz, A. Frequency–independent rules for the dielectric susceptibility derived from two forms of self–similar dynamical behavior of dipolar system. *J. Stat. Phys.* **1995**, *79*, 993–1003. [CrossRef]
100. Jurlewicz, A.; Weron, A.; Weron, K. Asymptotic behavior of stochastic systems with conditionally exponential decay property. *Appl. Math.* **1996**, *23*, 379–394.

101. Briane, V.; Kervrann, C.; Vimond, M. Statistical analysis of particle trajectories in living cells. *Phys. Rev. E* **2018**, *97*, 062121. [CrossRef]
102. Weron, A.; Janczura, J.; Boryczka, E.; Sungkaworn, T.; Calebiro, D. Statistical testing approach for fractional anomalous diffusion classification. *Phys. Rev. E* **2019**, *99*, 042149. [CrossRef]
103. Kowalek, P.; Loch-Olszewska, H.; Szwabinski, J. Classification of diffusion modes in single-particle tracking data: Feature-based versus deep-learning approach. *Phys. Rev. E* **2019**, *100*, 032410. [CrossRef]
104. Hubicka, K.; Janczura, J.; Time-dependent classification of protein diffusion types: A statistical detection of mean-squared-displacement exponent transitions. *Phys. Rev. E* **2020**, *101*, 022107. [CrossRef]
105. Wyłomańska, A.; Iskander, D.R.; Burnecki, K. Omnibus test for normality based on the Edgeworth expansion. *PLoS ONE* **2020**, *15*, e0233901. [CrossRef] [PubMed]
106. Kundu, D. Discriminating between normal and Laplace distributions. In *Advances in Ranking and Selection, Multiple Comparisons, and Reliability Statistics for Industry and Technology*; Balakrishnan, N.; Nagaraja, H.N., Kannan, N., Eds.; Birkhäuser: Boston, USA, 2005; pp. 65–79.
107. Lampo, T.J.; Stylianidou, S.; Backlund, M.P.; Wiggins, P.A.; Spakowitz, A.J. Cytoplasmic RNA-protein particles exhibit non-Gaussian subdiffusive behavior. *Biophys. J.* **2017**, *112*, 532–542. [CrossRef] [PubMed]

Article

Detection of Anomalous Diffusion with Deep Residual Networks

Miłosz Gajowczyk † and Janusz Szwabiński *,†

Faculty of Pure and Applied Mathematics, Hugo Steinhaus Center, Wrocław University of Science and Technology, 50-370 Wrocław, Poland; gajowczyk.milosz@gmail.com
* Correspondence: janusz.szwabinski@pwr.edu.pl
† These authors contributed equally to this work.

Abstract: Identification of the diffusion type of molecules in living cells is crucial to deduct their driving forces and hence to get insight into the characteristics of the cells. In this paper, deep residual networks have been used to classify the trajectories of molecules. We started from the well known ResNet architecture, developed for image classification, and carried out a series of numerical experiments to adapt it to detection of diffusion modes. We managed to find a model that has a better accuracy than the initial network, but contains only a small fraction of its parameters. The reduced size significantly shortened the training time of the model. Moreover, the resulting network has less tendency to overfitting and generalizes better to unseen data.

Keywords: SPT; anomalous diffusion; machine learning classification; deep learning; residual neural networks

Citation: Gajowczyk, M.; Szwabiński, J. Detection of Anomalous Diffusion with Deep Residual Networks. *Entropy* **2021**, *23*, 649. https://doi.org/10.3390/e23060649

Academic Editor: Alberto Guillén

Received: 6 April 2021
Accepted: 19 May 2021
Published: 22 May 2021

1. Introduction

Recent advances in single particle tracking (SPT) [1–4] have allowed to observe single molecules in living cells with remarkable spatio-temporal resolution. Monitoring the details of molecules' diffusion has become the key method for investigation of their complex environments.

The data collected in SPT experiments often reveal deviations from the Brownian motion [5], i.e., the normal diffusion governed by the Fick's laws [6] and characterized by a linear time-dependence of the mean square displacement (MSD) of the molecules. Those deviations are referred to as anomalous diffusion, a field intensively studied in the physical community [7–10]. Since Richardson found a cubic scaling of MSD for particles in turbulent flows [11], anomalous diffusion was observed in many processes including tracer particles in living cells [12–14], transport on fractal geometries [15], charge carrier transport in amorphous semiconductors [16], quantum optics [17], bacterial motion [18], foraging of animals [19], human travel patterns [20] and trends in financial markets [21]. Depending on the type of nonlinearity, the anomalous diffusion is further divided into sub- and superdiffusion—two categories corresponding to sub- and superlinear MSD, respectively.

Several analytical approaches have already been attempted to analyze mobility patterns of molecules. The most popular one is based on the mean square displacement [7,22–25]. The appeal of this method lies in its relative simplicity. However, it is known to have several limitations due to the finite precision of SPT setups [7,22,26,27] and the lack of significant statistics (short trajectories and/or very few ones). To overcome these problems, several other analytic methods have been proposed [27–38]. Most of them simply replace MSD by other features calculated from trajectories (e.g., radius of gyration [28] or velocity autocorrelation function [39]).

In the last few years, classification of diffusion modes utilizing machine learning (ML) algorithms is gaining on popularity. Bayesian approach [40–42], random forests [43–47], gradient boosting [44–47], neural networks [48], and deep neural networks [44,49–51] have already been used in an attempt to either just classify the trajectories or to extract quantitative information about them (e.g., the anomalous exponent [45,49,51]). The ML approach seems to be more powerful than the analytical one. However, the latter usually offers a deeper insight into the underlying processes governing the dynamics of molecules.

Despite the enormous progress in both the analytical and ML methods, the analysis of SPT data remains challenging. The classification results produced by different methods often do not agree with each other [27,38,46,47]. The reasons are similar to the ones limiting the applicability of MSD: localization errors, short trajectories, or irregular sampling. Thus, there is still need for new robust methods for anomalous diffusion. To catalog the already existing approaches, to assess their usability and to trigger the search for new ones, a challenge (called AnDi challenge) was launched last year by a team of international scientists [52].

In this paper, we are going to present a novel approach to anomalous diffusion based on deep residual networks (ResNets) [53]. In general, deep learning is quite interesting from the perspective of an end user, since it is able to extract features from raw data automatically, without any intervention by a human expert [54]. We already tested the applicability of convolutional neural networks (CNN) to SPT data [44]. They turned out to be very accurate. However, their architecture was quite complicated and the training times (including an automatic search for an optimal model) were of the order of days. Moreover, the resulting network had problems with the generalization to data coming from sources different than the ones used to generate the training set. Residual networks are a class of CNNs able to cure most of the problems the original CNN architecture is facing (i.e., vanishing and/or exploding gradients, saturiation of accuracy with increasing depth). They excel in image classification—a ResNet network won the ImageNet Large Scale Visual Recognition Challenge (ILSVRC) in 2015.

We will start from the smallest of the residual architectures, i.e., ResNet18, and then perform a series of numerical experiments in order to adopt it to characterization of anomalous diffusion. Our strategy for model tuning will be quite simple and focused mainly on the reduction of the parameters of the network. However, it should be noted here that there exist already sophisticated methods for designing small models with good performance [55–58]. The resulting network will then be applied to the G protein-coupled receptors and G proteins data set, already analyzed in Refs. [38,46,47]. Although our method is not a direct response to the AnDi Challenge [52] (e.g., we use different diffusion models for training), it is consistent with its goal to search for new robust algorithms for classification.

The paper is structured as follows. In Section 2, we briefly introduce the basics of MSD-based methods, the diffusion models we are interested in as well as the residual networks, which will be used for classification. In Section 3, data sets are briefly discussed. The search for the optimal architecture and the performance of the resulting model are presented in Section 4. The results are concluded in the last section.

2. Models and Methods

2.1. Traditional Analysis

A typical SPT experiment yields a series of coordinates (2D or 3D) over time for every observed particle. Those series have to be analyzed in order to find a relationship between the individual trajectories and the characteristics of the system at hand [59]. Typically, the first step of the analysis is the detection of the type of diffusion encoded in the trajectories.

The most common approach to classification of diffusion is based on the mean-square displacement (MSD) of particles [7,22–25]. The recorded time series is evaluated in terms of the time averaged MSD (TAMSD),

$$\overline{\delta_t^2(\Delta)} = \frac{1}{t-\Delta}\int_0^{\infty}\left[x(t'+\Delta)-x(t')\right]^2 \mathrm{d}t' \quad (1)$$

where $x(t)$ is the position of the particle at time t and Δ is the time lag separating the consequtive positions of the particle. Typically, $\overline{\delta_t^2(\Delta)}$ is calculated in the limit $\Delta \ll t$ to obtain good statistics, since the number of positions contributing to the average decreases with the increasing Δ.

The idea behind the MSD-based method is simply to evaluate the experimental MSD curves, i.e., $\overline{\delta_t^2(\Delta)}$ as a function of the varying time lag Δ and then to fit them with a theoretical model of the form

$$\overline{\delta_t^2(\Delta)} \simeq K_\alpha \Delta^\alpha, \tag{2}$$

where K_α is the generalized diffusion coefficient and α is the so-called anomalous exponent. The value of the latter one is used to discriminate between different diffusion types. The case $\alpha = 1$ corresponds to the normal diffusion (ND), also known as the Brownian motion [5]. In this physical scenario, a particle moves freely in its environment. In other words, it does not meet any obstacles in its path, and it also does not interact with other distant molecules. Any non-Brownian ($\alpha \neq 1$) emanation of particle transport is referred to as the anomalous diffusion. A sublinear MSD ($\alpha < 1$) stands for subdiffusion, which is appropriate to represent particles slowed down due to viscoelastic properties of their surroundings [60], particles colliding with obstacles [61,62] or trapped particles [63,64]. A superlinear case ($\alpha > 1$) indicates superdiffusion, which relates to a fast and usually directed motion of particles driven by molecular motors [65].

2.2. *Choice of Diffusion Models*

Many different theoretical models of diffusion may be used for analysis of experimental data (see Ref. [9] for a detailed overview). However, following Refs. [43,44], we decided to consider four models: normal diffusion [5], directed motion (DM) [22,66,67], fractional Brownian motion (FBM) in subdiffusive mode [68], and confined diffusion (CD) [40]. According to Saxton [7], for those basic models of diffusion in 2D, we have:

$$\begin{aligned}
\overline{\delta_{ND}^2(\Delta)} &= 4D\Delta, \\
\overline{\delta_{FBM}^2(\Delta)} &= 4D\Delta^\alpha, \\
\overline{\delta_{DM}^2(\Delta)} &= 4D\Delta + (v\Delta)^2, \\
\overline{\delta_{CD}^2(\Delta)} &\simeq r_c^2\left[1 - A_1 \exp\left(\frac{-4A_2 D\Delta}{r_c^2}\right)\right].
\end{aligned} \tag{3}$$

Here, v is the drift velocity in the directed motion, the constants A_1 and A_2 characterize the shape of the confinement, and r_c is the confinement radius.

2.3. *Deep Learning Classification Methods*

The above method has become very popular in the SPT community due to its simplicity. It should work flawlessly for pure long trajectories with no localization errors. However, real trajectories usually contain a lot of noise, which makes the fitting of mathematical models to MSD curves challenging, even in the case of normal diffusion [22]. Moreover, many experimental trajectories are short, limiting the evaluation of the MSD curves to just a few time lags. As a consequence, there is a need for methods going beyond MSD to provide a reliable information concerning the trajectories.

In a recent paper [44], we proposed two machine learning methods that outperform the MSD analysis in case of noisy data. The first one is perceived as traditional machine learning and utilizes a set of human-engineered features that should be extracted from trajectories to feed the classifiers (see also Refs. [46,47] for a more extensive analysis). The second one is based on deep neural networks, which constitute the state-of-the-art of the modern machine learning classification. We showed that both methods perform similarly on the synthetic test data. However, the deep learning approach may seem appealing to practitioners from the SPT community because it usually operates on raw trajectories as input data and does not require human intervention to create features for each trajectory.

A cascade of multiple layers of nonlinear processing units is used in this case for automatic feature identification, extraction, and transformation [69].

2.3.1. Convolutional Neural Networks

Convolutional neural networks (CNN) were used in Ref. [44] for classification purposes. This choice was triggered by the fact that those networks have already been successful in many tasks including time series analysis [70]. A CNN has usually two components. The first one consisting of hidden layers extracts features from raw data. The fully connected part of the network is responsible for classification (see Figure 1 for a schematic representation of a CNN). In order to detect features in the input data, the hidden layers perform a series of convolutions and pooling operations. Each convolution provides its own map of features (a 3D array) by utilizing a filter that is sliding over the input data. The size of the maps is reduced in the pooling elements.

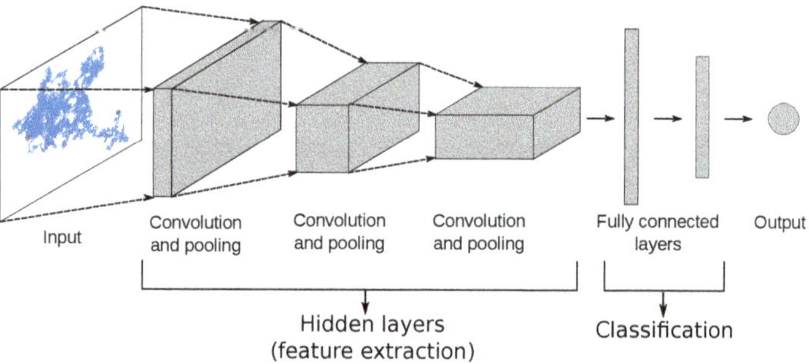

Figure 1. A schematic representation of a CNN network (source: Ref. [44]).

Choosing the right depth of the network is a challenging task. In Ref. [44], we assumed the architecture of the form (see also Ref. [71] for implementation details)

$$Batch - [Conv - Batch - ReLu] * N - Dense - ReLu - Dense - Batch - SoftMax, \quad (4)$$

and then performed a random search in the architecture and hyperparameter space in order to find the optimal model as well as other parameters required to initialize it. Here, $Batch$ is the batch normalization layer, i.e., a layer performing normalization of the data (not explicitly shown in Figure 1). $Conv$ and $Dense$ stand for convolution and dense layers, respectively. $ReLu$ is the abbreviation of the rectified linear unit, which is an activation function filtering out negative values from the output of the preceding layer. Finally, $SoftMax$ is the activation function determining the final output of the classifier. We haven't used the pooling layers in this model because reducing the spatial size of the 2D trajectories is usually not necessary. The procedure resulted in a network consisting of six convolutional layers and two dense ones.

2.3.2. ResNet Architecture

Although the model resulting from the above procedure performed well on our synthetic data (accuracy at the level of 97%), its architecture was quite complicated and the network itself was relatively deep, resulting in processing times of the order of days on a cluster of 24 CPUs with 50 GB total memory. However, long training times were not the only issue. It is known that with the increasing depth the problem of vanishing/exploding gradients may appear in the training phase of neural networks. Moreover, the training error may increase with the number of layers, resulting in a saturation of accuracy [53].

This is the reason why in this paper we decided to use the residual network (ResNet) [53]. It is a class of CNNs, which utilizes shortcuts (skip connections) to jump over several layers of the networks. Those shortcuts allow the network to make progress even if several layers have stopped learning because there is one blocking the backpropagation (Figure 2).

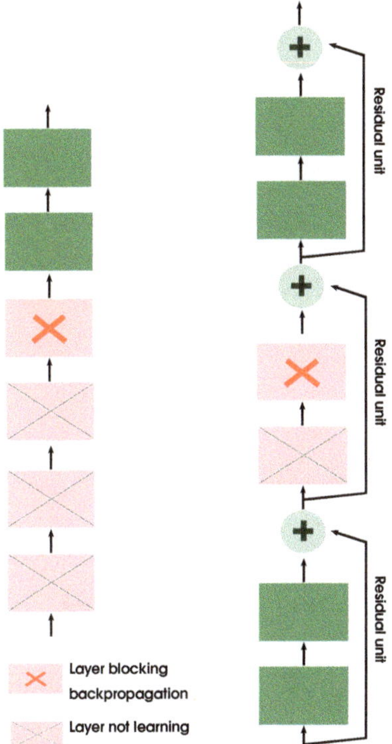

Figure 2. A regular CNN (left) versus a Resnet. Thanks to the skip connections in ResNet, the signal can easily pass a blocking layer in the backpropagation phase.

The residual network may be understood as a stack of residual units, where each unit is a small neural network with a skip connection. The outline of the unit is shown in Figure 3. For given input x, the desired mapping we want to obtain by learning is $H(x)$. Since the shortcut connection carries out the input layer to the addition operator shown in the figure, the rest of the unit needs only to learn the residual mapping $F(x) = H(x) - x$. When a regular CNN network is initialized, its weights are close to zero, so the network just outputs values close to zero. After adding the shortcuts, the network initially models the identity function. Therefore, if the target function is close to that function (which is often the case), the training phase will be significantly shorter than in the case of a regular CNN.

In Figure 4, the actual ResNet architecture is shown. We see that the core of the network is divided into four stages. Each of them contains, in addition to the residual units, a downsampling block. Its role is to reduce the information making its way across the network.

2.3.3. XResNet

In 2018, three modifications to the original ResNet architecture have been proposed under the common name XResNet [72]. Going into their details is beyond the scope of this paper. However, since they are known to have a non-negligible effect on the accuracy of the resulting model in some scenarios, we decided to include them in our search for the optimal architecture.

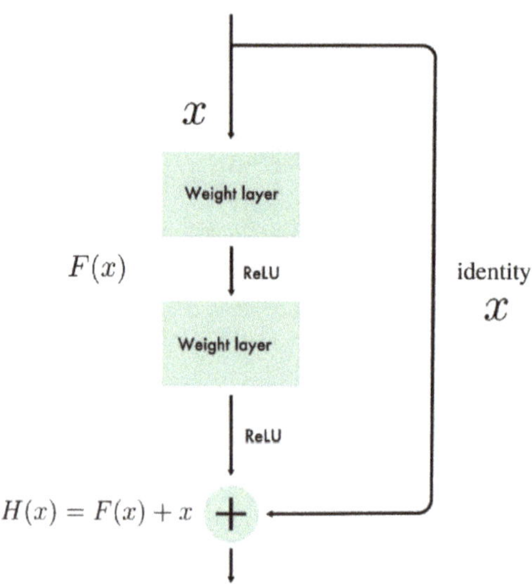

Figure 3. Residual unit in ResNet.

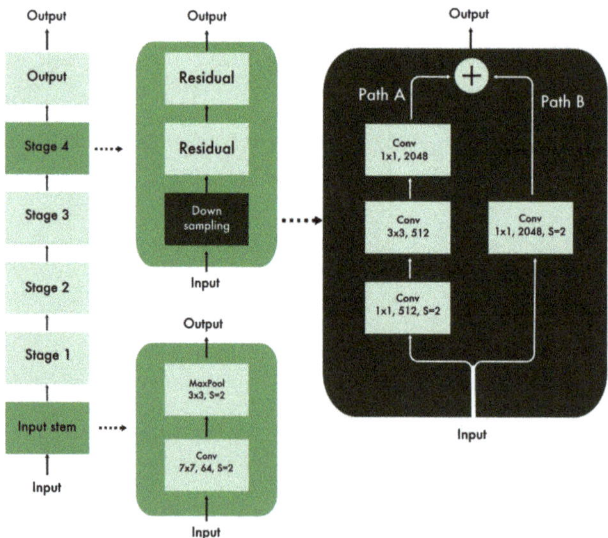

Figure 4. The architecture of ResNet. The downsampling block at the beginning of each stage help to reduce the amount of information in the case of deeper networks (path B is used in this case).

3. Synthetic and Experimental Data

3.1. Synthetic Training Data

The main factor limiting the deployment of machine learning to trajectory analysis is the availability of high-quality training data. It should contain a reasonable (i.e., large) amount of input data (trajectories) and corresponding desired output (their diffusion types). Since real data from experiments is not really provable (otherwise we would not need any new classification method), synthetic sets generated with computer simulations of different diffusion models are used for training. An ML algorithm uses the input–output pairs to learn the rules for data processing. Once trained, it is able to use those rules to classify new unseen trajectories.

As already mentioned in Section 2.2, we decided to follow Refs. [43,44] and use four basic models of diffusion to generate the training set of trajectories. The simulation methods will be briefly described in the remaining part of this section.

3.1.1. Normal Diffusion

Although several equivalent methods for simulation of Brownian motion exist, we will follow the approach presented by Michalet [22]. In case of normal diffusion, the probability distribution of the displacement's norm of a particle is given by the Rayleigh distribution

$$P(u) = \frac{2u}{4D\Delta t} \exp\left(\frac{-u^2}{4D\Delta t}\right), \; u \geq 0, \tag{5}$$

where u is the absolute distance traveled by the particle in time Δt. Thus, to simulate a trajectory, we have to randomly choose a start position of a particle and a random direction of the displacement φ and then pick a random step length u from the distribution (5). The new position of the particle is calculated,

$$\begin{aligned} x_{new} &= x_{old} + u \cos \varphi, \\ y_{new} &= y_{old} + u \sin \varphi, \end{aligned} \tag{6}$$

and taken as the starting point for the next move. The whole procedure is repeated till a trajectory of a desired length is generated.

3.1.2. Directed Motion

The simulation algorithm for the Brownian motion may be easily extended to generate a trajectory for diffusion with drift. All we have to do is simply to add a correction to the particle's position due to its active motion:

$$dx_i = v\Delta t \cos \beta, \tag{7}$$

$$dy_i = v\Delta t \sin \beta, \tag{8}$$

where v is the norm of the drift velocity and β its direction. Once we have the corrections, we add them to the new coordinates:

$$\begin{aligned} x_{new} &= x_{old} + u \cos \varphi + dx_i, \\ y_{new} &= y_{old} + u \sin \varphi + dy_i. \end{aligned} \tag{9}$$

The drift velocity is one of the parameters of the simulation. However, instead of setting its value directly, we will rather use an active-motion-to-diffusion ratio [43]:

$$R = \frac{v^2 T}{4D}, \tag{10}$$

where T is the time duration (i.e., the length of the trajectory). In our simulations, we will draw a random value of R from a given range and then calculate v for given D and T. In this way, it will be easier to generate similar trajectories with different values of v and D.

3.1.3. Confined Diffusion

Again, a small modification of the model for normal diffusion is needed to simulate a particle confined inside a reflective circular boundary. We simply divide every step of the simulation into 100 substeps with $\Delta t' = \Delta t/100$. Then, a normal diffusion move is carried out in every substep. The new position of the particle after all substeps will be updated only if the distance from the center of the boundary to new coordinates is smaller than the radius r_c of the boundary.

Following Wagner et al. [43], we will introduce a boundedness parameter B, defined as the area of the smallest ellipse enclosing a normal diffusion trajectory (with no confinement) divided by the area of the confinement,

$$B = \frac{A_{ellipse}}{\pi r_c^2} \simeq \frac{DN\Delta t}{r_c^2}. \tag{11}$$

It will help us to control the level of trapedness of particles in the simulations. B will be set randomly for each synthetic trajectory. Based on its value, the radius r_c will be calculated for given D, N, and Δt.

3.1.4. Fractional Brownian Motion

In addition to the confined diffusion, we will also use fractional Brownian motion to simulate the subdiffusive motion. FBM is the solution of the stochastic differential equation

$$dX_t^i = \sigma dB_t^{H,i}, i = 1, 2, \tag{12}$$

where $\sigma = \sqrt{2D}$ is the scale parameter related to the diffusion coefficient D, $H \in (0, 1)$ is the Hurst parameter and B_t^H is a continuous-time, zero-mean Gaussian process starting at zero, with the following covariance function

$$\mathrm{E}\left(B_t^H B_s^H\right) = \frac{1}{2}\left(|t|^{2H} + |s|^{2H} - |t - s|^{2H}\right). \tag{13}$$

The Hurst parameter H is connected with the anomalous exponent α via the relation

$$H = \frac{\alpha}{2}. \tag{14}$$

Since we want to use FBM for subdiffusion (i.e., $\alpha < 1$) only, the values of H will be restricted to the interval $(0, 1/2)$ in the simulations.

3.1.5. Creating Noisy Data

Real measurements of particles' positions are usually altered by noise from different sources including localization errors, vibrations of the sample, electronic noise or errors in the postprocessing phase [73]. Different methods of adding noise to synthetic trajectories are possible. One can, for instance, vary the diffusion coefficient of particles or simply add some disturbance to every point of a trajectory. We will go for the latter method and add normal Gaussian noise with zero mean and standard deviation σ to each simulated position.

To easily generate trajectories characterized by different levels of noise, we will proceed in the following way. We first introduce the signal-to-noise ratio:

$$Q = \begin{cases} \frac{\sqrt{D\Delta t}}{\sigma} & \text{for ND, CD, and FBM,} \\ \frac{\sqrt{D\Delta t + (v\Delta t)^2}}{\sigma} & \text{for DM.} \end{cases} \tag{15}$$

Then, we will randomly set Q and use the above formula to determine the standard deviation σ appropriate for given D, Δt, and v.

3.1.6. Simulation Details

For the sake of comparison, our synthetic data set should resemble all characteristics of the one used in Ref. [44]. To recap, we generated 20,000 trajectories, 5000 for each diffusion type. The time lag between consecutive points within a trajectory was set to $\Delta t = 1/30$ s, which is a typical value in experimental setups. All other parameters of the diffusion models were chosen randomly from the predefined ranges. Details can be found in Table 1.

Table 1. Parameters of the simulation and their values. All values except Δt were randomly chosen from given ranges.

Parameter	Meaning	Range of Values
Δt	timelag between steps	1/30 [s]
D	diffusion coefficient	0.1–20 [$\mu m^2/s$]
N	length of a trajectory	30–600
B	boundedness	1–6
R	active motion to diffusion ratio	1–17
α	anomalous exponent	0.3–0.7
SNR	signal to noise ratio	1–9

The data set was then divided into three subsets: the training set for fitting the machine learning models, the validation set used to estimate prediction errors for model selection and the test set for assessment of the final model. The stratified sampling method [74] was used for that purpose to guarantee a balanced representation of the diffusion modes in the subsets. Their sizes are presented in Table 2.

Table 2. Partition of the synthetic data set.

Subset Type	FBM	CD	DM	ND	Size	Share
Training	3500	3500	3500	3500	14,000	70%
Validation	750	750	750	750	3000	15%
Test	750	750	750	750	3000	15%

3.2. *Real Data*

We will apply our classifier to data from a single particle tracking experiment on G protein-coupled receptors and G proteins, already analyzed in Refs. [38,46,47]. The receptors mediate biological effects of many hormones and neourotransmitters and are also important as pharmacological targets [75]. Their signals are transmitted to the cell interior via interactions with G proteins. The analysis of the dynamics of these two types of molecules is extremely interesting because it may shed more light on how the receptors and G proteins meet, interact, and couple.

4. Results

The main goal of this work was to find a deep residual network with the simplest possible architecture, which is able to detect types of anomalous diffusion with satisfactory accuracy. In this section, we will first present a series of experiments that allowed us to significantly reduce the number of parameters of the original ResNet architecture. Then, we will apply the resulting model to classify both synthetic and real trajectories. All results were obtained with custom Python codes, available at https://github.com/milySW/NNResearchAPI, accessed on 20 May 2021. PyTorch library [76] was used to build the neural networks.

4.1. *Finding the Optimal Network Architecture*

We performed a series of computer experiments to find a reasonable ResNet architecture. Our goal was to keep the network as small as possible to reduce both the training times and the danger of overfitting. At the same time, we targeted the classification performance on synthetic data beyond the accuracy of 90%.

Before we dive into the results of the most important experiments, we would like to provide one important note. It is usually not worth investing effort and time in more complicated networks for tiny improvements of accuracy because, due to the stochastic nature of the networks, even different instances of the same model may yield slightly different results. Having that in mind, we introduced a (rather arbitrary) threshold equal to 0.2 percentage point as an indicator of improvements worth considering. All changes in accuracy smaller than the threshold were seen as irrelevant.

4.1.1. Impact of XResNet Modifications

Our first attempt was to check if the XResNet modifications [72] to the original architecture are worth considering. We took ResNet18, i.e., the smallest residual network with 18 layers, as the starting point. Results are shown in Table 3. Although the original architecture performs better on the training set, the modified one generalizes better to unseen data (i.e., has higher accuracy on the validation set). This may indicate the tendency of ResNet18 to overfit. The cost we have to pay for the improvement in validation accuracy by 0.34 percentage point is the increase in the number of parameters of the model (by 43,328) and a longer average time needed to complete one epoch (i.e., one cycle through the training data set). Despite the cost, we will keep the modifications in the model and try to reduce the number of parameters by other means.

Table 3. Impact of the XResNet modifications [72] on the accuracy of the model. Bold indicates the architecture we chose for further investigations.

Architecture	Number of Parameters	Accuracy (Training)	Accuracy (Validation)	Best Epoch	Epoch Time [s]
ResNet	11,177,092	93.16%	90.33%	12	40
XResNet	11,220,420	92.19%	90.67%	21	48

4.1.2. Depth of Neural Network

The baseline ResNet architecture consists of four stages, each of which is characterized by a different number of kernels that are convolved with the input [53]. However, ResNet was designed for classification of images, which are usually more complex than our trajectories. Thus, it will be interesting to check how a partial removal of those stages impacts the accuracy of the classifier. Results of our experiments are shown in Table 4. We see that reducing the depth of the network leads to a significant decrease in the number of the parameters in the model and improves its accuracy on the validation data.

Table 4. Relationship between the accuracy of the model and its depth. Depth equal to 3 was chosen for further investigations.

Depth	Number of Parameters	Accuracy (Training)	Accuracy (Validation)	Best Epoch	Epoch Time [s]
4	11,220,420	92.19%	90.67%	21	48
3	2,823,108	91.32%	91.10%	25	33
2	721,604	90.26%	90.80%	38	28

As expected, one does not need the full depth of the original ResNet architecture to classify the trajectories. Although the number of the parameters for two stages is very tempting, we decided to go further with depth 3 because it gives a slightly better performance.

4.1.3. Dimension and Size of Convolutions

The original Resnet architecture works with 2D objects and uses convolution kernels of size 3×3. It will be interesting to see how the model performs with smaller kernels. Although a 2×2 kernel is theoretically possible, one usually tries to avoid kernels of even sizes due to the lack of a well defined central pixel. Consequently, we will compare only 1×1 kernels with the baseline. As it follows from Table 5, the accuracy of the model declines significantly with the introduction of the smaller kernels.

Table 5. Relationship between the size of the 2D convolution kernels and the performance of the model.

Conv. Kernel	Number of Parameters	Accuracy (Training)	Accuracy (Validation)	Best Epoch	Epoch Time [s]
1×1	357,188	75.95%	76.93%	24	9
$\mathbf{3 \times 3}$	2,823,108	91.32%	91.10%	25	33

There is also a possibility of flattening the trajectories to 1D vectors and convolve them with $1 \times X$ kernels. We have checked the model for kernels with an odd X ranging from 3 to 11. Results are shown in Table 6. As we can see, those changes could slightly improve the performance of the model. Moreover, the size of the model was reduced by 44%. Thus, we will keep 1×5 kernels and work with 1D input for further investigations.

Table 6. Relationship between the size of the 1D convolution kernels and the performance of the model.

Conv. Kernel	Number of Parameters	Accuracy (Training)	Accuracy (Validation)	Best Epoch	Epoch Time [s]
1×3	973,668	92.69%	91.23%	25	39
$\mathbf{1 \times 5}$	1,590,148	92.77%	92.17%	30	24
1×7	2,206,628	90.91%	91.60%	14	28
1×9	2,823,108	93.59%	91.17%	20	30
1×11	3,439,588	92.71%	91.70%	15	32

4.1.4. Feature Maps

The number of parameters of the model may also be reduced by limiting its "breadth", understood here as the number of feature maps (convolution kernels) at each layer. The latter for the i-th block is given by the formula:

$$\begin{cases} x_0 = 64, \\ x_i = x_0 \cdot 2^{i-1}, \qquad \text{for } i = 1, 2, \ldots, n. \end{cases} \tag{16}$$

From Table 7, it follows that decreasing x_0 from 64 to 32 will not significantly decrease the accuracy of the model, but will reduce the number of parameters by a factor of 4. Moreover, the learning process of the network takes noticeably less time.

Table 7. Relationship between the number of feature maps and the accuracy of the model.

x_0	Number of Parameters	Accuracy (Training)	Accuracy (Validation)	Best Epoch	Epoch Time [s]
64	1,590,148	92.77%	92.17%	30	24
32	399,556	90.00%	92.00%	9	16
16	100,900	92.29%	91.10%	25	15

4.1.5. Additional Features

One of the advantages of deep networks, at least from the perspective of an end user, is the ability to work with raw experimental data. There is no need for human-engineered features as input because the network extracts its own features automatically from the data. While this is true for ResNet architecture as well, in principle, we could augment the input to the model by some additional attributes, including the ones tailor-made to the problem of diffusion.

A set of features with the potential of distinguishing different diffusion modes from each other was presented in Ref. [44]. Here, we would like to check if adding some of those attributes to the model will have a positive impact on accuracy. We decided to use asymmetry, efficiency, fractal dimension, and TAMSD at lag 20 as additional input (see Refs. [43,44] for definitions). For each trajectory, the values of the attributes were added to the network after the raw data went through all convolutional layers and was flattened.

Results of this series of experiments are shown in Table 8. Although the network was fed with additional information, its accuracy has not improved. To explain that, let us have a look at the distribution of asymmetry among trajectories in our data set. As it follows from Figure 5, its values for different types of diffusion overlap to some extent. Thus, classifying them based on the information encoded in asymmetry may be challenging. The same holds for the other attributes. Thus, we are not going to include them in our final model.

Table 8. Impact of additional attributes on the performance of the model.

Additional Features	Number of Parameters	Accuracy (Training)	Accuracy (Validation)	Best Epoch
None	399,556	90.00%	92.00%	9
Asymmetry	399,560	89.96%	91.03%	15
Efficiency	399,560	91.34%	91.90%	61
Fractal dimension	399,560	91.20%	91.17%	23
TAMSD	399,560	90.54%	91.03%	34
All	399,572	83.82%	83.97%	12

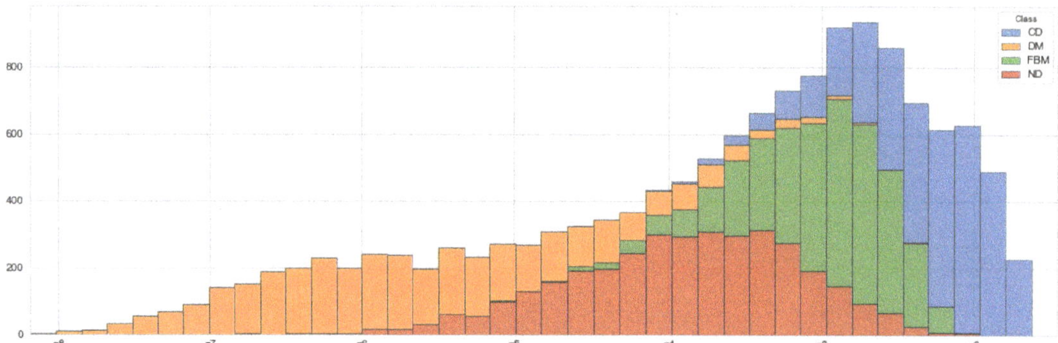

Figure 5. Distribution of asymmetry among trajectories in the synthetic data for different types of diffusion.

4.1.6. Impact of Autocorrelation

Following Ref. [77], we decided to check if the autocorrelation function taken as additional input improves the accuracy of the model. We combined the raw trajectories with their autocorrelations calculated at lags 8, 16, and 24 into a single tensor structure and used it as input to the model. Again, this measure did not improve the accuracy (Table 9).

Table 9. Using autocorrelation function as additional input to the model.

Autocorrelation	Number of Parameters	Accuracy (Training)	Accuracy (Validation)	Best Epoch
No	399,556	90.00%	92.00%	9
Yes	401,872	91.55%	91.93%	19

4.1.7. Selective Backprop

One of the interesting techniques to accelerate the training of deep neural networks is the selective backprop [78]. The idea behind this procedure is to prioritize samples with high loss at each iteration. It uses the output of sample's forward pass in the training phase to decide whether to use that sample to compute gradients and update parameters of the model or to skip immediately to other sample.

We carried out an experiment with two selective backprop scenarios. In the first one, a subset of training data covering 98% of the total loss was chosen for back-propagation. In this way, only 50–60% of trajectories were used in every epoch to update the network. In the second scenario, 50% of the training data were always taken, covering between 94% and 99% of the total loss in each epoch. It turned out that this method indeed shortens the training phase of the network (in particular average epoch time). However, it yields worse performance compared to the model utilizing the whole data set for back-propagation (Table 10).

Table 10. Different scenarios of selective backprop and their impact on the accuracy of the model.

Scenario	Number of Parameters	Accuracy (Training)	Accuracy (Validation)	Best Epoch	Epoch Time [s]
None	399,556	90.00%	92.00%	9	16
98% of cost	399,556	81.81%	90.67%	14	11
50% of cost	399,556	80.57%	90.97%	19	10

4.1.8. Choice of Hyperparameters

In the last series of experiments, we tried to find optimal values of some hyperparameters of the model. First, we looked at the cost function. Its choice allows us to control the focus in the training phase. Cross entropy for instance strongly penalizes misclassification, as it grows exponentially while approaching a wrong prediction [79]. Mean squared error (MSE) is usually used for regression problems. It does not punish wrong classifications enough, but rather promotes being close to a desired value. Although the cross entropy is the natural choice in classification tasks, the choice of the cost function seems to have no significant impact on the model's validation accuracy (Table 11). We kept MSE for shorter training times.

Table 11. Impact of cost function on the accuracy of the model.

Cost Function	Accuracy (Training)	Accuracy (Validation)	Best Epoch
Cross-entropy	91.91%	91.97%	26
MSE	90.00%	92.00%	9

An activation function defines the output of a node for the given input. It usually introduces some nonlinearity to the model. We checked four different functions. Sigmoid [80] is one of the most widely used activation functions today. It nicely mimics the behavior of real neurons; however, it may suffer from vanishing/exploding gradients. ReLU [81] is computationally very cheap, but it is also known to "die" in some situations (weights may update in such a way that the neuron never activates). Leaky ReLU [82] and ELU [83] are modifications of ReLU that mitigate that problem.

According to Table 12, ReLU activation function offers the highest accuracy on the validation set.

Table 12. Accuracy of the model for different choices of the activation function.

Activation Function	Accuracy (Training)	Accuracy (Validation)	Best Epoch
Sigmoid	87.65%	85.13%	10
ReLU	90.00%	92.00%	9
LeakReLU	91.50%	91.53%	24
ELU	85.15%	87.20%	3

The batch size is another important hyperparameter in the model. It defines the number of samples to work through before the model's internal parameters are updated. Larger batches should allow for more efficient computation, but may not generalize well to unseen data [84]. Small batches, on the other hand, are known to sometimes have problems with arriving at local minima [79].

Results for three different batch sizes are shown in Table 13—512 turned out to be the best one in our model.

Table 13. Accuracy of the model for different batch sizes.

Batch Size	Accuracy (Training)	Accuracy (Validation)	Best Epoch
256	89.15%	91.63%	7
512	90.00%	92.00%	9
1024	94.75%	91.37%	23

4.1.9. Resulting Model

Based on the results of the above experiments, we were able to reduce the number of parameters in the model from 11,220,420 in Resnet18 with XResNet modifications to 399,556. In the same time, the accuracy of the model on validation data increased by 1.33 percentage points.

The architecture of the final model is summarized in Table 14. Besides the already mentioned parameters and hyperparameters, there are two others that have not been discussed yet. The activation threshold is a boolean flag telling the model whether it should automatically estimate the threshold value, above which the neurons become active. In addition, the learning rate is a tuning parameter that determines the step size at each iteration while moving toward a minimum of the loss function. To find its value, we used a finder algorithm proposed in Ref. [85] and implemented in a PyTorch Lightning module [86].

Table 14. Details of the optimal architecture.

Category	Feature	Value
Architecture	XResNet	Yes
	Dimension	1D
	Depth	3
	Feature map number	32
Modifications	Additional attributes	No
	Autocorrelation	No
	Filtering	No
hyperparameters	Conv. kernel	1×5
	Cost function	MSE
	Activation function	ReLU
	Batch size	512
	Activation threshold	Yes
	Learning rate	0.0003

4.2. *Performance of the Model*

A test set consisting of 3000 samples (750 for each diffusion type) was used to assess the performance of the final model (see Section 3.1.6 for details). In Figure 6, the confusion matrix of the classifier is shown. By definition, an element C_{ij} of the matrix is equal to the number of observations known to be in class i (true labels) and predicted to be in class j [87].

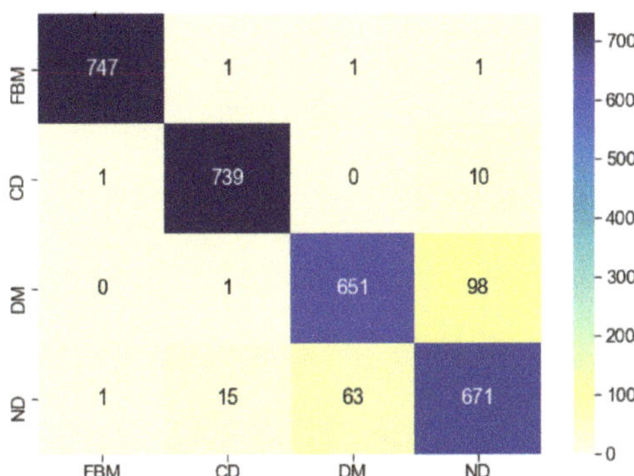

Figure 6. Confusion matrix of the model. Rows correspond to the true labels and columns to the predicted ones.

The model achieves the best performance for subdiffusion. Only 12 out of 750 trajectories have been wrongly classified in case of FBM and 25 out of 750 in case of CD. The other two modes are more challenging for the classifier. As for DM, 136 trajectories are misclassified, most of them as normal diffusion. The performance for the latter is slightly better—109 trajectories got wrong labels.

In Section 4.1.5, we tried to improve the performance of the model with some additional human-engineered features, which were motivated by the characteristics of diffusion itself. We were not really successful because it turned out that the distributions of those features overlap with each other, particularly for DM and ND, contributing to the confusion of the classifier. We guess that the same holds for features extracted automatically by the ResNet model—they are not specific enough to better distinguish DM from ND.

The confusion matrix may be used to calculate the basic performance metrics of the classifier. They are summarized in Table 15. Accuracy is defined as the number of correct predictions divided by the total number of predictions. Precision is the fraction of correct predictions of a class among all predictions of that class. It indicates how often a classifier is correct if it predicts a given class. Recall is the fraction of correct predictions of a given class over the total number of samples in that class. It measures the number of relevant results within a predicted class. Finally, F1 score is the harmonic mean of precision and recall.

Table 15. Basic performance metrics of the model on test data.

	Accuracy	Precision	Recall	F1 Score
FBM	-	96.98%	98.40%	97.68%
CD	-	91.77%	96.67%	94.16%
DM	-	92.33%	81.87%	86.78%
ND	-	81.76%	85.47%	83.57%
Total/Average	90.6%	90.71%	90.6%	90.55%

Even though the model has apparently some problems with DM and ND classes, its overall accuracy on test data are high. It returns much more relevant results than the irrelevant ones (high average precision), and it is able to yield most of the relevant results (high average recall). The F1 score simply confirms that.

It could be also interesting to check how the performance metrics of the classifier evolve with the training time (i.e., with the number of epochs). The results are presented in Figure 7. To generate the plots, we trained 50 instances of the model and then averaged the metrics. In this way, we could also estimate the 95% confidence levels. We see that all metrics reach a satisfactory level already in the third epoch. Further training improves the performance of the model only slightly.

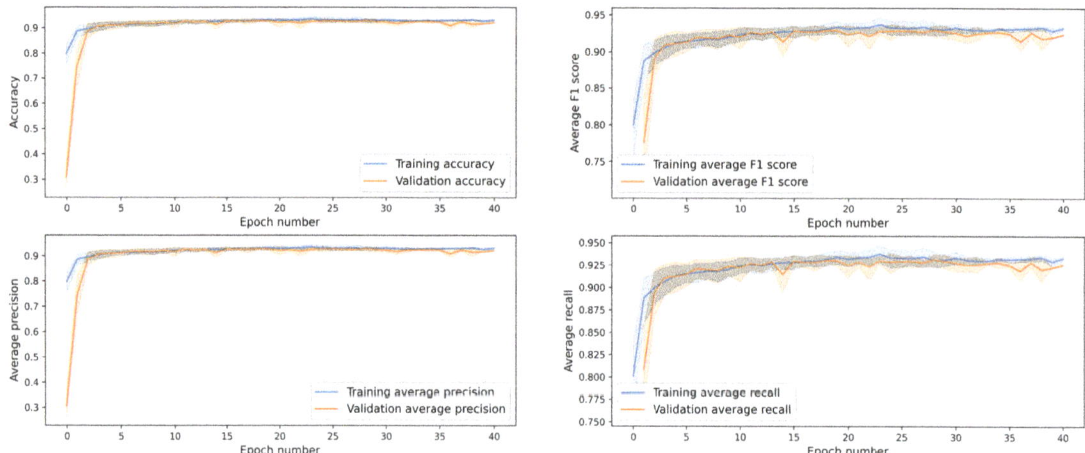

Figure 7. Performance metrics (on validation data) of the model as functions of the training time.

The same results, but this time broken down into separate diffusion modes, are shown in Figure 8. The measures for DM and ND are not only smaller than the ones for subdiffusion, but they also fluctuate to a higher extent when we look at values after the early epochs. This is due to the fact that these two classes are often confused with each other.

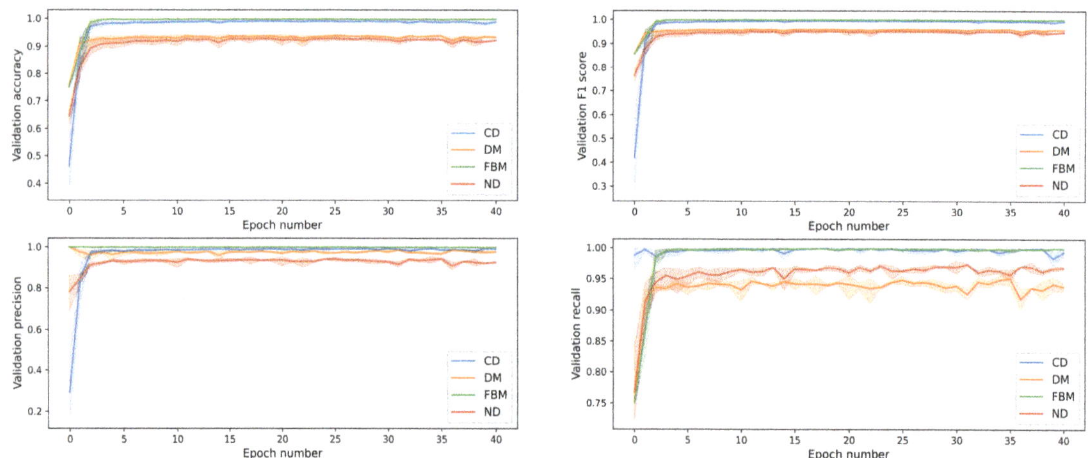

Figure 8. Performance metrics (on validation data) for each diffusion mode as functions of the training time.

The metrics for individual classes in the best epoch are shown in Figure 9. Again, we see a small gap between the subdiffusive classes on one hand and the problematic ones (i.e., DM and ND) on the other. However, even in the worst case, the metrics are above 80% indicating a good performance of the classifier.

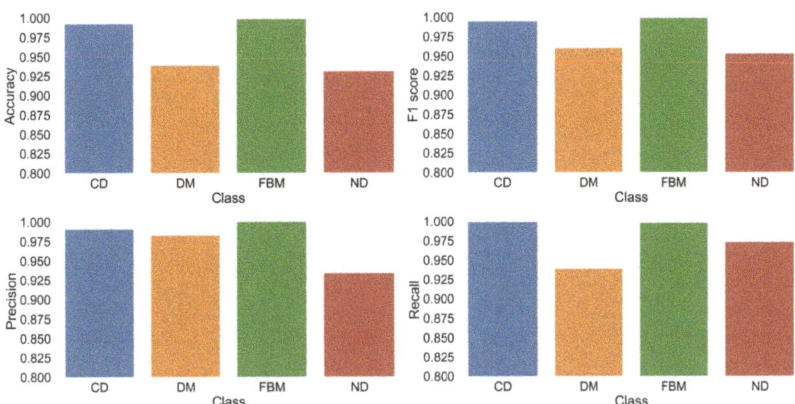

Figure 9. Performance metrics in the best epoch for each diffusion mode.

4.3. Classification of Real Data

From the available data on G protein-coupled receptors and G proteins, we took into account only trajectories with at least 50 steps. In this way, the data set was reduced to 1029 G proteins and 1218 receptors. Classification results are shown in Table 16. For the sake of comparison, two other predictions are reported in the table: a gradient boosting method utilizing noisy training data and a set of human-engineered features (reduced Set A trained with noise, see Table 15 in Ref. [47] for details) and a statistical testing procedure based on the maximum distance traveled by the particle (MAX method, see Refs. [38,46] for details).

Table 16. Classification of real data: comparison of our model with the feature based ML method from Ref. [47] (Set A with noise) and the statistical hypothesis testing from Ref. [38,46] (MAX method). "Rec." and "G Prot." stand for G protein-coupled receptors and G proteins, respectively. Due to rounding, the numbers may not add up precisely to 100%.

	Our Model		**Set A with Noise**		**MAX Method**	
	Rec.	**G Prot.**	**Rec.**	**G Prot.**	**Rec.**	**G Prot.**
Subdiffusion	0%	0.6%	25%	34%	21%	24%
Normal diffusion	70%	65%	72%	58%	79%	76%
Superdiffusion	30%	34.4%	1%	6%	0%	1%

Despite some differences in the absolute numbers, all three methods classify most of the trajectories as normal diffusion. However, there are significant discrepancies between them in the classification of the remaining time series. While our method labels almost all of them as superdiffusion, the other two ones predict subdiffusion in most of the cases. Unfortunately, the ground-truth for real data are missing and the results cannot be proven. However, it was already pointed out in Ref. [38] that different classification algorithms may provide substantially different results for the same data sets. Averaging of the results from all available methods has been proposed to mitigate the risk of large classification errors.

5. Discussion and Conclusions

Identifying the type of motion of particles in living cells is crucial to deduct their driving forces and hence to get insight into the mechano-structural characteristics of the cells. With the development of advanced AI methods in the last decades, there is an increasing interest to use them for that purpose. These methods are expected to outperform the well established statistical approach, in particular for noisy and small data sets.

In this paper, deep residual networks have been used to classify the SPT trajectories. We started from the well-known ResNet architecture [72], which excels in image classification, and carried out a series of numerical experiments to adapt it to detection of diffusion modes. We managed to find a model that has a better accuracy than the initial network, but contains only a small fraction of its parameters (399,556 vs. 11,177,092 in ResNet18, i.e., the smallest among ResNet networks). The reduced number of parameters had a huge positive impact on the training time of the model. Moreover, the resulting network has less tendency to overfitting and generalizes better to unseen data.

The overall accuracy of our model on the synthetic test data with noise is pretty good (90.6%). Breaking down the predictions into individual classes reveals that the model is able to recognize FBM and confined diffusion with a remarkable accuracy (99.6% and 98.53%, respectively). The detection of normal diffusion and directed motion seems to be more challenging and the model mixes up those two categories with each other from time to time.

Regarding the classification of real data, the predictions of our model are a little bit confusing. Compared to two other methods, i.e., a statistical testing procedure based on the maximum distance traveled by the particle [38,46] and gradient boosting methods with a set of tailor-made features characterizing the trajectories [47], it gives a similar fraction of normal diffusion (the majority class) among the trajectories. However, while our model classifies the remaining data as superdiffusion, the other ones assign most of those trajectories to the subdiffusive class. Moreover, it should be mentioned that some other classifiers provide results different from the ones in Table 16 [38,46]. In light of the above, the authors in Ref. [38] suggested taking a mean of the results of all available methods to minimize the risk of large errors. Therefore, there is still need to search for new classification methods for SPT data.

Author Contributions: Conceptualization, J.S.; methodology, J.S.; software, M.G.; validation, M.G.; investigation, M.G. and J.S.; writing—original draft preparation, J.S.; writing—review and editing, M.G. and J.S.; supervision, J.S. Both authors have read and agreed to the published version of the manuscript.

Funding: This work was partially supported by core funding for statutory R&D activities. J.S. was also funded by NCN Beethoven Grant No. 2016/23/G/ST1/04083.

Data Availability Statement: Codes required to generate training datasets may be found at https://github.com/milySW/NNResearchAPI (accessed on 20 May 2021).

Conflicts of Interest: The authors declare no conflict of interest.

References

1. Geerts, H.; Brabander, M.D.; Nuydens, R.; Geuens, S.; Moeremans, M.; Mey, J.D.D.; Hollenbeck, P. Nanovid tracking: A new automatic method for the study of mobility in living cells based on colloidal gold and video microscopy. *Biophys. J.* **1987**, *52*, 775–782. [CrossRef]
2. Barak, L.; Webb, W. Diffusion of low density lipoprotein-receptor complex on human fibroblasts . *J. Cell Biol.* **1982**, *95*, 846–852. [CrossRef] [PubMed]
3. Kusumi, A.; Sako, Y.; Yamamoto, M. Confined Lateral Diffusion of Membrane Receptors as Studied by Single Particle Tracking (Nanovid Microscopy). Effects of Calcium-induced Differentiation in Cultured Epithelial Cells. *Biophys. J.* **1993**, *65*, 2021–2040. [CrossRef]
4. Xie, X.S.; Choi, P.J.; Li, G.W.; Lee, N.K.; Lia, G. Single-Molecule Approach to Molecular Biology in Living Bacterial Cells. *Annu. Rev. Biophys.* **2008**, *37*, 417–444. [CrossRef]
5. Alves, S.B.; Oliveira, G.F., Jr.; Oliveira, L.C.; de Silansa, T.P.; Chevrollier, M.; Oriá, M.; Cavalcante, H.L.S. Characterization of diffusion processes: Normal and anomalous regimes. *Physica A* **2016**, *447*, 392–401. [CrossRef]
6. Fick, A. Ueber Diffusion (On Diffusion). *Ann. Phys. Chem.* **1855**, *170*, 59–86. [CrossRef]
7. Saxton, M.J.; Jacobson, K. Single-Particle Tracking: Applications to Membrane Dynamics. *Annu. Rev. Biophys. Biomol. Struct.* **1997**, *26*, 373–399. [CrossRef] [PubMed]
8. Barkai, E.; Garini, Y.; Metzler, R. Strange kinetics of single molecules in living cells. *Phys. Today* **2012**, *65*, 29. [CrossRef]
9. Metzler, R.; Jeon, J.H.; Cherstvy, A.G.; Barkai, E. Anomalous diffusion models and their properties: Non-stationarity, non-ergodicity, and ageing at the centenary of single particle tracking. *Phys. Chem. Chem. Phys.* **2014**, *16*, 24128–24164. [CrossRef]

10. Metzler, R.; Jeon, J.H.; Cherstvy, A. Non-Brownian diffusion in lipid membranes: Experiments and simulations. *Biochim. Biophys. Acta BBA Biomembr.* **2016**, *1858*, 2451–2467. [CrossRef]
11. Richardson, L.F. Atmospheric Diffusion Shown on a Distance-Neighbour Graph. *Proc. R. Soc. Lond. Ser. A* **1926**, *110*, 709–737. [CrossRef]
12. Goldberg, Y. Primer on Neural Network Models for Natural Language Processing. *J. Artif. Intell. Res.* **2016**, *57*, 345–420. [CrossRef]
13. Bronstein, I.; Israel, Y.; Kepten, E.; Mai, S.; Shav-Tal, Y.; Barkai, E.; Garini, Y. Transient Anomalous Diffusion of Telomeres in the Nucleus of Mammalian Cells. *Phys. Rev. Lett.* **2009**, *103*, 018102. [CrossRef] [PubMed]
14. Jeon, J.H.; Tejedor, V.; Burov, S.; Barkai, E.; Selhuber-Unkel, C.; Berg-Sørensen, K.; Oddershede, L.; Metzler, R. In Vivo Anomalous Diffusion and Weak Ergodicity Breaking of Lipid Granules. *Phys. Rev. Lett.* **2011**, *106*, 048103. [CrossRef]
15. Porto, M.; Bunde, A.; Havlin, S.; Roman, H.E. Structural and dynamical properties of the percolation backbone in two and three dimensions. *Phys. Rev. E* **1997**, *56*, 1667–1675. [CrossRef]
16. Zumofen, G.; Blumen, A.; Klafter, J. Current flow under anomalous-diffusion conditions: Lévy walks. *Phys. Rev. A* **1990**, *41*, 4558–4561. [CrossRef]
17. Schaufler, S.; Schleich, W.P.; Yakovlev, V.P. Keyhole Look at Lévy Flights in Subrecoil Laser Cooling. *Phys. Rev. Lett.* **1999**, *83*, 3162–3165. [CrossRef]
18. Klafter, J.; White, B.S.; Levandowsky, M. Microzooplankton Feeding Behavior and the Levy Walk. In *Biological Motion. Lecture Notes in Biomathematics*; Alt, W., Hoffmann, G., Eds.; Springer: Berlin/Heidelberg, Germnay, 1990; Volume 89.
19. Viswanathan, G.M.; da Luz, M.G.E.; Raposo, E.P.; Stanley, H.E. *The Physics of Foraging: An Introduction to Random Searches and Biological Encounters*; Cambridge University Press: Cambridge, UK, 2011. [CrossRef]
20. González, M.C.; Hidalgo, C.A.; Barabási, A.L. Understanding individual human mobility patterns. *Nature* **2008**, *453*, 779–782. [CrossRef]
21. Michael, F.; Johnson, M. Financial market dynamics. *Phys. A Stat. Mech. Its Appl.* **2003**, *320*, 525–534. [CrossRef]
22. Michalet, X. Mean square displacement analysis of single-particle trajectories with localization error: Brownian motion in an isotropic medium. *Phys. Rev. E* **2010**, *82*, 041914. [CrossRef]
23. Kneller, G.R. Communication: A scaling approach to anomalous diffusion. *J. Chem. Phys.* **2014**, *141*, 041105. [CrossRef] [PubMed]
24. Qian, H.; Sheetz, M.P.; Elson, E.L. Single particle tracking. Analysis of diffusion and flow in two-dimensional systems. *Biophys. J.* **1991**, *60*, 910–921. [CrossRef]
25. Gal, N.; Lechtman-Goldstein, D.; Weihs, D. Particle tracking in living cells: A review of the mean square displacement method and beyond. *Rheol. Acta* **2013**, *52*, 425–443. [CrossRef]
26. Kepten, E.; Weron, A.; Sikora, G.; Burnecki, K.; Garini, Y. Guidelines for the Fitting of Anomalous Diffusion Mean Square Displacement Graphs from Single Particle Tracking Experiments. *PLoS ONE* **2015**, *10*, e0117722. [CrossRef] [PubMed]
27. Briane, V.; Kervrann, C.; Vimond, M. Statistical analysis of particle trajectories in living cells. *Phys. Rev. E* **2018**, *97*, 062121. [CrossRef] [PubMed]
28. Saxton, M.J. Lateral diffusion in an archipelago. Single-particle diffusion. *Biophys. J.* **1993**, *64*, 1766–1780. [CrossRef]
29. Valentine, M.T.; Kaplan, P.D.; Thota, D.; Crocker, J.C.; Gisler, T.; Prud'homme, R.K.; Beck, M.; Weitz, D.A. Investigating the microenvironments of inhomogeneous soft materials with multiple particle tracking. *Phys. Rev. E* **2001**, *64*, 061506. [CrossRef]
30. Gal, N.; Weihs, D. Experimental evidence of strong anomalous diffusion in living cells. *Phys. Rev. E* **2010**, *81*, 020903. [CrossRef]
31. Raupach, C.; Zitterbart, D.P.; Mierke, C.T.; Metzner, C.; Müller, F.A.; Fabry, B. Stress fluctuations and motion of cytoskeletal-bound markers. *Phys. Rev. E* **2007**, *76*, 011918. [CrossRef]
32. Burov, S.; Tabei, S.M.A.; Huynh, T.; Murrell, M.P.; Philipson, L.H.; Rice, S.A.; Gardel, M.L.; Scherer, N.F.; Dinner, A.R. Distribution of directional change as a signature of complex dynamics. *Proc. Natl. Acad. Sci. USA* **2013**, *110*, 19689–19694. [CrossRef] [PubMed]
33. Tejedor, V.; Bénichou, O.; Voituriez, R.; Jungmann, R.; Simmel, F.; Selhuber-Unkel, C.; Oddershede, L.B.; Metzler, R. Quantitative Analysis of Single Particle Trajectories: Mean Maximal Excursion Method. *Biophys. J.* **2010**, *98*, 1364–1372. [CrossRef] [PubMed]
34. Burnecki, K.; Kepten, E.; Garini, Y.; Sikora, G.; Weron, A. Estimating the anomalous diffusion exponent for single particle tracking data with measurement errors—An alternative approach. *Sci. Rep.* **2015**, *5*, 11306. [CrossRef] [PubMed]
35. Das, R.; Cairo, C.W.; Coombs, D. A Hidden Markov Model for Single Particle Tracks Quantifies Dynamic Interactions between LFA-1 and the Actin Cytoskeleton. *PLoS Comput. Biol.* **2009**, *5*. [CrossRef] [PubMed]
36. Slator, P.J.; Cairo, C.W.; Burroughs, N.J. Detection of Diffusion Heterogeneity in Single Particle Tracking Trajectories Using a Hidden Markov Model with Measurement Noise Propagation. *PLoS ONE* **2015**, *10*, e0140759. [CrossRef] [PubMed]
37. Slator, P.J.; Burroughs, N.J. A Hidden Markov Model for Detecting Confinement in Single-Particle Tracking Trajectories. *Biophys. J.* **2018**, *115*, 1741–1754. [CrossRef] [PubMed]
38. Weron, A.; Janczura, J.; Boryczka, E.; Sungkaworn, T.; Calebiro, D. Statistical testing approach for fractional anomalous diffusion classification. *Phys. Rev. E* **2019**, *99*, 042149. [CrossRef] [PubMed]
39. Grebenkov, D.S. Optimal and suboptimal quadratic forms for noncentered Gaussian processes. *Phys. Rev. E* **2013**, *88*, 032140. [CrossRef] [PubMed]
40. Monnier, N.; Guo, S.M.; Mori, M.; He, J.; Lénárt, P.; Bathe, M. Bayesian Approach to MSD-Based Analysis of Particle Motion in Live Cells. *Biophys. J.* **2012**, *103*, 616–626. [CrossRef]

41. Thapa, S.; Lomholt, M.A.; Krog, J.; Cherstvy, A.G.; Metzler, R. Bayesian analysis of single-particle tracking data using the nested-sampling algorithm: Maximum-likelihood model selection applied to stochastic-diffusivity data. *Phys. Chem. Chem. Phys.* **2018**, *20*, 29018–29037. [CrossRef]
42. Cherstvy, A.G.; Thapa, S.; Wagner, C.E.; Metzler, R. Non-Gaussian, non-ergodic, and non-Fickian diffusion of tracers in mucin hydrogels. *Soft Matter* **2019**, *15*, 2526–2551. [CrossRef]
43. Wagner, T.; Kroll, A.; Haramagatti, C.R.; Lipinski, H.G.; Wiemann, M. Classification and Segmentation of Nanoparticle Diffusion Trajectories in Cellular Micro Environments. *PLoS ONE* **2017**, *12*, e0170165. [CrossRef]
44. Kowalek, P.; Loch-Olszewska, H.; Szwabiński, J. Classification of diffusion modes in single-particle tracking data: Feature-based versus deep-learning approach. *Phys. Rev. E* **2019**, *100*, 032410. [CrossRef]
45. Muñoz-Gil, G.; Garcia-March, M.A.; Manzo, C.; Martín-Guerrero, J.D.; Lewenstein, M. Single trajectory characterization via machine learning. *New J. Phys.* **2020**, *22*, 013010. [CrossRef]
46. Janczura, J.; Kowalek, P.; Loch-Olszewska, H.; Szwabiński, J.; Weron, A. Classification of particle trajectories in living cells: Machine learning versus statistical testing hypothesis for fractional anomalous diffusion. *Phys. Rev. E* **2020**, *102*, 032402. [CrossRef]
47. Loch-Olszewska, H.; Szwabiński, J. Impact of feature choice on machine learning classification of fractional anomalous diffusion. *Entropy* **2020**, *22*, 1436. [CrossRef]
48. Dosset, P.; Rassam, P.; Fernandez, L.; Espenel, C.; Rubinstein, E.; Margeat, E.; Milhiet, P.E. Automatic detection of diffusion modes within biological membranes using backpropagation neural network. *BMC Bioinform.* **2016**, *17*, 197. [CrossRef]
49. Granik, N.; Weiss, L.E.; Nehme, E.; Levin, M.; Chein, M.; Perlson, E.; Roichman, Y.; Shechtman, Y. Single-Particle Diffusion Characterization by Deep Learning. *Biophys. J.* **2019**, *117*, 185–192. [CrossRef]
50. Bo, S.; Schmidt, F.; Eichhorn, R.; Volpe, G. Measurement of anomalous diffusion using recurrent neural networks. *Phys. Rev. E* **2019**, *100*, 010102. [CrossRef]
51. Gentili, A.; Volpe, G. Characterization of anomalous diffusion statistics powered by deep learning. *arXiv* **2021**, arXiv:2102.07605.
52. Muñoz-Gil, G.; Volpe, G.; García-March, M.A.; Metzler, R.; Lewenstein, M.; Manzo, C. The anomalous diffusion challenge: Single trajectory characterisation as a competition. In Proceedings of the Emerging Topics in Artificial Intelligence 2020, Halkidiki, Greece, 5–7 June 2020. [CrossRef]
53. He, K.; Zhang, X.; Ren, S.; Sun, J. Deep Residual Learning for Image Recognition. In Proceedings of the 2016 IEEE Conference on Computer Vision and Pattern Recognition (CVPR), Las Vegas, NV, USA, 27–30 June 2016; pp. 770–778. [CrossRef]
54. Hatami, N.; Gavet, Y.; Debayle, J. Classification of Time-Series Images Using Deep Convolu- tional Neural Networks. In *Proceedings of SPIE, Tenth International Conference on Machine Vision (ICMV 2017)*; Verikas, A., Radeva, P., Nikolaev, D., Zhou, J., Eds.; SPIE Publications: Bellingham WA, USA, 2018; p. 10696.
55. Ye, K.; Kovashka, A.; Sandler, M.; Zhu, M.; Howard, A.; Fornoni, M. SpotPatch: Parameter-Efficient Transfer Learning for Mobile Object Detection. In *Computer Vision—ACCV 2020*; Ishikawa, H., Liu, C.L., Pajdla, T., Shi, J., Eds.; Springer International Publishing: Cham, Switzerland, 2021; pp. 239–256.
56. Guo, Y.; Li, Y.; Wang, L.; Rosing, T. Depthwise Convolution Is All You Need for Learning Multiple Visual Domains. In Proceedings of the AAAI Conference on Artificial Intelligence, Honolulu, HI, USA, 27 January–1 February 2019; Volume 33, pp. 8368–8375. [CrossRef]
57. Guo, Y.; Shi, H.; Kumar, A.; Grauman, K.; Rosing, T.; Feris, R. SpotTune: Transfer Learning Through Adaptive Fine-Tuning. In Proceedings of the IEEE/CVF Conference on Computer Vision and Pattern Recognition (CVPR), Long Beach, CA, USA, 15–20 June 2019.
58. Mudrakarta, P.K.; Sandler, M.; Zhmoginov, A.; Howard, A.G. K For The Price Of 1: Parameter Efficient Multi-task In addition, Transfer Learning. *arXiv* **2018**, arXiv:1810.10703.
59. Bressloff, P.C. *Stochastic Processes in Cell Biology*; Literaturverz, S., Ed.; Interdisciplinary Applied Mathematics; Springer: Cham, Switzerland, 2014; pp. 645–672.
60. Weiss, M.; Elsner, M.; Kartberg, F.; Nilsson, T. Anomalous Subdiffusion Is a Measure for Cytoplasmic Crowding in Living Cells. *Biophys. J.* **2004**, *87*, 3518–3524. [CrossRef] [PubMed]
61. Saxton, M.J. Single-particle tracking: Models of directed transport. *Biophys. J.* **1994**, *67*, 2110–2119. [CrossRef]
62. Berry, H.; Chaté, H. Anomalous diffusion due to hindering by mobile obstacles undergoing Brownian motion or Orstein-Ulhenbeck processes. *Phys. Rev. E* **2014**, *89*, 022708. [CrossRef] [PubMed]
63. Metzler, R.; Klafter, J. The random walk's guide to anomalous diffusion: A fractional dynamics approach. *Phys. Rep.* **2000**, *339*, 1–77. [CrossRef]
64. Hoze, N.; Nair, D.; Hosy, E.; Sieben, C.; Manley, S.; Herrmann, A.; Sibarita, J.B.; Choquet, D.; Holcman, D. Heterogeneity of AMPA receptor trafficking and molecular interactions revealed by superresolution analysis of live cell imaging. *Proc. Natl. Acad. Sci. USA* **2012**, *109*, 17052–17057. [CrossRef]
65. Arcizet, D.; Meier, B.; Sackmann, E.; Rädler, J.O.; Heinrich, D. Temporal Analysis of Active and Passive Transport in Living Cells. *Phys. Rev. Lett.* **2008**, *101*, 248103. [CrossRef] [PubMed]
66. Ruan, G.; Agrawal, A.; Marcus, A.I.; Nie, S. Imaging and Tracking of Tat Peptide-Conjugated Quantum Dots in Living Cells: New Insights into Nanoparticle Uptake, Intracellular Transport, and Vesicle Shedding. *J. Am. Chem. Soc.* **2007**, *129*, 14759–14766. [CrossRef]

67. Bannunah, A.M.; Vllasaliu, D.; Lord, J.; Stolnik, S. Mechanisms of Nanoparticle Internalization and Transport Across an Intestinal Epithelial Cell Model: Effect of Size and Surface Charge. *Mol. Pharm.* **2014**, *11*, 4363–4373. [CrossRef]
68. Mandelbrot, B.B.; Ness, J.W.V. Fractional Brownian Motions, Fractional Noises and Applications. *SIAM Rev.* **1968**, *10*, 422–437. [CrossRef]
69. Deng, L.; You, D. Deep Learning: Methods and Applications. *Found. Trends Signal Process.* **2014**, *7*, 1–199. [CrossRef]
70. Yang, J.B.; Nguyen, M.N.; San, P.P.; Li, X.L.; Krishnaswamy, S. Deep Convolutional Neural Networks on Multichannel Time Series for Human Activity Recognition. In Proceedings of the 24th International Conference on Artificial Intelligence, IJCAI'15, Buenos Aires, Argentina, 25–31 July 2015; AAAI Press: Buenos Aires, Argentina, 2015; pp. 3995–4001.
71. van Kuppevelt, D.; Meijer, C.; Huber, F.; van der Ploeg, A.; Georgievska, S.; van Hees V.T. Mcfly: Automated deep learning on time series. *SoftwareX* **2020**, *12*. [CrossRef]
72. He, T.; Zhang, Z.; Zhang, H.; Zhang, Z.; Xie, J.; Li, M. Bag of Tricks for Image Classification with Convolutional Neural Networks. In Proceedings of the IEEE/CVF Conference on Computer Vision and Pattern Recognition, Salt Lake City, UT, USA, 18–23 June 2018.
73. Lanoiselée, Y.; Briand, G.; Dauchot, O.; Grebenkov, D.S. Statistical analysis of random trajectories of vibrated disks: Towards a macroscopic realization of Brownian motion. *Phys. Rev. E* **2018**, *98*, 062112. [CrossRef]
74. Shahrokh Esfahani, M.; Dougherty, E.R. Effect of separate sampling on classification accuracy. *Bioinformatics* **2013**, *30*, 242–250. [CrossRef] [PubMed]
75. Pierce, K.L.; Premont, R.T.; Lefkowitz, R.J. Seven-transmembrane receptors. *Nat. Rev. Mol. Cell Biol.* **2002**, *3*, 639. [CrossRef]
76. Paszke, A.; Gross, S.; Massa, F.; Lerer, A.; Bradbury, J.; Chanan, G.; Killeen, T.; Lin, Z.; Gimelshein, N.; Antiga, L.; et al. PyTorch: An Imperative Style, High-Performance Deep Learning Library. In *Advances in Neural Information Processing Systems 32*; Wallach, H., Larochelle, H., Beygelzimer, A., d'Alché-Buc, F., Fox, E., Garnett, R., Eds.; Curran Associates, Inc.: Red Hook, NY, USA, 2019; pp. 8024–8035.
77. Szarek, D.; Sikora, G.; Balcerek, M.; Jabłoński, I.; Wyłomańska, A. Fractional Dynamics Identification via Intelligent Unpacking of the Sample Autocovariance Function by Neural Networks. *Entropy* **2020**, *22*, 1322. [CrossRef]
78. Jiang, A.H.; Wong, D.L.; Zhou, G.; Andersen, D.G.; Dean, J.; Ganger, G.R.; Joshi, G.; Kaminsky, M.; Kozuch, M.; Lipton, Z.C.; et al. Accelerating Deep Learning by Focusing on the Biggest Losers. *arXiv* **2019**, arXiv:1910.00762.
79. Géron, A. *Hands-On Machine Learning with Scikit-Learn and TensorFlow: Concepts, Tools, and Techniques to Build Intelligent Systems*; O'Reilly Media: Sebastopol, CA, USA, 2017.
80. Han, J.; Moraga, C. The influence of the sigmoid function parameters on the speed of backpropagation learning. In *From Natural to Artificial Neural Computation*; Mira, J., Sandoval, F., Eds.; Springer: Berlin/Heidelberg, Germany, 1995; pp. 195–201.
81. Nair, V.; Hinton, G.E. Rectified Linear Units Improve Restricted Boltzmann Machines. In Proceedings of the 27th International Conference on International Conference on Machine Learning, ICML'10, Haifa, Israel, 21–24 June 2010; Omnipress: Madison, WI, USA, 2010; pp. 807–814.
82. Maas, A.L.; Hannun, A.Y.; Ng, A.Y. Rectifier nonlinearities improve neural network acoustic models. In Proceedings of the ICML Workshop on Deep Learning for Audio, Speech and Language Processing, JMLR.org, Atlanta, GA, USA, 16 June 2013.
83. Clevert, D.A.; Unterthiner, T.; Hochreiter, S. Fast and Accurate Deep Network Learning by Exponential Linear Units (ELUs). *arXiv* **2016**, arXiv:1511.07289.
84. Keskar, N.S.; Mudigere, D.; Nocedal, J.; Smelyanskiy, M.; Tang, P.T.P. On Large-Batch Training for Deep Learning: Generalization Gap and Sharp Minima. *arXiv* **2017**, arXiv:1609.04836.
85. Smith, L.N. No More Pesky Learning Rate Guessing Games. *arXiv* **2017**, arXiv:1506.01186.
86. Falcon, W. PyTorch Lightning. GitHub. Note. 2019; p. 3. Available online: https://github.com/PyTorchLightning/pytorch-lightning (accessed on 20 October 2020).
87. Raschka, S. *Python Machine Learning*; Packt Publishing: Birmingham, UK, 2015.

Article

Impact of Feature Choice on Machine Learning Classification of Fractional Anomalous Diffusion

Hanna Loch-Olszewska *,† **and Janusz Szwabiński** *,†

Faculty of Pure and Applied Mathematics, Hugo Steinhaus Center, Wrocław University of Science and Technology, 50-370 Wrocław, Poland

* Correspondence: hanna.loch@pwr.edu.pl (H.L.-O.); janusz.szwabinski@pwr.edu.pl (J.S.)

† These authors contributed equally to this work.

Received: 12 November 2020; Accepted: 12 December 2020; Published: 19 December 2020

Abstract: The growing interest in machine learning methods has raised the need for a careful study of their application to the experimental single-particle tracking data. In this paper, we present the differences in the classification of the fractional anomalous diffusion trajectories that arise from the selection of the features used in random forest and gradient boosting algorithms. Comparing two recently used sets of human-engineered attributes with a new one, which was tailor-made for the problem, we show the importance of a thoughtful choice of the features and parameters. We also analyse the influence of alterations of synthetic training data set on the classification results. The trained classifiers are tested on real trajectories of G proteins and their receptors on a plasma membrane.

Keywords: anomalous diffusion; machine learning classification; feature engineering

---

1. Introduction

Starting with the pioneering experiment performed by Perrin [1], the quantitative analysis of microscopy images has become an important technique for various disciplines ranging from physics to biology. Over the last century, it has evolved to what is now known as single-particle tracking (SPT) [2–4]. In recent years, SPT has gained popularity in the biophysical community. The method serves as a powerful tool to study the dynamics of a wide range of particles including small fluorophores, single molecules, macromolecular complexes, viruses, organelles and microspheres [5,6]. Processes such as microtubule assembly and disassembly [7], cell migration [8], intracellular transport [9,10] and virus trafficking [11] have been already successfully studied with this technique.

A typical SPT experiment results in a series of coordinates over time (also known as "trajectory") for every single particle, but it does not provide any directed insight into the dynamics of the investigated process by itself. Mobility patterns of particles encoded in their trajectories have to be extracted in order to relate individual trajectories to the behavior of the system at hand and the associated biological process [12]. The analysis of SPT trajectories usually starts with the detection of a corresponding motion type of a particle, because this information may already provide insights into mechanical properties of the particle's surrounding [13]. However, this initial task usually constitutes a challenge due to the stochastic nature of the particles' movement.

There are already several approaches to analyse the mobility patterns of particles. The most commonly used one is based on the mean square displacement (MSD) of particles [10,14–17]. The idea behind this method is quite simple: a MSD curve (i.e., an average square displacement as a function of the time lag) is quantified from a single experimental trajectory and then fitted with a theoretical expression [18]. A linear best fit indicates normal diffusion (Brownian motion) [19], which corresponds to a particle moving freely in its environment. Such a particle neither interacts with other distant particles nor is hindered by any obstacles. If the fit is sublinear, the particle's movement is referred to as subdiffusion. It is appriopriate to represent particles moderated by viscoelastic properties of the environment [20], particles which hit upon obstacles [21,22] or trapped particles [9,23]. Finally, a superlinear MSD curve means superdiffusion, which relates to the motion of particles driven by molecular motors. This type of motion is faster than the linear case and usually in a specific direction [24].

Although popular in the SPT community, the MSD approach has several drawbacks. First of all, experimental uncertainties introduce a great amount of noise into the data, making the fitting of mathematical models challenging [10,14,25,26]. Moreover, the observed trajectories are often short, limiting the MSD curves to just a few first time lags. In this case, distinguishing between different theoretical models may not be feasible. To overcome these problems, several analytical methods that improve or go beyond MSD have already been proposed. The optimal least-square fit method [10], the trajectory spread in space measured with the radius of gyration [27], the van Hove displacements distributions [28], self-similarity of trajectory using different powers of the displacement [29] or the time-dependent directional persistence of trajectories [30] are examples of methods belonging to the first category. They may be combined with the results of the pure MSD analysis to improve the outcome of classification. The distribution of directional changes [31], the mean maximum excursion method [32] and the fractionally integrated moving average (FIMA) framework [33] belong to the other class. They allow efficient replacement of the MSD estimator for classification purposes. Hidden Markov models (HMM) turned out to be quite useful in heterogeneity checking within single trajectories [34,35] and in the detection of confinement [36]. Classification based on hypothesis testing, both relying on MSD and going beyond this statistics, has been shown to be quite successful as well [26,37].

In the last few years, machine learning (ML) has started to be employed for the analysis of single-particle tracking data. In contrast to standard algorithms, where the user is required to explicitly define the rules of data processing, ML algorithms can learn those rules directly from series of data. Thus, the principle of ML-based classification of trajectories is simple: an algorithm learns by adjusting its behavior to a set of input data (trajectories) and corresponding desired outputs (real motion types, called the ground truth). These input–output pairs constitute the training set. A classifier is nothing but a mapping between the inputs and the outputs. Once trained, it may be used to predict the motion type of a previously unseen sample.

The main factor limiting the deployment of ML to trajectory analysis is the availability of high-quality training data. Since the data collected in the experiments is not really provable (otherwise, we would not need any new classification method), synthetic sets generated with computer simulations of different diffusion models are usually used for training.

Despite the data-related limitations, several attempts at ML-based analysis of SPT experiments have been already carried out. The applicability of the Bayesian approach [18,38,39], random forests [40–43], neural networks [44] and deep neural networks [41,45,46] was extensively studied. The ultimate goal of those works was the determination of the diffusion modes. However, some of them went beyond the pure classification and focused on extraction of quantitative information about the trajectories (e.g., the anomalous exponent [42,45]).

In one of our previous papers, we compared two different ML approaches to classification [41]. Feature-based methods do not use raw trajectories as input for the classifiers. Instead, they require a set of human-engineered features, which are then used to feed the algorithms. In contrast, deep learning (DL) methods extract features directly from raw data without any effort from human experts. In this case, the representation of data is constructed automatically and there is no need for complex data preprocessing. Deep learning is currently treated as the state-of-the-art technology for automatic data classification and slightly overshadows the feature-based methods. However, from our results, it follows that the latter are still worth to consider. Compared to DL, they may arrive at similar accuracies in much shorter training times, are usually easier to interpret, allow to work with trajectories of different lengths in a natural way and often do not require any normalisation of data. The only drawback of those methods is that there is not a universal set of features that works well for trajectories of any type. Choosing the features is challenging and may have an impact on the classification results.

In this paper, we would like to elaborate on the choice of proper features to represent trajectories. Comparing classifiers trained on the same set of trajectories, but with slightly different features, we will address some of the challenges of feature-based classification.

The paper is structured as follows. In Section 2, we briefly introduce the concept of anomalous diffusion and present the stochastic models that we chose to model it. In Section 3, methods and data sets used in this work are discussed. The results of classification are extensively analysed in Section 4. In the last section, we summarise our findings.

2. Anomalous Diffusion and Its Stochastic Models

Non-Brownian movements that exhibit non-linear mean squared displacement can be described by multiple models, depending on some specific properties of the corresponding trajectories. The most popular models are the continuous-time random walk (CTRW) [9], random walks on percolating clusters (RWPC) [47,48], fractional Brownian motion (FBM) [49–51], fractional Lévy α-stable motion (FLSM) [52], fractional Langevin equation (FLE) [53] and autoregressive fractionally integrated moving average (ARFIMA) [54].

In this paper, we follow the model choice described in [26,37,43]—namely, we use FBM, the directed Brownian motion (DBM) [55] and Ornstein–Uhlenbeck (OU) processes [56]. With the particular choice of the parameters, all these models simplify to the classical Brownian motion (i.e., normal diffusion).

The FBM is the solution of the stochastic differential equation

$$dX_t^i = \sigma dB_t^{H,i}, \ \ i = 1,2, \tag{1}$$

where $\sigma > 0$ is the scale coefficient, which relates to the diffusion coefficient D via $\sigma = \sqrt{2D}$, $H \in (0,1)$ is the Hurst parameter and B_t^H is a continuous-time, zero-mean Gaussian process starting at zero, with the following covariance function

$$\mathrm{E}\left(B_t^H B_s^H\right) = \frac{1}{2}\left(|t|^{2H} + |s|^{2H} - |t-s|^{2H}\right). \tag{2}$$

The value of H determines the type of diffusion in the process. For $H < \frac{1}{2}$, FBM produces subdiffusion. It corresponds to a movement of a particle hindered by mobile or immobile obstacles [57]. For $H > \frac{1}{2}$, FBM generates superdiffusive motion. It reduces to the free diffusion at $H = \frac{1}{2}$.

The directed Brownian motion, also known as the diffusion with drift, is the solution to

$$dX_t^i = v_i dt + \sigma dB_t^{1/2,i}, \ \ i = 1,2, \tag{3}$$

where $v = (v_1, v_2) \in \mathbf{R}^2$ is the drift parameter and σ is again the scale parameter. For $v = 0$, it reduces to normal diffusion. For other choices of v, it generates superdiffusion related to an active transport of particles driven by molecular motors.

The Ornstein–Uhlenbeck process is often used as a model of a confined diffusion (a subclass of subdiffusion). It describes the movement of a particle inside a potential well and can be determined as the solution to the following stochastic differential equation:

$$dX_t^i = -\lambda_i(X_t^i - \theta_i)dt + \sigma dB_t^{1/2,i}, \ \ i = 1,2, \ \ \theta_i \in \mathbf{R}. \tag{4}$$

The parameter $\theta = (\theta_1, \theta_2)$ is the long-term mean of the process (i.e., the equilibrium position of a particle), $\lambda = (\lambda_1, \lambda_2)$ is the value of a mean-reverting speed and and σ is again the scale parameter. If there is no mean reversion effect, i.e., $\lambda_i = 0$, OU reduces to normal diffusion.

3. Methods and Used Data Sets

In this paper, we discuss two feature-based classifiers: random forest (RF) and gradient boosting (GB) [58]. The term feature-based relates to the fact that the corresponding algorithms do not operate on raw trajectories of a process. Instead, for each trajectory a vector of human-engineered features is calculated and then used as input for the classifier. This approach for the diffusion mode classification has already been used in [41–43,45], but here, we propose a new set of features, which gives better results on synthetic data sets.

Both RF and GB are examples of ensemble methods, which combine multiple classifiers to obtain better predictive performance. They use decision trees [59] as base classifiers. A single decision tree is fairly simple to build. The original data set is split into smaller subsets based on values of a given feature. The process is recursively repeated until the resulting subsets are homogeneous (all samples from the same class) or further splitting does not improve the classification performance. A splitting feature for each step is chosen according to Gini impurity or information gain measures [58].

A single decision tree is popular among ML methods due to the ease of its interpretation. However, it has several drawbacks that disqualify it as a reliable classifier: it is sensitive to even small variations of data and prone to overfitting. Ensemble methods combining many decision trees help to overcome those drawbacks while maintaining most of the advantages of the trees. A multitude of independent decision trees is constructed by making use of the bagging idea with the random subspace method [60–62] to form a random forest. Their prediction is aggregated and the mode of the classes of the individual trees is taken as the final output. In contrast, the trees in gradient boosting are built in a stage-wise fashion. At every step, a new tree learns from mistakes committed by the ensemble. GB is usually expected to perform better than RF, but the latter one may be a better choice in case of noisy data.

In this work, we used implementations of RF and GB provided by the scikit-learn Python library [63]. The performance of the classifiers was evaluated with the common measures including accuracy, precision, recall, F1 score and confusion matrices (although the information given by those measures is to some extent redundant, we decided to use all of them due to their popularity). The accuracy is a percentage of correct predictions among all predictions, that is a general information about the performance of a classifier (reliable in case of the balanced data set). The precision and recall give us a bit more detailed information for each class. The precision is a ratio of the correct predictions to all predictions in that class (including the cases falsely assigned to this class). On the other hand, the recall (also called sensitivity or true positive rate) is the ratio of correct predictions of that class to all members of that class (including the ones that were falsely assigned to another class). The F1 score is a harmonic mean of precision and recall, resulting in high value only if both precision and recall are high. Finally, the confusion matrices show detailed results of classification: element $c_{i,j}$ of matrix C is the percentage of the observations from class i assigned to class j (a row presents actual class, while the column presents predicted class).

The Python codes for the data simulation, features calculation, models preparation and performance calculation are available at Zenodo (see Supplementary Materials).

3.1. Features Used for Classification

As already mentioned above, both ensemble methods require vectors of human-engineered features representing the trajectories as input. In some sense, those methods may be treated as a kind of extension to the statistical methods usually used for classification purposes. Instead of conducting a statistical testing procedure of diffusion based on one statistic, what is often the case, we can combine several statistics with each other bu turning them into features, which are then used to train a classifier. This could be of particular importance in situations, when single statistics yield results differing from each other (cf. [43]). It should be mentioned, however, that choosing the right features is a challenging task. For instance, we have already shown in [41] that classifiers trained with a popular set of features do not generalise well beyond the situations encoutered in the training set. Thus, great attention needs to be paid to the choice of the input features to machine learning classifiers as well. They ought to cover all the important characteristics of the process, but at the same time, they should contain the minimal amount of unnecessary information, as each redundant piece of data causes noise in the classification or may lead to overfitting, for example (for a general discussion concerning a choice of features, see, for instance, [64]).

Based on the results in [41,43], we decided to use the following features in our analysis, hereinafter referred to as Set A:

- Anomalous exponent α, fitted to the time-averaged mean square displacement (TAMSD). This exponent relates to the Hurst parameter in Equation (1) via $\alpha = 2H$.
- Diffusion coefficient D, fitted to TAMSD.
- Mean squared displacement ratio, characterising the shape of a MSD curve. In general, it is given with the formula

$$\kappa(n_1, n_2) = \frac{\frac{1}{N-n_1}\sum_{i=1}^{N-n_1}\left|X_{i+n_1} - X_i\right|^2}{\frac{1}{N-n_2}\sum_{i=1}^{N-n_2}\left|X_{i+n_2} - X_i\right|^2} - \frac{n_1}{n_2},$$

where $n_1 < n_2$. In this work, we set $n_2 = n_1 + 1$ and averaged the output over n_1. In other words, we used (n_1 replaced by n for convenience):

$$\kappa = \frac{1}{N-1}\sum_{n=1}^{N-1}\kappa(n, n+1). \tag{5}$$

- Efficiency, calculated as

$$E = \frac{|X_{N-1} - X_0|}{(N-1)\sum_{i=1}^{N-1}|X_i - X_{i-1}|^2}, \tag{6}$$

which measures the linearity of a trajectory.
- Straightness, a measure of the average direction change between subsequent steps, calculated as:

$$S = \frac{|X_{N-1} - X_0|}{(N-1)\sum_{i=1}^{N-1}|X_i - X_{i-1}|}. \tag{7}$$

- The value of empirical velocity autocorrelation function [65] of lag 1 in point $n = 1$, that is

$$\chi = \frac{1}{N-2}\sum_{i=1}^{N-2}(X_{i+2} - X_{i+1}) \cdot (X_{i+1} - X_i).$$

- Maximal excursion, given by the formula

$$ME = \frac{\max(X_{i+1} - X_i)}{X_{N-1} - X_0}. \tag{8}$$

It is inspired by the mean maximal excursion (MME) [32], detecting the jumps that are long as compared to the overall displacement.
- The statistics based on p-variation [52]:

$$V_m^{(p)} = \sum_{i=0}^{N/m-1}|X_{(i+1)m} - X_{im}|^p.$$

The usefulness of this statistic to recognition of the fractional Lévy stable motion (including fractional Brownian motion) was shown in [52]. We introduce a quantity that verifies if for any p the function $V_m^{(p)}$ of the variable m changes the monotonicity. We provide the information if for the highest value of p such that $V_m^{(p)}$ does change the monotonicity, it is convex or concave. In short, we analyse $V_m^{(p)}$ as a function of m to provide one the following values:

$$P = \begin{cases} 0 & \text{if it does not change the monotonicity,} \\ 1 & \text{if it is convex for the highest } p \text{ for which it is not mononononuous,} \\ -1 & \text{if it is concave for the highest } p \text{ for which it is not mononononuous.} \end{cases} \quad (9)$$

The first five features were already used in [41]. It should also be mentioned here that three of them are based on MSD curves. There is one important point to consider while calculating the curves, namely the maximum time lag. If not specified otherwise, we will use the lag equal to 10% of each trajectory's length. Since this choice is not obvious and may impact the classification performance, we will discuss the sensitivity of classifiers' accuracies to different choices of the lag in Section 4.5.

Apart from the set of features presented above, denoted Set A, we are going to analyse two other sets: the one used in [40,41], referred as Set B, and the one proposed in [43] (set C). The lists of features used in each set are given in Table 1 (for their exact definition, please see the mentioned references). Sets A and B have several features in common. The link between sets A and C is not so apparent, but the maximal excursion and p-variation-based statistics play in the description of trajectories a role similar to the standardised maximum distance and the exponent of power function fitted to p-variation, respectively.

Following [41], we consider four classifiers for each set of features: RF and GB classifiers built with the full set (labelled as "with D") and with a reduced one after the removal of the diffusion constant D ("no D").

Table 1. Features used for classification purposes in each of analysed sets.

Set A	**Set B (from [41])**	**Set C (from [43])**
Anomalous exponent α	Anomalous exponent α	Anomalous exponent α
Diffusion coefficient D	Diffusion coefficient D	Diffusion coefficient D
MSD ratio	MSD ratio	—
Efficiency	Efficiency	—
Straightness	Straightness	—
VAC (for lag 1)	—	—
Maximal excursion	—	—
p-variation-based statistics	—	—
—	Asymmetry	—
—	Fractal dimension	—
—	Gaussianity	—
—	Kurtosis	—
—	Trappedness	—
—	—	Standardised maximum distance
—	—	Exponent of power function fitted to p-variation (for $p = 1, 2, \ldots, 5$)

3.2. Synthetic Data

Unlike the explicitly programmed methods, machine learning algorithms are not ready-made solutions for arbitrary data. Instead, an algorithm needs to be firstly fed with a reasonable amount of data (so-called training data) that should contain the main characteristics of the process under investigation in order to find and learn some hidden patterns. As the classifier is not able to extract any additional patterns from previously unseen samples after this stage, its performance is highly dependent on the quality of the training data. Hence, the training set needs to be complete in some sense.

First, we created our main data set, which will be referred to as the base data set for the remainder of this paper. It is analogous to the one used in [43]. We generated a number of 2D trajectories according to the three diffusion models described in Section 2, with no correlations between the coordinates. A single trajectory can be denoted as

$$X_n = (X_{t_0}, X_{t_1}, \ldots, X_{t_N}), \tag{10}$$

where $X_{t_i} = \left(X^1_{t_i}, X^2_{t_i}\right) \in \mathbf{R}^2$ is the position of the particle at time $t_i = t_0 + i\Delta t$, $i = 0, 1, \ldots, N$. We kept the lag Δt between two consecutive observations constant.

The details of our simulations are summarised in Table 2. In total, 120,000 trajectories have been produced, 40,000 for each diffusion mode, in order to balance the data set. The length of the trajectories was randomly chosen from the range between 50 and 500 steps to mimic typical observations in experiments. We set $\sigma = 1\ \mu\mathrm{m\,s}^{-1/2}$ and $\Delta t = 1$ s.

Table 2. Characteristics of the simulated trajectories used to train the classifiers. For the base training set, the following values were used: $c = 0.1$, $\sigma = 1\ \mu\mathrm{m\,s}^{-1/2}$ and $\Delta t = 1$ s.

Diffusion Class	Model	Parameter Ranges	Number of Trajectories
Normal diffusion	FBM	$H \in [0.5 - c, 0.5 + c]$	20,000
	DBM	$v = (v_1, v_2),\ v_1, v_2 \in [0, c]$	10,000
	OU	$\theta = 0,\ \lambda = (\lambda_1, \lambda_2),\ \lambda_1, \lambda_2 \in [0, c]$	10,000
Subdiffusion	FBM	$H \in [0.1, 0.5 - c)$	20,000
	OU	$\theta = 0,\ \lambda = (\lambda_1, \lambda_2),\ \lambda_1, \lambda_2 \in (c, 1]$	20,000
Superdiffusion	FBM	$H \in (0.5 + c, 0.9]$	20,000
	DBM	$v = (v_1, v_2),\ v_1, v_2 \in (c, 1]$	20,000

Since the normal diffusion can be generated by a particular choice of the models' parameters ($H = 0.5$ for FBM, $v = 0$ for DBM and $\lambda = 0$ for OU), it is almost indistinguishable from the anomalous diffusion generated with the parameters in the vicinity of those special values. The addition of the noise complicates the problem even more. Thus, following [43], we introduced a parameter c that defines a range in which a weak sub- or superdiffusion should be treated as a normal one. Although introduced here at a different level, it bears resemblance to the cutoff c used in [37].

Apart from the base data set, we are going to use several auxiliary ones to elaborate on different aspects of the feature choice. In Section 4.3, we will work with a training set, in which the trajectories from the base one are disturbed with a Gaussian noise to resemble experimental uncertainties. In Section 4.4, we will analyse the performance of classifiers trained on synthetic data generated with $\sigma = 0.38$, corresponding to the diffusion coefficient $D = 0.0715\ \mu\mathrm{m}^2\,\mathrm{s}^{-1}$, which is adequate for the analysis of real data samples. To study the sensitivity of the classifiers to the value of the cutoff c in Section 4.6, we will use three further sets with $c = 0$, $c = 0.001$ and $c = 0.01$. In Section 4.7, a synthetic set with $\sigma = 2D$, where D is drawn from the uniform distribution on $[1, 9]$ will be used to check how the classifiers cope with the trajectories characterised by heterogeneous mobilities.

For all data sets, the training and testing subset were randomly selected with a 70%/30% ratio.

3.3. Empirical Data

To check how our classifiers work on unseen data, we will apply them to some real data. We decided to use the trajectories of G proteins and G-protein-coupled receptors already analysed in [37,43,66]. To avoid some issues related with short time series, we limited ourselves to trajectories with at least 50 steps only, obtaining 1037 G proteins' and 1218 receptors' trajectories. They are visualised in Figure 1.

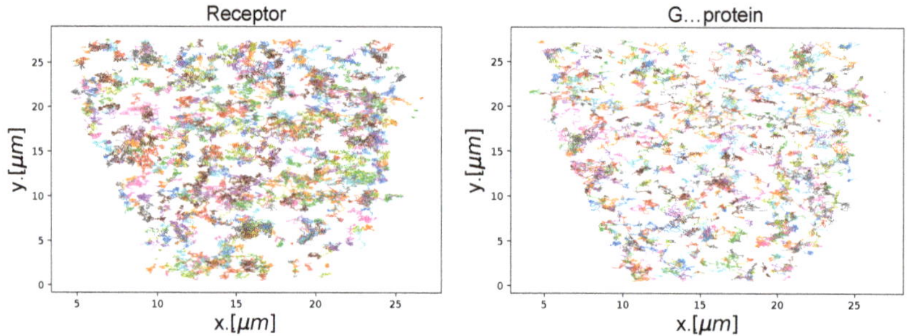

Figure 1. Trajectories of the receptors (**left**) and G proteins (**right**) used as input for the classifiers. Different colors are introduced to indicate different trajectories. The set of the receptors contains 1218 trajectories and the one of G proteins—1037 trajectories. The lengths of the trajectories are from range $[50, 401]$, the time step is equal to 28.4 ms and recorded positions are given in μm.

4. Results

The main goal of our work is a comparative analysis of classifiers trained using different sets of features (see Table 1 for their definition). The classifiers were trained and tested on our base data set and the auxiliary data sets, for comparison.

In order to optimise both classification algorithms, we looked for their hyperparameters using the `RandomisedSearchCV` method from scikit-learn library. It performs a search over values of hyperparameters generated from their distributions (in our case, discrete uniform ones). The term hyperparameter in this context means a parameter required for the construction of the classifier, which has to be set by a human expert before the learning process starts. In general, it influences the performance of the classifier, hence its choice is essential.

4.1. Classification Results on Base Data Set Using Proposed Set of Features

We start with the classifiers trained on the base set (see Table 2 for details). We trained four different classifiers: RF and GB for both the full set of attributes ("with D") and a reduced one ("no D"). Set A of features was used for representation of trajectories. The performance of these classifiers will be treated as a benchmark in our further analysis.

The hyperparameters of the classifiers are presented in Table 3 (for the detailed explanation of each of these parameters, please see [43,58]). It is worth noticing a difference in the ensemble sizes between the full set and the reduced one—in case of the gradient boosting, we observe a ninefold reduction of the number of trees. However, this difference does not reflect in the performance of the classifiers. Taking the number of features into account, the value of the `max_depth` hyperparameter for RF with D is surprisingly high. It seems to be an artifact of the hyperparameter tuning procedure via random grid search. From our analysis (not included in this paper), it follows that this value can be set to 20 without a negative impact on accuracy. Nevertheless, we decided to keep the original result of the automatic hyperparameter tuning in order to treat all of the classifiers on the same footing. We should probably add that the largest tree in RF was 38 levels deep, despite such a high value of the maximum depth.

We begin the analysis of the classifiers by inspecting their accuracies. The results are shown in Table 4. As we can see, both classifiers perform excellently, with more than 95% of correct predictions for the test set. In the case of the training data, GB performs better than RF. However, RF is slightly more accurate on the test set, indicating a small tendency of GB to overfit.

Table 3. Hyperparameters of the optimal classifiers built on base data set with Set A of features. The full set of features is labelled as "with *D*". The "no *D*" columns stand for the reduced set of features after the removal of the diffusion coefficient *D*. N/A (i.e., "Not Applicable") indicates hyperparameters specific for random forest.

	Random Forest		Gradient Boosting	
Hyperpareameters	**With *D***	**No *D***	**With *D***	**No *D***
`bootstrap`	*True*	*True*	N/A	N/A
`criterion`	*gini*	*entropy*	N/A	N/A
`max_depth`	80	10	50	10
`max_features`	*sqrt*	*sqrt*	*sqrt*	log_2
`min_samples_leaf`	4	2	4	2
`min_samples_split`	2	10	10	2
`n_estimators`	800	600	900	100

Table 4. Accuracy of the best classifiers trained on the base data set (see Table 2) with Set A of features. The "with *D*" and "no *D*" columns refer to the full and reduced (after removal of *D*) sets of features, respectively. The results are rounded to three decimal digits.

	Random Forest		Gradient Boosting	
Data Set	**With *D***	**No *D***	**With *D***	**No *D***
Training	0.979	0.962	1.0	0.993
Test	0.957	0.955	0.956	0.955

To explain the relatively small differences in the performance between the "with D" and "no D" versions of the classifiers, we may want to look at the importances of features. There are several ways to calculate those importances. We used a method which defines the importance as the decrease in accuracy after a random shuffling of values of one of the features. Results are given in Table 5. Just to recall, features with high importances are the drivers of the outcome. The last important ones might often be omitted, making the classification model faster to fit and predict. The results of the node impurity importances (the total decrease in node impurity caused by a given feature, averaged over all trees in the ensemble [67]) are similar.

Table 5. Permutation feature importances of the classifiers built on base data set with Set A of features. The "with *D*" and "no *D*" columns refer to the full and reduced (after removal of *D*) sets of features, respectively. The rows are sorted according to the decreasing importances for random forest with *D*. The most and least important features are indicated with bold or underlining, respectively.

	Random Forest		Gradient Boosting	
Feature	**With *D***	**No *D***	**With *D***	**No *D***
χ - VAC for $\delta = 1, n = 1$	**0.1428**	**0.0812**	**0.1612**	**0.2292**
Anomalous exponent α	0.0212	0.0436	0.0244	0.0204
MSD ratio	0.0128	0.0194	0.0080	0.0168
Efficiency	0.0118	0.0074	0.0030	<u>0.0046</u>
Straightness	0.0110	0.0062	0.0064	0.0048
p-variation statistic *P*	0.0104	0.0090	<u>0.0024</u>	0.0060
Maximal excursion	0.0080	<u>0.0046</u>	0.0068	0.0094
D	<u>0.0074</u>	–	0.0056	–

It turns out that *D* is the least important feature for RF classifier trained on the full set and the third one with the smallest importance for GB classifier. That is why its removal has a small impact on the accuracy of prediction and why the classifiers trained on the reduced set of features with no *D* are worth considering—we expect them to work better on unseen data having diffusion coefficients different from the one used in the base set. Indeed, its removal does not change the performance of the classifier on the test set (see Table 4). Later in Section 4.7, we will show that in case of the training set with varying D, the situation is different: D will become more important and excluding it from the set will reduce the accuracy.

The most informative feature in all cases is the velocity autocorrelation function for lag $\delta = 1$ at point $n = 1$. It is worth mentioning that this quantity has been already successfully used for the distinction of subdiffusion models [68], but not in the ML context. The anomalous exponent α, which is a standard method for the diffusion mode classification, is the second most important feature for all models, with a significant influence on the results. Thus, it seems that the classifiers distinguish between the models first and then assess the mode of diffusion.

To get more insight into the detailed performance of the classifiers, their normalised confusion matrices are shown in Figure 2. Please note that the percentages may not sum to 1.0 due to rounding. We see that all models have the biggest problems with the classification of normal diffusion. This is simply due to the fact that the differences between normal diffusion and realizations of weak sub- or superdiffusion are negligible and it is challenging to classify it properly even after introduction of the parameter c (the role of which will be studied in more detail in Section 4.6).

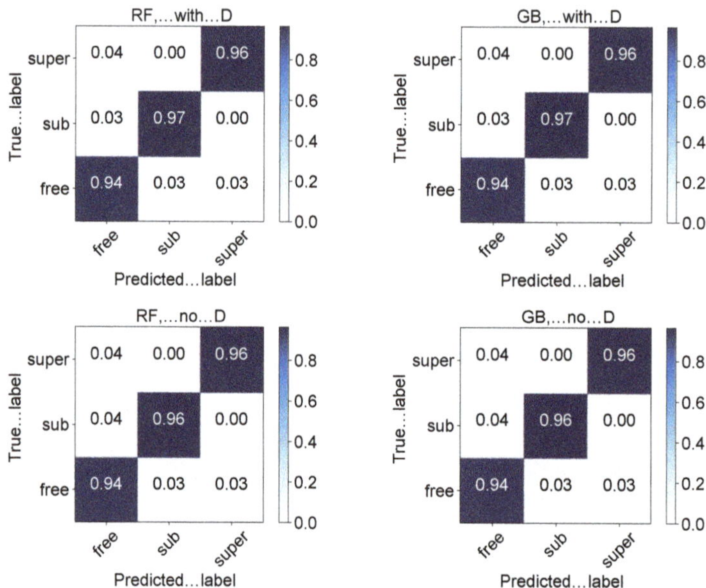

Figure 2. Normalised confusion matrices for classifiers built on base training data (see Table 2) with Set A of features. The "with D" (top row) and "no D" (bottom row) labels refer to the full and reduced (after removal of D) sets of features, respectively. All results are rounded to two decimal digits.

The values presented in Figure 2 may be used to calculate the other popular measures of performance: precision, recall and F1 score (see Section 3). The results, rounded to three decimal digits, are summarised in Table 6. Again, we see that the measures point to the highest error rate for the normal diffusion: for the random forest model with D as one of the features, only 92.9% of the trajectories classified as normal diffusion were in fact in this class (precision), whereas 94.4% of freely diffusing trajectories were correctly classified (recall). Such a high error rate is related to the mentioned lack of distinctions between the nodes—the normal diffusion is some kind of buffer between subdiffusion and superdiffusion, thus it can be incorrectly classified as one of these two.

Table 6. Precision, recall and F1 scores of the classifiers trained on base synthetic data with Set A of features. For each classifier, the testing set consists of 12,000 trajectories per diffusion mode—that is, 36,000 in total. All classifiers were built on base data set with Set A of features.

Method	Variant	Measure	Normal Diffusion	Subdiffusion	Superdiffusion	Total/Average
RF	with D	Precision	0.929	0.973	0.970	0.957
		Recall	0.944	0.966	0.962	0.957
		F1	0.936	0.969	0.966	0.957
	no D	Precision	0.922	0.971	0.971	0.955
		Recall	0.943	0.963	0.958	0.955
		F1	0.933	0.967	0.964	0.955
GB	with D	Precision	0.928	0.972	0.970	0.956
		Recall	0.942	0.966	0.961	0.956
		F1	0.935	0.969	0.965	0.956
	no D	Precision	0.925	0.970	0.969	0.955
		Recall	0.940	0.964	0.960	0.955
		F1	0.932	0.967	0.965	0.955

4.2. Comparison with Other Sets of Features

Below, we show the comparison of the classification results with all considered classifiers (based on three different set of features) on our base data set (Table 2).

In Table 7, the accuracies on the test set are shown, calculated using the tenfold cross-validation method [58]. As the calculation of the accuracy of the classifier is based on the single train/test split, in an unfortunate case, the test set can contain the data with characteristics that have not been seen by classifier during training, and thus the accuracy would be falsely low. The k-fold cross-validation is a technique that helps to reduce that bias. The data is randomly split into k folds (without replacement) and the model is trained and tested k times—each time one fold is the test set, whereas the remaining ones create the training set. The overall accuracy is the mean of the accuracies of each run. The hyperparameters of the particular models are summarised in Table 8 and they were established using the `RandomisedSearchCV` method again.

Table 7. Accuracy of the classifiers built on the base data set using different sets of features, measured using tenfold cross-validation method. All results are rounded to three decimal digits.

	Random Forest		Gradient Boosting	
Data Set	With D	No D	With D	No D
Set A	0.957	0.955	0.956	0.953
Set B	0.946	0.928	0.945	0.928
Set C	0.948	0.946	0.948	0.944

In the comparison of all these classifiers, the ones based on the set of features proposed in this article provide the best results on our base synthetic data set. Actually, the choice of features was inspired by two of our previous articles [41,43]. The new set combines the attributes used in those papers: it contains the anomalous exponent α, diffusion coefficient D, efficiency, straightness and mean squared displacement ratio that have been used in [41], and the normalised maximal excursion and p-variation-based features used in [43].

Nevertheless, we need to underline here that it does not mean that this set of features is the solution for all the classification problems—it simply seems to be the best choice for such synthetic data set. The lack of universality of feature-based methods was already presented in [41]: the classifiers did not generalise well to samples generated with slightly altered models.

Table 8. Hyperparameters of the optimal classifiers built on base data set used for the calculation of tenfold cross-validation accuracy in Table 7. The "with D" and "no D" columns refer to the full and reduced (after removal of D) sets of features, respectively. N/A stands for "Not Applicable" (the first two parameters are random forest specific). The definitions of the feature sets are given in Table 1.

Features	Model	Variant	Bootstrap	Criterion	max_depth	max_features	min_samples_leaf	min_samples_split	n_estimators
Set A	RF	with D	True	gini	80	sqrt	4	2	800
		no D	True	entropy	10	sqrt	2	10	600
	GB	with D	N/A	N/A	50	sqrt	4	10	900
		no D	N/A	N/A	10	log2	2	2	100
Set B	RF	with D	True	entropy	None	None	2	5	1000
		no D	True	entropy	None	log2	1	10	600
	GB	with D	N/A	N/A	110	log2	2	10	400
		no D	N/A	N/A	10	log2	4	5	100
Set C	RF	with D	True	entropy	60	log2	4	2	900
		no D	True	entropy	10	sqrt	2	10	600
	GB	with D	N/A	N/A	10	log2	2	2	100
		no D	N/A	N/A	10	log2	2	2	100

To compare the performance of these models in more details, the values of recall, precision and F1 score are given in Table 9. For the sake of clarity, we only compare the random forest classifiers built on the complete features' sets (with the diffusion coefficient D). For the remaining cases, the behaviour is alike, except for the fact that all measures for classifiers with features as in Set B but without diffusion coefficient D are significantly lower than for other classifiers. We would like to underline here that the set of features proposed in Section 3.1 provides the best results in all measures used here. For all classifiers, the results for superdiffusion and subdiffusion are better than for normal diffusion class, what is understandable, as the only kind of error that occurs is the misclassification of anomalous diffusion trajectories as the normal diffusion. In case of normal diffusion, a part of misclassified trajectories is labelled as superdiffusion, and another part is labelled as subdiffusion.

Table 9. Detailed performance comparison of random forest classifiers based on three sets of features, built on the base data set. Metrics are calculated on the test data. All results are rounded to three decimal digits. For each classifier, the test set consists of 12,000 trajectories per diffusion mode—that is, 36,000 in total.

Set of Features	Measure	Normal Diffusion	Subdiffusion	Superdiffusion	Total/Average
Set A	Precision	0.929	0.973	0.970	0.957
	Recall	0.944	0.966	0.962	0.957
	F1	0.936	0.969	0.966	0.957
Set B	Precision	0.910	0.970	0.963	0.948
	Recall	0.934	0.957	0.950	0.947
	F1	0.922	0.964	0.956	0.947
Set C	Precision	0.912	0.969	0.966	0.949
	Recall	0.935	0.958	0.951	0.948
	F1	0.923	0.963	0.959	0.948

4.3. Adding Noise

The results on our base data set are promising, but, unfortunately, real data are more challenging to classify, as they usually contain some noise and/or measurement error. Thus, we added a random Gaussian noise with zero mean and standard deviation σ_{Gn} to our trajectories. In order to control the noise amplitude with respect to standard deviation of a process, we followed the idea used in [40,41,43], namely setting a random signal-to-noise ratio instead of σ_{Gn}. The signal-to-noise ratio is defined as

$$Q = \begin{cases} \frac{\sqrt{D\Delta t + v^2 \Delta t^2}}{\sigma_{Gn}} & \text{for DBM,} \\ \frac{\sqrt{D\Delta t}}{\sigma_{Gn}} & \text{otherwise,} \end{cases} \tag{11}$$

where $v = \sqrt{v_1^2 + v_2^2}$. The value of σ_{Gn} was calculated for each trajectory separately, based on the random value of Q drawn from the uniform distribution on interval $[1, 9]$.

The accuracies of the classifiers trained on the data set with noise are given in Table 10. It is worth comparing the results with Table 4—there is a decrease of the accuracy, especially in case of the reduced set of features ("no D"), but both methods still classify the diffusion modes well. Nevertheless, in this case, it turns out that the inclusion of the diffusion coefficient D as one of the features is important. Still, for our synthetic data set with noise, the features in Set A seem to describe the characteristics of the used processes most precisely.

Table 10. Performance of the classifiers trained on data with random Gaussian noise. Accuracies (for test data only) are rounded to three decimal digits.

	Random Forest		Gradient Boosting	
Features	**With D**	**No D**	**With D**	**No D**
Set A	0.950	0.937	0.949	0.937
Set B	0.941	0.918	0.941	0.918
Set C	0.944	0.932	0.943	0.930

4.4. Empirical Data

In order to present the methods in a practical context, we are going to apply the classifiers from Sections 4.1 and 4.3 to real G protein data (see Section 3.3). Additionaly, to follow the approach from [43], we will consider additional classifiers fed with the data set similar to the base one, but with $\sigma = 0.38$, since this value corresponds to the mean diffusion coefficient of the real data sample ($D = 0.0715\ \mu\text{m}^2\text{s}^{-1}$). Accuracies of the additional classifiers are shown in Table 11. Interestingly, they are slightly better than the ones for the base set. It seems that the change of the scale parameter positively influenced the ranges of other characteristics, resulting in an increased accuracy (it worked as implicit feature engineering in the absence of data normalization).

Table 11. Performance of the classifiers trained on data with $\sigma = 0.38$. Accuracies (for test data only) are rounded to three decimal digits.

	Random Forest		Gradient Boosting	
Features	**With D**	**No D**	**With D**	**No D**
Set A	0.961	0.959	0.960	0.958
Set B	0.949	0.927	0.948	0.928
Set C	0.953	0.951	0.952	0.949

Before we start to analyse the results for real data, there are several points to consider. First, it should be emphasised once again that the data collected in experiments is not provable. Since the ground truth is missing, we cannot really choose the best among the classifiers. We just could use some additional information about the G proteins in order to indicate if the classifiers work reasonably or not. Second, real trajectories are often heterogeneous, meaning that a particle may change its type of motion within a single trajectory [69]. Thus the classifiers fed with homogeneous synthetic data may be not the best choice to work with such data.

In Tables 12–14, we show the results of classification of real data with the base classifiers, the ones with the noise and the ones with $\sigma = 0.38$, respectively. In all three cases, we considered only the "with D" classifiers (for the justification, see Section 4.7). The results obtained with the classifiers trained on different data sets vary slightly, but they agree on a small percentage of superdiffusive trajectories. This is somehow expected from the biological background: during their movement, the G proteins and G-protein-coupled receptors pair, spending some amount of time immobilised. In the same time, there is no evidence of any other force that can accelerate the movement.

Table 12. Classification results for real trajectories. The base data set ($\sigma = 1$, no noise; see Section 4.1) with the full sets features (labelled as "with D" in the previous sections) was used for training. The numbers may not add up precisely to 100% due to rounding.

		Random Forest		Gradient Boosting	
Features	**Classified Mode**	**Receptor**	**G Protein**	**Receptor**	**G Protein**
Set A	Free diffusion	22%	26%	12%	17%
	Subdiffusion	76%	64%	84%	70%
	Superdiffusion	1%	9%	2%	12%
Set B	Free diffusion	2%	7%	0%	0%
	Subdiffusion	97%	90%	99%	97%
	Superdiffusion	0%	1%	0%	2%
Set C	Free diffusion	40%	45%	41%	40%
	Subdiffusion	59%	52%	57%	54%
	Superdiffusion	0%	1%	1%	5%

Table 13. Classification results for real trajectories. The noisy data set ($\sigma = 1$, see Section 4.3) with the full sets of features (labelled as "with D" in the previous sections) was used for training. The numbers may not add up precisely to 100% due to rounding.

		Random Forest		Gradient Boosting	
Features	**Classified Mode**	**Receptor**	**G Protein**	**Receptor**	**G Protein**
Set A	Free diffusion	28%	31%	27%	28%
	Subdiffusion	70%	61%	70%	61%
	Superdiffusion	1%	6%	2%	9%
Set B	Free diffusion	3%	11%	2%	9%
	Subdiffusion	96%	86%	96%	87%
	Superdiffusion	0%	1%	0%	3%
Set C	Free diffusion	45%	48%	41%	41%
	Subdiffusion	54%	49%	58%	53%
	Superdiffusion	0%	1%	0%	5%

On our base data set, the classifiers based on Set A label most of both G proteins' and G protein-coupled receptors' trajectories as subdiffusion (64–84%, depending on particle type and method). This is somewhat in between the results of classifiers based on Set B and Set C, where the former point to subdiffusion more frequently, while the latter apply only in 52–59% of cases.

Comparing the behaviour of the classifiers based on the different data sets used for training, we can see that the classifiers built on the Set C are the most stable in some sense—they yield similar results independently of the training data, indicating to a significant fraction of subdiffusive and freely diffusing trajectories. For the new proposed set of features, Set A, as well as for Set B, the introduction of noise does not alter the classification significantly, but the decrease of the scale of the trajectories in data set (setting $\sigma = 0.38$) leads to recognition of more trajectories as the normal diffusion, similarly to the p-variation-based statistical test proposed in [37]. Alternately, the GB classifier based on Set B and scaled data set classifies a significant percentage of trajectories as superdiffusive, which is rather unexpected.

Table 14. Classification results for real trajectories. The data set with $\sigma = 0.38$ (no noise) and with the full sets of features was used for training. The numbers may not add up precisely to 100% due to rounding.

		Random Forest		Gradient Boosting	
Features	**Classified Mode**	**Receptor**	**G Protein**	**Receptor**	**G Protein**
Set A	Free diffusion	42%	40%	36%	35%
	Subdiffusion	56%	54%	61%	58%
	Superdiffusion	1%	5%	1%	5%
Set B	Free diffusion	51%	38%	44%	24%
	Subdiffusion	44%	50%	37%	44%
	Superdiffusion	3%	10%	17%	30%
Set C	Free diffusion	54%	51%	54%	51%
	Subdiffusion	45%	47%	45%	46%
	Superdiffusion	0%	1%	0%	1%

For the full picture, in Table 15, we also include the results for the classifiers built with the reduced Set A—that is, without diffusion coefficient D ("no D"). Following the results for the synthetic trajectories, where on the noisy data set the accuracy for the classifiers based on the reduced set of features is smaller (see Table 10), we acknowledge that the results on that data set can be biased. Indeed, such classifiers claim that most of the trajectories exhibit the normal diffusion, whereas the classifiers built on the base and the scaled data set classify them as subdiffusion.

Table 15. Classification results for real trajectories. The classifiers were trained with the reduced Set A (labelled as "no D"). The numbers may not add up precisely to 100% due to rounding.

		Random Forest		Gradient Boosting	
Classifier	**Classified Mode**	**Receptor**	**G Protein**	**Receptor**	**G Protein**
Base classifier	Free diffusion	33%	35%	32%	30%
	Subdiffusion	65%	59%	65%	59%
	Superdiffusion	0%	5%	2%	9%
Trained with noise	Free diffusion	72%	58%	77%	60%
	Subdiffusion	25%	34%	18%	29%
	Superdiffusion	1%	6%	3%	10%
Trained with $\sigma = 0.38$	Free diffusion	34%	34%	28%	30%
	Subdiffusion	63%	58%	69%	59%
	Superdiffusion	1%	7%	2%	10%

To sum up, all the classifiers identify most trajectories as normal or subdiffusive, but the fraction of both diffusion modes varies between classifiers. The scaling of trajectories in the training data set has introduced significant changes in the results (please compare Tables 12 and 14), thus the properties of particular features should be further examined (for example, their normalisation). Moreover, in [69], the authors showed that the trajectories in the analysed data set change their character during the time evolution. Different features used in the classifiers probably capture slightly different characteristics of the trajectories; thus, the sensitivity of features for the heterogeneity of movement should be verified.

4.5. Influence of MSD Calculation Methods

Some of the features used in our set—that is, the diffusion coefficient D, the anomalous exponent α and the mean displacement ratio κ, are based on the time-averaged MSD. This quantity can be highly biased for large lags, as then only a few displacements are included in the calculation of the mean value. Alternately, if we choose to fit the diffusion coefficient or the anomalous exponent to only a few data points (to MSD calculated for a few lags only), the estimation could be biased. This is a known problem in the analysis of the biological data and has already been discussed in [26,70,71].

We have considered the influence of the number of lags on the accuracy of the classifiers and trained them on the base data set with the values of features calculated using 50% or 10% of available TAMSD length. In Table 16, the comparison of these accuracies on the test set is shown, using all three sets of features. For each set, only the "with D" variant has been considered. The better results are obtained with the shorter TAMSD curve, but the differences are only slight. Thus, we have set the 10% as the fixed value for all our considerations.

Table 16. Accuracies on test sets for the classifiers built with the features' sets with 10% or 50% of MSD curve length used for calculation of the MSD-based features. All results are rounded to three decimal digits.

	Random Forest		Gradient Boosting	
Features	**10%**	**50%**	**10%**	**50%**
Set A	0.957	0.956	0.956	0.955
Set B	0.947	0.942	0.947	0.942
Set C	0.948	0.947	0.947	0.946

4.6. Sensitivity of the Model to Parameter C

Up to this point, we used set of synthetic data generated with $c = 0.1$ (see Table 2 for the meaning of c). This parameter was used to define ranges, outside of which weak sub- or superdiffusion should be distinguished from the normal one. It is time to analyse the impact of c on the prediction performance of our classification models.

In Table 17, the accuracies on the test set of the particular classifiers are presented. The highest value of this metrics for $c = 0.1$ could suggest that it is is the best choice, but there is the other side of a coin—the highest c means that more trajectories in the data set were falsely labelled as normal diffusion on the data set simulation stage, despite the fact that they were generated from models with the parameters corresponding to the anomalous diffusion. In Table 18, the values of precision, recall and F1 are shown for the random forest classifier ("with D") trained on each of the analysed sets. Although the precision for the normal diffusion grows with the increasing value of c, there is a drop in the recall value between $c = 0.01$ and $c = 0.1$. Inversely, for both modes of anomalous diffusion, the precision drops when changing from $c = 0.01$ and $c = 0.1$. It means that we not only make a base mistake in labelling, falsely labelling some normal trajectories as anomalous ones at the data set generation stage (what is not visible here), but also setting too high value of c parameter adds some confusion.

The issue is visualised in Figure 3, where the histograms of predicted labels are shown (please mind the logarithmic scale on y-axis). The ranges defined by the parameter C are indicated with black dashed lines. All observations between the dashed lines were treated as normal diffusion by the classifiers (such label was assigned at the data set generation stage as ground truth). Although for $c = 0.1$ and all diffusion models, the major part of trajectories was classified correctly, the distribution of the normal diffusion label assigned is wider than, for example, $c = 0.01$, especially in the case of fractional Brownian motion. Thus, to diminish the error (understood as an incorrect label in comparison to real diffusion mode, not assigned ground truth label), a smaller value of c should be taken—for example, the mentioned $c = 0.01$.

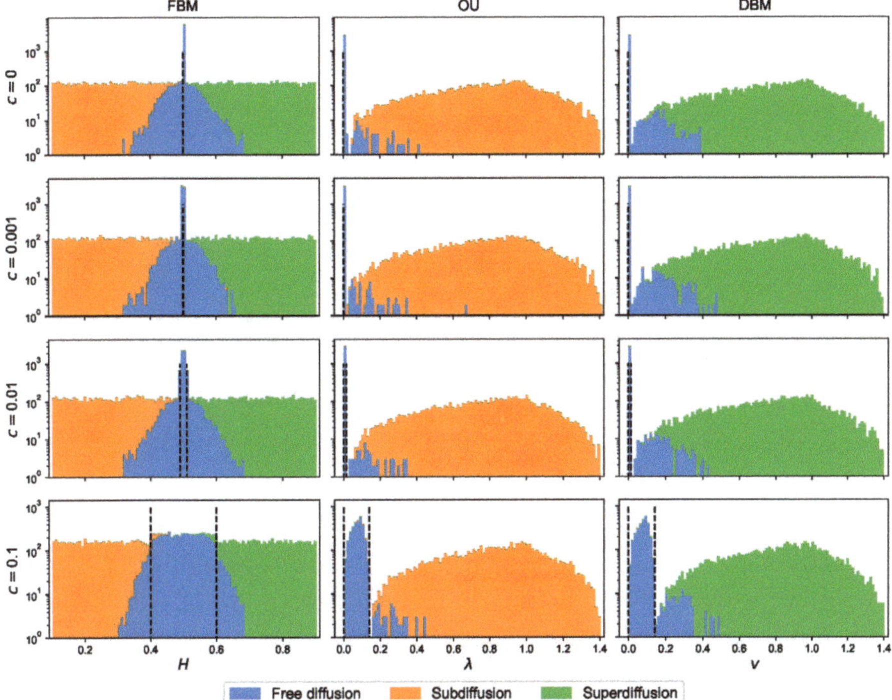

Figure 3. The histograms of assigned labels for different diffusion models, as predicted for the test sets by classifiers built on data sets with different values of parameter c with Set A of features. Please mind the logarithmic scale on y-axis. The dashed lines bounds the regions for which the normal diffusion was assigned as ground truth despite the real character of trajectories.

Table 17. Accuracies on test set of the optimal classifiers built on data sets with different values of parameter c and Set A of features. All results are rounded to three decimal digits.

	Random Forest		Gradient Boosting	
Data Set	**With D**	**No D**	**With D**	**No D**
$c = 0$	0.920	0.919	0.920	0.919
$c = 0.001$	0.924	0.924	0.923	0.923
$c = 0.01$	0.929	0.929	0.928	0.926
$c = 0.1$ (base)	0.957	0.955	0.956	0.955

4.7. Role of Diffusion Coefficient D

Finally, we move to the case in which parameter σ varies between trajectories. The data set for the classification was prepared according to Table 2, but each trajectory was characterised by a random σ value equal to $\sqrt{2D}$, where D was drawn from the uniform distribution on the interval $[1, 9]$. The same set of features was used and an additional regularisation was performed in the classifier training procedure.

The accuracy results for such classifiers are shown in Table 19. As one can see, the classifiers are still correct in more than 90% of cases and we can still consider them as useful. Interestingly, the changes in D have bigger influence to values than adding noise, introduced in Section 4.3. Thus, our classifiers work better in case of homogeneous environment with a constant diffusion coefficient, and as could be somehow expected, the difference between the classifiers with the diffusion coefficient D as a feature and the ones without it is visible, in favour of the all features' set. Thus, there is no reason to consider the reduced set of features in future research.

Table 18. Precision, recall and F1 scores for classifiers trained on data with different values of the cutoff c. Set A of features was used. All results are rounded to three decimal digits. For each data set, the support of the testing set is 12,000 trajectories per diffusion mode, giving 36,000 in total.

c Value	Measure	Normal Diffusion	Subdiffusion	Superdiffusion	Total/Average
	Precision	0.835	0.972	0.974	0.927
$c = 0$	Recall	0.950	0.910	0.900	0.920
	F1	0.889	0.940	0.936	0.921
	Precision	0.842	0.975	0.972	0.930
$c = 0.001$	Recall	0.952	0.915	0.906	0.924
	F1	0.894	0.944	0.938	0.925
	Precision	0.850	0.976	0.976	0.934
$c = 0.01$	Recall	0.955	0.918	0.913	0.929
	F1	0.900	0.946	0.943	0.930
	Precision	0.929	0.973	0.970	0.957
$c = 0.1$	Recall	0.944	0.966	0.962	0.957
	F1	0.936	0.969	0.966	0.957

Table 19. Performance of the best classifiers trained on the data set with varying diffusion coefficient D and Set A of features. Accuracies are rounded to three decimal digits.

	Random Forest		Gradient Boosting	
Data Set	**With D**	**No D**	**With D**	**No D**
Training	0.971	0.921	0.979	0.966
Test	0.919	0.912	0.920	0.909

In Figure 4, the confusion matrices of the analysed classifiers are shown. There is definitely more confusion between superdiffusion and free diffusion, in both directions, but still there is no misclassification between super- and subdiffusion (what would point to more serious problems with the classification). We think that these results can be even improved with the revision of the diffusion coefficient estimation method.

4.8. Beyond Multi-Class Classification

Up to this point, the classifiers were set to output only one among three available classes. However, both RF and GB classifiers are ensemble methods that determine the final output through voting of their base learners (decision trees). That voting can be exploited to provide probabilities of being assigned to each class. Their analysis can help in understanding the classifiers' behaviour and sources of misclassifications.

In Figure 5, ternary plots for both random forest and gradient boosting classifiers based on full Set A of features are shown. They complement the results shown in Table 4 and Figure 2. As we can see, the majority of the points is concentrated at the edges of the plots, corresponding to a situation with at most two non-vanishing class probabilities for given trajectories. The points located near the vertices depict the trajectories with one dominant class. There is much less of a burden in case of the gradient boosting classifier—the probability of assigning a trajectory to a finally claimed class is much higher and there are almost no trajectories with non-zero probabilities for all classes. This is clearly linked to the construction of both these classifiers. In random forest, each base classifier independently returns a predicted class and the final output is the most frequent class returned. Thus, the spread of the predictions can be high. In gradient boosting, the trees are constructed sequentially: each new one is supposed to correct the predictions of the ensemble and its results have a higher weight in the final aggregation. Thus, the final trees are having the greatest impact on the outcome and we expect GB to produce output with one dominant probability in most of the cases.

In Figure 6, predicted class probabilities for sample trajectories are shown, for random forest (left graph) and gradient boosting (right graph). Indeed, the gradient boosting classifier was more decisive, producing more univocal results, even if they were incorrect (please see the first trajectory from the top and the second trajectory form the bottom).

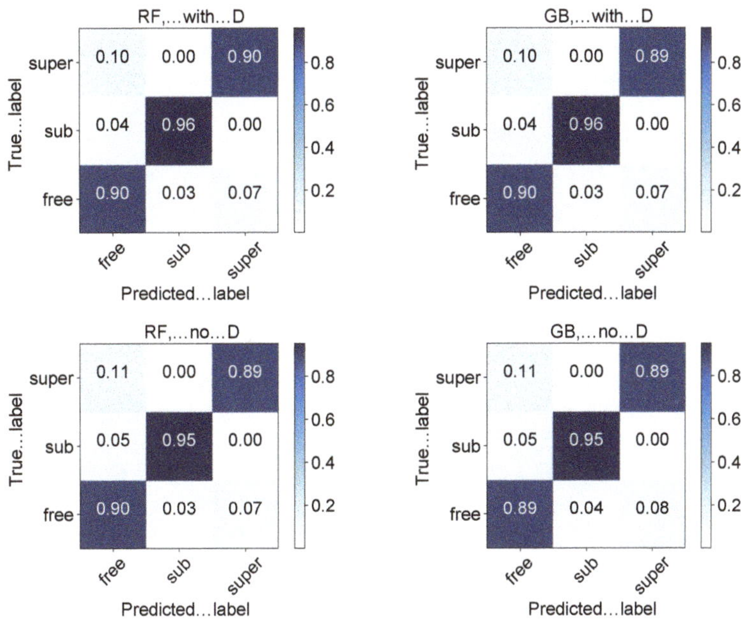

Figure 4. Normalised confusion matrices for classifiers built on training data with varying D and Set A of features. All results are rounded to two decimal digits.

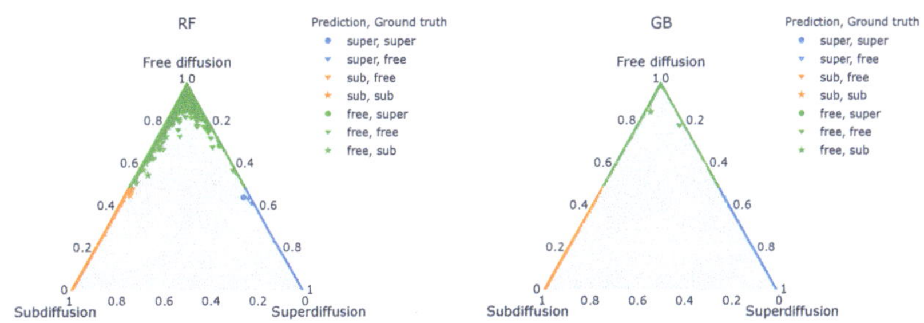

Figure 5. Ternary plots of the class probabilities assigned to the testing data by the classifiers trained on the base data set with Set A of features.

Finally, we can verify the distribution of the class probabilities for our experimental data (see Section 3.3 and 4.4), where the ground truth for the diffusion type is not known. In Figure 7, the corresponding ternary plots for empirical data are presented, for random forest and gradient boosting classifiers (left and right column, respectively) and for both G-protein-coupled receptors and G proteins (top and bottom row, respectively). These graphs can clearly show us the trajectories for which the classifiers' decisions were the most vague—all points near the center of the triangle correspond to trajectories with significant probabilities of all of three diffusion types. Moreover, we can see that in case of random forest, the trajectories classified as superdiffusion had also a significant probability of being a normal diffusion, whereas the gradient boosting classifier undoubtedly returned high probability of them belonging to superdiffusion.

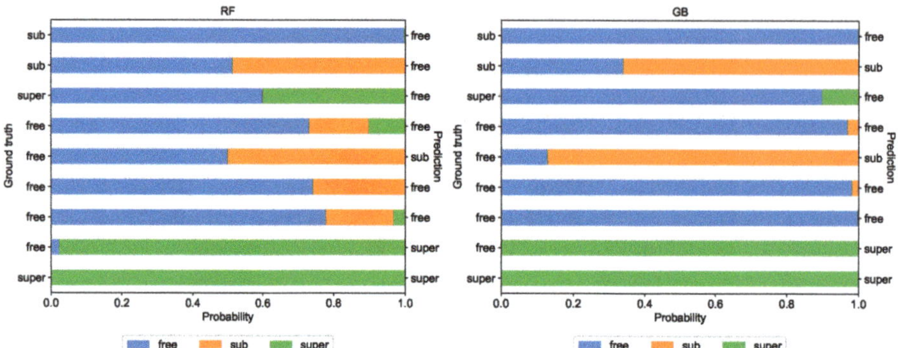

Figure 6. The class probabilities for exemplary trajectories from the testing set, based on the classifiers trained on the base data set and constructed with Set A of features.

In Figure 8, the predicted class probabilities for several interesting trajectories are shown, for both random forest (left graph) and gradient boosting (right graph). Again, the gradient boosting algorithm is more firm, but in cases of misclassification, it also claims the incorrect diffusion type with less doubt. Such an analysis of the classifiers decisions is a great starting point for further research—the output classifiers build on different data sets and with different sets of features can be examined in detail to find the exact source of a given prediction. That can also lead us to a reasonable model for the anomaly detection in the trajectories.

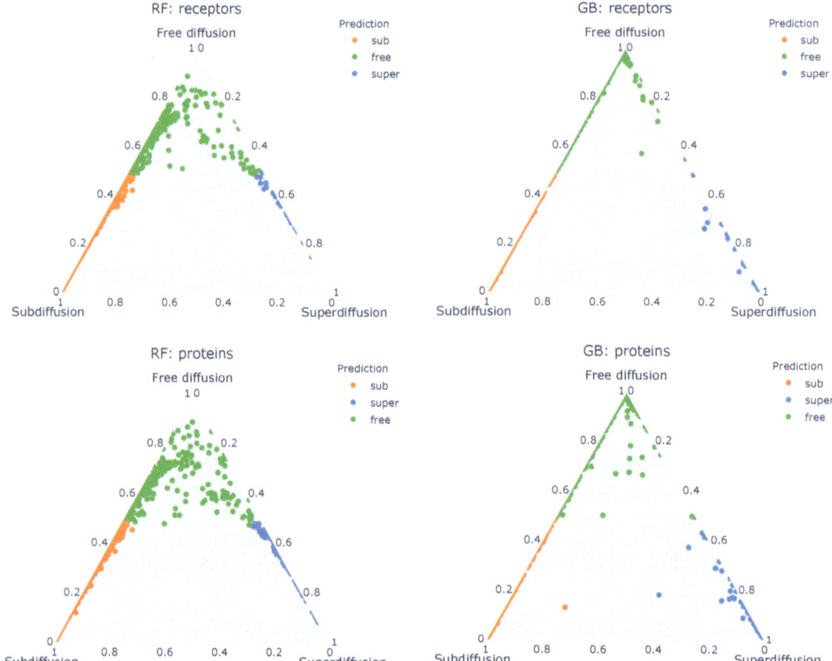

Figure 7. Ternary plots of the class probabilities assigned to empirical data by the classifiers trained on the base data set with Set A of features.

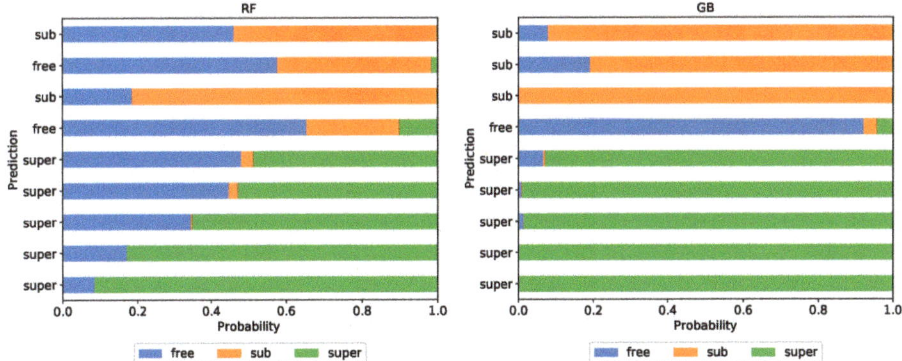

Figure 8. The class probabilities for exemplary trajectories from the empirical data set, based on the classifiers trained on the base data set and constructed with Set A of features.

5. Conclusions

In this paper, we presented a new set of features (referred to as Set A, see Table 1) for the two types of machine learning classifiers, random forest and gradient boosting, that on the synthetic data set gives good results, better than the set used previously in [43]. We have analysed the performance of our classifier trained and tested on the multiple versions of the synthetic data set, allowing us to assess its usefulness, flexibility and robustness. Moreover, we compared the proposed set with the ones already used in this problem, from [40,41,43]. Our set gives the best results in terms of the most common metrics.

Although the results on the synthetic data set are promising, we acknowledge the challenge with the application of the classifiers to real data. As discussed in [41], the classifiers trained on particular models for given diffusion modes do not generalise well. In Section 4.4, we show that even the classifiers with good accuracy return not clear result when used with the data of potentially different characteristics. To some extent, it can be improved by including more models in the training data set.

Thus, we would like to underline the importance of the features' selection for a given problem—even for the same task (e.g., diffusion mode classification), both models chosen for the training data generation and features chosen for their characterisation have a great influence on the performance of classifiers. Moreover, the assumptions made in constructions of the classifiers, such as hyperparameters' values or simply the choice of classifier type, are also highly important.

Supplementary Materials: Python codes for every stage of the classification procedure, together with a short documentation, are publicly available at Zenodo (https://doi.org/10.5281/zenodo.4317214).

Author Contributions: Conceptualization, H.L.-O. and J.S.; methodology, J.S.; software, H.L.-O.; validation, H.L.-O. and J.S.; investigation, H.L.-O. and J.S.; writing—original draft preparation, H.L.-O.; writing—review and editing, J.S.; supervision, J.S. All authors have read and agreed to the published version of the manuscript.

Funding: This research was funded by NCN-DFG Beethoven Grant No. 2016/23/G/ST1/04083.

Acknowledgments: Calculations were carried out using resources provided by the Wroclaw Centre for Networking and Supercomputing (http://wcss.pl).

Conflicts of Interest: The authors declare no conflict of interest.

Abbreviations

The following abbreviations are used in this manuscript:

DBM	directed Brownian motion
DL	deep learning

FBM	fractional Brownian
GB	gradient boosting
ML	machine learning
MSD	mean square displacement
OU	Ornstein–Uhlenbeck process
RF	random forest
SPT	single-particle tracking
TAMSD	time-averaged mean square displacement

References

1. Perrin, J. Mouvement brownien et molécules. *J. Phys. Theor. Appl.* **1910**, *9*, 5–39. [CrossRef]
2. Geerts, H.; Brabander, M.D.; Nuydens, R.; Geuens, S.; Moeremans, M.; Mey, J.D.D.; Hollenbeck, P. Nanovid tracking: A new automatic method for the study of mobility in living cells based on colloidal gold and video microscopy. *Biophys. J.* **1987**, *52 5*, 775–782. [CrossRef]
3. Barak, L.; Webb, W. Diffusion of low density lipoprotein-receptor complex on human fibroblasts. *J. Cell Biol.* **1982**, *95*, 846–852. [CrossRef] [PubMed]
4. Kusumi, A.; Sako, Y.; Yamamoto, M. Confined Lateral Diffusion of Membrane Receptors as Studied by Single Particle Tracking (Nanovid Microscopy). Effects of Calcium-induced Differentiation in Cultured Epithelial Cells. *Biophys. J.* **1993**, *65*, 2021–2040. [CrossRef]
5. Chenouard, N.; Smal, I.; de Chaumont, F.; Maška, M.; Sbalzarini, I.F.; Gong, Y.; Cardinale, J.; Carthel, C.; Coraluppi, S.; Winter, M.; et al. Objective comparison of particle tracking methods. *Nat. Methods* **2014**, *11*, 281–289. [CrossRef]
6. Saxton, M.J. Single-particle tracking: Connecting the dots. *Nat. Methods* **2008**, *5*, 671–672. [CrossRef]
7. Akhmanova, A.; Steinmetz, M.O. Tracking the ends: A dynamic protein network controls the fate of microtubule tips. *Nat. Rev. Mol. Cell Biol.* **2008**, *9*, 309–322. [CrossRef]
8. Berginski, M.E.; Vitriol, E.A.; Hahn, K.M.; Gomez, S.M. High-Resolution Quantification of Focal Adhesion Spatiotemporal Dynamics in Living Cells. *PLoS ONE* **2011**, *6*, e22025. [CrossRef] [PubMed]
9. Metzler, R.; Klafter, J. The random walk's guide to anomalous diffusion: A fractional dynamics approach. *Phys. Rep.* **2000**, *339*, 1–77. [CrossRef]
10. Michalet, X. Mean square displacement analysis of single-particle trajectories with localization error: Brownian motion in an isotropic medium. *Phys. Rev. E* **2010**, *82*, 041914. [CrossRef]
11. Brandenburg, B.; Zhuang, X. Virus trafficking—Learning from single-virus tracking. *Nat. Rev. Microbiol.* **2007**, *5*, 197–208. [CrossRef] [PubMed]
12. Bressloff, P.C. *Stochastic Processes in Cell Biology*; Interdisciplinary Applied Mathematics; Springer: Cham, Switzerland, 2014; pp. 645–672.
13. Mahowald, J.; Arcizet, D.; Heinrich, D. Impact of External Stimuli and Cell Micro-Architecture on Intracellular Transport States. *ChemPhysChem* **2009**, *10*, 1559–1566. [CrossRef] [PubMed]
14. Saxton, M.J.; Jacobson, K. Single-Particle Tracking: Applications to Membrane Dynamics. *Annu. Rev. Biophys. Biomol. Struct.* **1997**, *26*, 373–399. [CrossRef] [PubMed]
15. Kneller, G.R. Communication: A scaling approach to anomalous diffusion. *J. Chem. Phys.* **2014**, *141*, 041105. [CrossRef] [PubMed]
16. Qian, H.; Sheetz, M.P.; Elson, E.L. Single particle tracking. Analysis of diffusion and flow in two- dimensional systems. *Biophys. J.* **1991**, *60*, 910–921. [CrossRef]
17. Gal, N.; Lechtman-Goldstein, D.; Weihs, D. Particle tracking in living cells: A review of the mean square displacement method and beyond. *Rheol. Acta* **2013**, *52*, 425–443. [CrossRef]
18. Monnier, N.; Guo, S.M.; Mori, M.; He, J.; Lénárt, P.; Bathe, M. Bayesian Approach to MSD-Based Analysis of Particle Motion in Live Cells. *Biophys. J.* **2012**, *103*, 616–626. [CrossRef]
19. Alves, S.B.; de Oliveira, G.F., Jr.; Oliveira, L.C.; de Silansa, T.P.; Chevrollier, M.; Oriá, M.; Cavalcante, H.L.S. Characterization of diffusion processes: Normal and anomalous regimes. *Physica A* **2016**, *447*, 392–401. [CrossRef]
20. Weiss, M.; Elsner, M.; Kartberg, F.; Nilsson, T. Anomalous Subdiffusion Is a Measure for Cytoplasmic Crowding in Living Cells. *Biophys. J.* **2004**, *87*, 3518–3524. [CrossRef]

21. Saxton, M.J. Single-particle tracking: Models of directed transport. *Biophys. J.* **1994**, *67*, 2110–2119. [CrossRef]
22. Berry, H.; Chaté, H. Anomalous diffusion due to hindering by mobile obstacles undergoing Brownian motion or Orstein-Ulhenbeck processes. *Phys. Rev. E* **2014**, *89*, 022708. [CrossRef] [PubMed]
23. Hoze, N.; Nair, D.; Hosy, E.; Sieben, C.; Manley, S.; Herrmann, A.; Sibarita, J.B.; Choquet, D.; Holcman, D. Heterogeneity of AMPA receptor trafficking and molecular interactions revealed by superresolution analysis of live cell imaging. *Proc. Natl. Acad. Sci. USA* **2012**, *109*, 17052–17057. [CrossRef] [PubMed]
24. Arcizet, D.; Meier, B.; Sackmann, E.; Rädler, J.O.; Heinrich, D. Temporal Analysis of Active and Passive Transport in Living Cells. *Phys. Rev. Lett.* **2008**, *101*, 248103. [CrossRef] [PubMed]
25. Kepten, E.; Weron, A.; Sikora, G.; Burnecki, K.; Garini, Y. Guidelines for the Fitting of Anomalous Diffusion Mean Square Displacement Graphs from Single Particle Tracking Experiments. *PLoS ONE* **2015**, *10*, e0117722. [CrossRef]
26. Briane, V.; Kervrann, C.; Vimond, M. Statistical analysis of particle trajectories in living cells. *Phys. Rev. E* **2018**, *97*, 062121. [CrossRef]
27. Saxton, M.J. Lateral diffusion in an archipelago. Single-particle diffusion. *Biophys. J.* **1993**, *64*, 1766–1780. [CrossRef]
28. Valentine, M.T.; Kaplan, P.D.; Thota, D.; Crocker, J.C.; Gisler, T.; Prud'homme, R.K.; Beck, M.; Weitz, D.A. Investigating the microenvironments of inhomogeneous soft materials with multiple particle tracking. *Phys. Rev. E* **2001**, *64*, 061506. [CrossRef]
29. Gal, N.; Weihs, D. Experimental evidence of strong anomalous diffusion in living cells. *Phys. Rev. E* **2010**, *81*, 020903. [CrossRef]
30. Raupach, C.; Zitterbart, D.P.; Mierke, C.T.; Metzner, C.; Müller, F.A.; Fabry, B. Stress fluctuations and motion of cytoskeletal-bound markers. *Phys. Rev. E* **2007**, *76*, 011918. [CrossRef]
31. Burov, S.; Tabei, S.M.A.; Huynh, T.; Murrell, M.P.; Philipson, L.H.; Rice, S.A.; Gardel, M.L.; Scherer, N.F.; Dinner, A.R. Distribution of directional change as a signature of complex dynamics. *Proc. Natl. Acad. Sci. USA* **2013**, *110*, 19689–19694. [CrossRef]
32. Tejedor, V.; Bénichou, O.; Voituriez, R.; Jungmann, R.; Simmel, F.; Selhuber-Unkel, C.; Oddershede, L.B.; Metzler, R. Quantitative Analysis of Single Particle Trajectories: Mean Maximal Excursion Method. *Biophys. J.* **2010**, *98*, 1364–1372. [CrossRef] [PubMed]
33. Burnecki, K.; Kepten, E.; Garini, Y.; Sikora, G.; Weron, A. Estimating the anomalous diffusion exponent for single particle tracking data with measurement errors—An alternative approach. *Sci. Rep.* **2015**, *5*, 11306. [CrossRef] [PubMed]
34. Das, R.; Cairo, C.W.; Coombs, D. A Hidden Markov Model for Single Particle Tracks Quantifies Dynamic Interactions between LFA-1 and the Actin Cytoskeleton. *PLoS Comput. Biol.* **2009**, *5*, 1–16. [CrossRef] [PubMed]
35. Slator, P.J.; Cairo, C.W.; Burroughs, N.J. Detection of Diffusion Heterogeneity in Single Particle Tracking Trajectories Using a Hidden Markov Model with Measurement Noise Propagation. *PLoS ONE* **2015**, *10*. [CrossRef] [PubMed]
36. Slator, P.J.; Burroughs, N.J. A Hidden Markov Model for Detecting Confinement in Single-Particle Tracking Trajectories. *Biophys. J.* **2018**, *115*, 1741–1754. [CrossRef] [PubMed]
37. Weron, A.; Janczura, J.; Boryczka, E.; Sungkaworn, T.; Calebiro, D. Statistical testing approach for fractional anomalous diffusion classification. *Phys. Rev. E* **2019**, *99*, 042149. [CrossRef]
38. Thapa, S.; Lomholt, M.A.; Krog, J.; Cherstvy, A.G.; Metzler, R. Bayesian analysis of single-particle tracking data using the nested-sampling algorithm: Maximum-likelihood model selection applied to stochastic-diffusivity data. *Phys. Chem. Chem. Phys.* **2018**, *20*, 29018–29037. [CrossRef]
39. Cherstvy, A.G.; Thapa, S.; Wagner, C.E.; Metzler, R. Non-Gaussian, non-ergodic, and non-Fickian diffusion of tracers in mucin hydrogels. *Soft Matter* **2019**, *15*, 2526–2551. [CrossRef]
40. Wagner, T.; Kroll, A.; Haramagatti, C.R.; Lipinski, H.G.; Wiemann, M. Classification and Segmentation of Nanoparticle Diffusion Trajectories in Cellular Micro Environments. *PLoS ONE* **2017**, *12*, e0170165. [CrossRef]
41. Kowalek, P.; Loch-Olszewska, H.; Szwabiński, J. Classification of diffusion modes in single-particle tracking data: Feature-based versus deep-learning approach. *Phys. Rev. E* **2019**, *100*, 032410. [CrossRef]
42. Muñoz-Gil, G.; Garcia-March, M.A.; Manzo, C.; Martín-Guerrero, J.D.; Lewenstein, M. Single trajectory characterization via machine learning. *New J. Phys.* **2020**, *22*, 013010. [CrossRef]

43. Janczura, J.; Kowalek, P.; Loch-Olszewska, H.; Szwabiński, J.; Weron, A. Classification of particle trajectories in living cells: Machine learning versus statistical testing hypothesis for fractional anomalous diffusion. *Phys. Rev. E* **2020**, *102*, 032402. [CrossRef] [PubMed]
44. Dosset, P.; Rassam, P.; Fernandez, L.; Espenel, C.; Rubinstein, E.; Margeat, E.; Milhiet, P.E. Automatic detection of diffusion modes within biological membranes using backpropagation neural network. *BMC Bioinform.* **2016**, *17*, 197. [CrossRef] [PubMed]
45. Granik, N.; Weiss, L.E.; Nehme, E.; Levin, M.; Chein, M.; Perlson, E.; Roichman, Y.; Shechtman, Y. Single-Particle Diffusion Characterization by Deep Learning. *Biophys. J.* **2019**, *117*, 185–192. [CrossRef]
46. Bo, S.; Schmidt, F.; Eichhorn, R.; Volpe, G. Measurement of anomalous diffusion using recurrent neural networks. *Phys. Rev. E* **2019**, *100*, 010102. [CrossRef]
47. Straley, J.P. The ant in the labyrinth: Diffusion in random networks near the percolation threshold. *J. Phys. C Solid State Phys.* **1980**, *13*, 2991–3002. [CrossRef]
48. Metzler, R.; Jeon, J.H.; Cherstvy, A.G.; Barkai, E. Anomalous diffusion models and their properties: Non-stationarity, non-ergodicity, and ageing at the centenary of single particle tracking. *Phys. Chem. Chem. Phys.* **2014**, *16*, 24128–24164. [CrossRef]
49. Mandelbrot, B.B.; Ness, J.W.V. Fractional Brownian Motions, Fractional Noises and Applications. *SIAM Rev.* **1968**, *10*, 422–437. [CrossRef]
50. Guigas, G.; Kalla, C.; Weiss, M. Probing the Nanoscale Viscoelasticity of Intracellular Fluids in Living Cells. *Biophys. J.* **2007**, *93*, 316–323. [CrossRef]
51. Burnecki, K.; Kepten, E.; Janczura, J.; Bronshtein, I.; Garini, Y.; Weron, A. Universal Algorithm for Identification of Fractional Brownian Motion. A Case of Telomere Subdiffusion. *Biophys. J.* **2012**, *103*, 1839–1847. [CrossRef]
52. Burnecki, K.; Weron, A. Fractional Lévy stable motion can model subdiffusive dynamics. *Phys. Rev. E* **2010**, *82*, 021130. [CrossRef] [PubMed]
53. Kou, S.C.; Xie, X.S. Generalized Langevin Equation with Fractional Gaussian Noise: Subdiffusion within a Single Protein Molecule. *Phys. Rev. Lett.* **2004**, *93*, 180603. [CrossRef] [PubMed]
54. Burnecki, K.; Weron, A. Algorithms for testing of fractional dynamics: A practical guide to ARFIMA modelling. *J. Stat. Mech. Theory Exp.* **2014**, *2014*, P10036. [CrossRef]
55. Elston, T.C. A macroscopic description of biomolecular transport. *J. Math. Biol.* **2000**, *41*, 189. [CrossRef]
56. MacLeod, C.L.; Ivezi, Z.; Kochanek, C.S.; Kozłowski, S.; Kelly, B.; Bullock, E.; Kimball, A.; Sesar, B.; et al. Modeling the time variability of SDSS stripe 82 quasars as a damped random walk. *Astrophys. J.* **2010**, *721*, 1014. [CrossRef]
57. Jeon, J.H.; Tejedor, V.; Burov, S.; Barkai, E.; Selhuber-Unkel, C.; Berg-Sørensen, K.; Oddershede, L.; Metzler, R. In Vivo Anomalous Diffusion and Weak Ergodicity Breaking of Lipid Granules. *Phys. Rev. Lett.* **2011**, *106*, 048103. [CrossRef]
58. Raschka, S. *Python Machine Learning*; Packt Publishing: Birmingham, UK, 2015.
59. Song, Y.Y.; LU, Y. Decision tree methods: Applications for classification and prediction. *Shanghai Arch. Psychiatry* **2015**, *27*, 130–135.
60. Breiman, L. Random Forests. *Mach. Learn.* **2001**, *45*, 5–32. [CrossRef]
61. Ho, T.K. Random Decision Forests. In Proceedings of the Third International Conference on Document Analysis and Recognition, Montreal, QC, Canada, 14–16 August 1995; IEEE Computer Society: Washington, DC, USA, 1995; Volume 1.
62. Ho, T.K. The Random Subspace Method for Constructing Decision Forests. *IEEE Trans. Pattern Anal. Mach. Intell.* **1998**, *20*, 832–844. [CrossRef]
63. Pedregosa, F.; Varoquaux, G.; Gramfort, A.; Michel, V.; Thirion, B.; Grisel, O.; Blondel, M.; Prettenhofer, P.; Weiss, R.; Dubourg, V.; et al. Scikit-learn: Machine Learning in Python. *J. Mach. Learn. Res.* **2011**, *12*, 2825–2830.
64. James, G.; Witten, D.; Hastie, T.; Tibshirani, R. *An Introduction to Statistical Learning with Applications in R*; Springer: New York, NY, USA, 2013.
65. Grebenkov, D.S. Optimal and suboptimal quadratic forms for noncentered Gaussian processes. *Phys. Rev. E* **2013**, *88*, 032140. [CrossRef] [PubMed]
66. Sungkaworn, T.; Jobin, M.L.; Burnecki, K.; Weron, A.; Lohse, M.J.; Calebiro, D. Single-molecule imaging reveals receptor–G protein interactions at cell surface hot spots. *Nature* **2017**, *550*, 543. [CrossRef] [PubMed]

67. Breiman, L.; Friedman, J.H.; Olshen, R.A.; Stone, C.J. *Classification and Regression Trees*; Wadsworth and Brooks: Monterey, CA, USA, 1984.
68. Weber, S.C.; Spakowitz, A.J.; Theriot, J.A. Bacterial Chromosomal Loci Move Subdiffusively through a Viscoelastic Cytoplasm. *Phys. Rev. Lett.* **2010**, *104*, 238102. [CrossRef] [PubMed]
69. Hubicka, K.; Janczura, J. Time-dependent classification of protein diffusion types: A statistical detection of mean-squared-displacement exponent transitions. *Phys. Rev. E* **2020**, *101*, 022107. [CrossRef] [PubMed]
70. Vega, A.R.; Freeman, S.A.; Grinstein, S.; Jaqaman, K. Multistep Track Segmentation and Motion Classification for Transient Mobility Analysis. *Biophys. J.* **2018**, *114*, 1018. [CrossRef] [PubMed]
71. Lanoiselée, Y.; Sikora, G.; Grzesiek, A.; Grebenkov, D.S.; Wyłomańska, A. Optimal parameters for anomalous-diffusion-exponent estimation from noisy data. *Phys. Rev. E* **2018**, *98*, 062139. [CrossRef]

Article

Fractional Dynamics Identification via Intelligent Unpacking of the Sample Autocovariance Function by Neural Networks

Dawid Szarek [1], Grzegorz Sikora [1], Michał Balcerek [1], Ireneusz Jabłoński [2] and Agnieszka Wyłomańska [1,*]

1 Faculty of Pure and Applied Mathematics, Hugo Steinhaus Center, Wroclaw University of Science and Technology, Wybrzeże Wyspiańskiego 27, 50-370 Wroclaw, Poland; 234856@student.pwr.edu.pl (D.S.); grzegorz.sikora@pwr.edu.pl (G.S.); michal.balcerek@pwr.edu.pl (M.B.)

2 Department of Electronics, Wroclaw University of Science and Technology, B. Prusa 53/55, 50-317 Wroclaw, Poland; ireneusz.jablonski@pwr.edu.pl

* Correspondence: agnieszka.wylomanska@pwr.edu.pl

Received: 31 October 2020; Accepted: 18 November 2020; Published: 20 November 2020

Abstract: Many single-particle tracking data related to the motion in crowded environments exhibit anomalous diffusion behavior. This phenomenon can be described by different theoretical models. In this paper, fractional Brownian motion (FBM) was examined as the exemplary Gaussian process with fractional dynamics. The autocovariance function (ACVF) is a function that determines completely the Gaussian process. In the case of experimental data with anomalous dynamics, the main problem is first to recognize the type of anomaly and then to reconstruct properly the physical rules governing such a phenomenon. The challenge is to identify the process from short trajectory inputs. Various approaches to address this problem can be found in the literature, e.g., theoretical properties of the sample ACVF for a given process. This method is effective; however, it does not utilize all of the information contained in the sample ACVF for a given trajectory, i.e., only values of statistics for selected lags are used for identification. An evolution of this approach is proposed in this paper, where the process is determined based on the knowledge extracted from the ACVF. The designed method is intuitive and it uses information directly available in a new fashion. Moreover, the knowledge retrieval from the sample ACVF vector is enhanced with a learning-based scheme operating on the most informative subset of available lags, which is proven to be an effective encoder of the properties inherited in complex data. Finally, the robustness of the proposed algorithm for FBM is demonstrated with the use of Monte Carlo simulations.

Keywords: anomalous diffusion; fractional Brownian motion; estimation; autocovariance function; neural network; Monte Carlo simulations

---

1. Introduction

Many single-particle tracking data related to the motion in crowded environments exhibit anomalous diffusion behavior [1,2]. This behavior is also visible in various phenomena such as finance [3,4], ecology [5], hydrology [6], and biology [7], as well as meteorology and geophysics [8,9]. Anomalous diffusion behavior is manifested by deviations from the laws of Brownian motion (BM). One of the most common definitions of the anomalous diffusion process is expressed in the nonlinear behavior of its second moment:

$$\mathbb{E}\left[X^2(t)\right] \sim t^{\alpha}, \tag{1}$$

where α is the so-called anomalous diffusion exponent. When $\alpha = 1$, the process is classified as diffusion, while for $\alpha \neq 1$, we call it anomalous diffusion. More precisely, for $\alpha < 1$, it is called subdiffusion, while for $\alpha > 1$, it is superdiffusion. It should be mentioned that anomalous diffusion can be also related to the non-Gaussian probability density function of the corresponding process; for instance, see [10–13].

The class of anomalous diffusion processes is very rich. The most classical anomalous diffusion models are fractional Brownian motion (FBM) [8,14], Lévy stable motion [15], continuous-time random walk [16,17], and the subordinated processes (also called time-changed processes) [18–25]. We also mention here the processes with time- or position-dependent diffusion coefficients such as scaled Brownian motion [26,27] or heterogeneous diffusion models [28], as well as the superstatistical process [29] or diffusing diffusivity models (also called Brownian yet non-Gaussian diffusion process) [30]. We also refer the readers to the articles [31–34] and the references therein.

In the case of experimental data with anomalous dynamics, the main problem is first to recognize the type of anomaly and then to reconstruct properly the physical rules governing such a phenomenon. The main challenge is to identify the process from short trajectory inputs. Various approaches to address this problem can be found in the literature; for instance, see [35–42]. One of the simplest and most efficient approaches is based on the theoretical properties of the sample autocovariance function (ACVF) [43]. It is known that the ACVF is the characteristic that determines completely the centered Gaussian process. Thus, its sample version is a proper tool for the testing and estimation of the parameters of this process. The approach presented in the literature [43] is effective; however, it does not utilize all of the information contained in the sample ACVF for a given trajectory, i.e., only values of statistics for selected lags are utilized. Therefore, an evolution of this approach is proposed in this paper.

Herein, we compared three approaches that apply the ACVF for estimation of the anomalous diffusion exponent. The first one, the so-called *naive* method, uses the ACVF in one specific lag to estimate the anomalous diffusion exponent of the given process. The second algorithm is based on the ACVF corresponding to the vector of the selected lags. The last technique is based on the sample ACVF information extracted with the most informative subset of available lags. The designed novel method in the third approach is intuitive and it uses information directly available in a new fashion. Information retrieval from the sample ACVF vector is performed here with a learning-based scheme operating on the most informative lags, i.e., a feedforward neural network (FNN) [44] is designed and applied for solving the regression task. This approach has been proven to be an effective encoder of the properties inherited in complex data [45–47]. The goal is to preliminarily assess (using computer simulations) the predictive properties of an FNN for the estimation of the anomalous diffusion exponent based on a short data set. This exercise provides evidence of the performance of the simple version of the neural network (NN) adapted for the defined regimes (anomalous diffusion) and the ACVF vector, which is a projection of a valid complex and real trajectory, e.g., for particle movement in a solution. The reported results can be further enhanced by the exploitation of adaptive mechanisms inserted into recurrent neural networks (RNNs) and/or the detailed and proper inferences of multiscale pattern(s) for deep learning [1–5]. The advantage of the application of an FNN to the task defined above is that a trained neural network model can be a robust and efficient estimator for anomalous diffusion exponents, e.g., complex relations hidden in ACVF data can be extracted within one-step, concluding in a trained feedforward neural network.

The robustness of the introduced algorithm based on the ACVF and NN methods in comparison to the known ACVF-based techniques is demonstrated herein for the exemplary Gaussian process using Monte Carlo simulations. We considered the FBM as an exemplary model with fractional dynamics; that is, the Gaussian process with stationary increments and the so-called self-similar property parametrized by the Hurst exponent $H = 0.5\alpha$, where α is the anomalous diffusion exponent given in (1).

The main goal of the paper was to prove that the incorporation of intelligent-based algorithms into classical estimation schemes can shed new light on the investigation of the anomalous diffusion phenomenon. Moreover, the classical tools enhanced by artificial intelligence (AI) methods are more effective in comparison to the known statistical algorithms used for anomalous diffusion parametrization.

The rest of the paper is organized as follows: In Section 2, we outline the definition of the fractional Brownian motion and the exemplary Gaussian process with anomalous diffusion behavior. Next, in Section 3, we discuss two of the estimation methods for the Hurst exponent based on the ACVF that are commonly used in various applications. In the next section, we present a new approach for the estimation of the H index. Namely, in the new algorithm, we combined the ACVF and NN methods. To demonstrate the effectiveness of the new approach, in Section 5, we present a simulation study where we compare the three considered algorithms for the estimation of the Hurst exponent. The last section concludes the paper and presents a future study.

2. Fractional Brownian Motion

Fractional Brownian motion (FBM) $\{X_H(t),\ t \geq 0\}$ with the Hurst index $H \in (0,1)$ is a continuous and centered Gaussian process defined through the following Langevin equation [14,48–50]:

$$\frac{dX_H(t)}{dt} = D\xi_H(t), \tag{2}$$

where the parameter D is the diffusion coefficient. In Equation (2), $\{\xi_H(t),\ t \geq 0\}$ is the fractional Gaussian noise process with the autocorrelation function satisfying the following:

$$\mathrm{Corr}(\xi_H(0), \xi_H(t)) \sim 2H(2H-1)Dt^{2(H-1)},\ t \geq 0. \tag{3}$$

The FBM was introduced by Kolmogorov in 1940 (see [8]). FBM is the only Gaussian process with the self-similar property. Because the FBM is a centered Gaussian process, it can be also defined through the ACVF that, in this case, is given by [14]:

$$\mathbb{E}\left[X_H(t)X_H(s)\right] = \frac{1}{2}D\left(t^{2H} + s^{2H} - |t-s|^{2H}\right), \qquad \text{where } t,s \geq 0. \tag{4}$$

Thus, for the given $t \geq 0$, $X_H(t) \sim \mathcal{N}\left(0, Dt^{2H}\right)$. The FBM has stationary increments. Moreover, if $H > 1/2$, then the increments of the process are positively correlated, while for $H < 1/2$, they are negatively correlated. Moreover, for $H > 1/2$, the FBM exhibits the so-called long range dependence, which means that the following property is satisfied:

$$\sum_{n=1}^{\infty} \mathbb{E}\left[X_H(1)(X_H(n+1) - X_H(n))\right] = \infty. \tag{5}$$

As can be seen, for $H = 1/2$, the FBM reduces to the ordinary Brownian motion (BM). It should be mentioned that the FBM is considered one of the classical processes used to describe the anomalous diffusion phenomenon. Indeed, for $H < 1/2$, it exhibits subdiffusion behavior, while for $H > 1/2$, it shows superdiffusion behavior. To see the differences between the behavior of the trajectories corresponding to different anomalous types, in Figure 1, we demonstrate the exemplary trajectories of FBM for $H = 0.3$ (subdiffusion), $H = 0.5$ (diffusion), and $H = 0.7$ (superdiffusion).

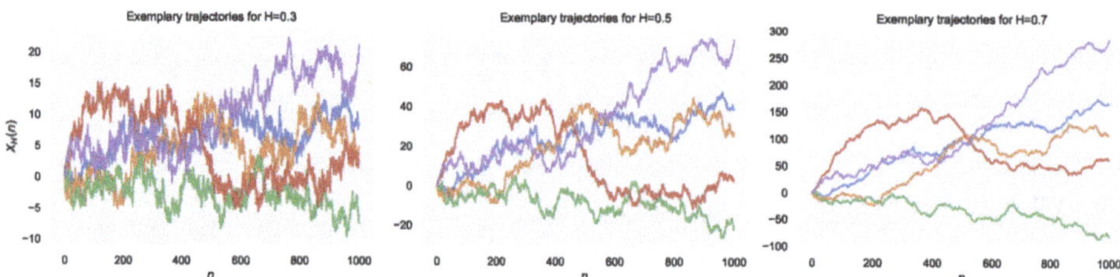

Figure 1. Exemplary trajectories for $H \in \{0.3, 0.5, 0.7\}$.

3. ACVF-Based Methods for the Estimation of the Hurst Exponent

In this article, we consider three approaches that utilize the ACVF in the estimation of the Hurst parameter H. Here, we depict two approaches known from the literature. The last technique is described in detail in the next section.

The methods presented in this section are based on the sample version of the ACVF. Let us consider the trajectory of FBM, $\mathbf{X}_H = \{X_H(1), X_H(2), \cdots, X_H(n)\}$, and the corresponding sample of increments, $\xi_H = \{\xi_H(1), \xi_H(2), \cdots, \xi_H(n)\}$, where $\xi_H(t) = X_H(t) - X_H(t-1)$ for $t = 1, 2, \ldots, n$. The sample ACVF for $\boldsymbol{\xi}_H$ is given by [51]:

$$\hat{\gamma}_\xi(\tau) = \frac{1}{n}\sum_{t=1}^{n-\tau} \xi_H(t)\xi_H(t+\tau), \qquad \tau = 0, 1, \ldots, n-1. \tag{6}$$

The statistic $\hat{\gamma}_\xi(\tau)$ is a rescaled estimator of the theoretical autocovariance function for $\gamma_\xi(\tau)$, corresponding to lag τ, where $\gamma_\xi(\tau) = \mathbb{E}\left(\xi_H(1)\xi_H(1+\tau)\right)$. One can easily show that the statistic (6) is a biased estimator of $\gamma_\xi(\tau)$, namely:

$$\begin{aligned}
\mathbb{E}\left[\hat{\gamma}_\xi(\tau)\right] &= \frac{1}{n}\mathbb{E}\left[\sum_{t=1}^{n-\tau} \xi_H(t)\xi_H(t+\tau)\right] = \\
&= \frac{n-\tau}{n}\mathbb{E}\left[\xi_H(1)\xi_H(1+\tau)\right] = \frac{D(n-\tau)}{2n}\left(|\tau+1|^{2H} + |\tau-1|^{2H} - 2|\tau|^{2H}\right) = \frac{n-\tau}{n}\gamma_\xi(\tau).
\end{aligned}$$

However, in our considerations, we used the biased version of the $\gamma_\xi(\tau)$ estimator due to its lower variance in comparison to the unbiased one.

In the literature, a few possibilities of estimating the Hurst parameter H using the ACVF have been presented. In the simplest approach, we considered only the first lag $\tau = 1$, and compared the statistic $\hat{\gamma}_\xi(1)$ with the desirable theoretical value of $\gamma_\xi(1) = \frac{D}{2}(2^{2H} - 2)$. Thus, using the following relation:

$$\hat{\gamma}_\xi(1) = \frac{D}{2}\left(2^{2H} - 2\right),$$

one can obtain the simple estimator of H:

$$\hat{H} = \frac{1}{2}\log_2\left(\hat{\gamma}_\xi(1)\frac{2}{D} + 2\right). \tag{7}$$

As such an approach is simple, we refer to it as *naive* (the M1 method) in the following analyses. In this approach, the diffusion coefficient D is assumed to be known. However, in real applications, the D parameter can also be estimated as a sample variance of the vector $\boldsymbol{\xi}$. As one can see, the ACVF for lag

$\tau = 1$ includes the necessary information about the FBM process. Unfortunately, if the corresponding measurements are burdened by an additive error (i.e., some measurement noise), $\gamma(1)$ changes accordingly, resulting in the need to consider more advanced techniques for the estimation of the H parameter; for instance, see [52,53] and the discussion therein.

In an alternative approach for estimating the H parameter, one can simultaneously use more lags τ. Thus, we can fit the function $\tau \mapsto \frac{D}{2}\left(|\tau+1|^{2H} + |\tau-1|^{2H} - 2|\tau|^{2H}\right)$ to the empirical ACVF $\hat{\gamma}_{\xi}(\tau)$ for the corresponding lags $\tau = 1, 2, \ldots, \tau_{max}$ in the least squares sense, i.e., the estimator is calculated as follows:

$$\hat{H} = \arg\min_{H\in(0,1)} \sum_{\tau=1}^{\tau_{max}} \left[\hat{\gamma}_{\xi}(\tau) - \frac{D}{2}\left(|\tau+1|^{2H} + |\tau-1|^{2H} - 2|\tau|^{2H}\right)\right]^2, \quad (8)$$

for some maximum lag τ_{max}. Again, if the diffusion coefficient is unknown, we can estimate it by considering arg min over $(H, D) \in (0,1) \times (0,\infty)$. This approach is unfortunately much more complex, as it requires nonlinear regression methods. In the further analysis, we refer to this approach as the M2 method.

4. ACVF and NN-Based Methods for the Estimation of the Hurst Exponent

As mentioned above, three methods for the estimation of the Hurst exponent H are compared in this article. All of the identification algorithms are based on the ACVF. The first (M1) and the second (M2) methods were described in the previous section. In this section, we present the algorithm based on the ACVF and NN methods, denoted as the M3 method in the simulation study.

One might expect that the two proposed methods provide the best estimation results when the data follows the "pure" theoretical model, i.e., FBM. In reality, this is rarely the case—often, the observed trajectories are biased by a measurement error and/or various interleaving processes cover the anomalous diffusion component in the acquired signal. A purely statistical approach, such as in the M1 and M2 methods, can bring about limitations in real-world data applications, whereas artificial intelligence has shown its potential to overcome parasitic conditions in data from numerous fields of applications [45,54–56]. This triggers applications of learning-based schemes that enable weakening of the initial assumptions related to data properties and model building, providing good estimators for a wide range of anomalous diffusion regimes (i.e., for a wide range of H values). The assumed architecture for the neural network model and the training process are crucial aspects for efficient data exploration in artificial intelligent schemes, with the latter being of particular importance for NN model performance [44]. As a consequence, the input data properties used for the NN model training condition the reliability of the observed outputs—i.e., the value of the estimated H and its uncertainty in the reported study.

In the previous section, we assumed that the relationship between the data and the estimated parameter is known and given by Equations (7) and (8) for the M1 and M2 methods, respectively. Now, we propose to use a feedforward neural network as the predictor of this theoretical relationship, i.e., $\mathbb{E}[H|\{\hat{\gamma}_{\xi}(\tau)\}]$. To be more precise, the FNN is proposed as the model of the hidden relationships in the experimental data. The last one means that obtaining a formal expression for the rules governing the phenomena in a real-world system is not the main subject of interest here, but encoded in the FNN model, these rules are used to enhance the reliability of the Hurst exponent estimation from the data that correspond to the FBM, according to our assumption. It is worth noting that the modeling of $\mathbb{E}[H|\boldsymbol{X}_H]$ or $\mathbb{E}[H|\boldsymbol{\xi}_H]$ is also possible with the NN-based approach; however, more sophisticated NN topologies are required to reconstruct the long dependency valid for the FBM model (for $H \gg 0.5$). Thus, dealing with long and varying input vector lengths is required to realize this task; models based on RNNs, long short-term memory (LSTM) neural networks [57], or other forms of intelligent recurrence should be used, which implies higher requirements regarding their training [58].

The information about the underlying process is concentrated in the first couple of lags of the sample ACVF. In this paper, 32 lags (including the 0th lag) were used as the input for the Hurst exponent estimation. In the next section, it is shown that this amount was sufficient for our study.

The proposed architecture of the neural network model consists of three hidden layers of consecutive sizes of 64, 64, and 32. The Swish-1 [59] activation function, defined as:

$$\text{Swish-}\beta(x) := x \times \sigma(\beta x) = \frac{x}{1+e^{-\beta x}}, \tag{9}$$

was used for the neurons in each layer—in many applications, this expression outperforms the other activation functions [59].

In the designed FNN predictor, the size of the first layer was conditioned by the number of lags corresponding to the ACVF used during the experiment (i.e., 32 neurons were inserted into the first layer) and the output layer consisted of one neuron, which produced the H estimator as the model response. Since the estimated value of the Hurst exponent was within the range of $H \in (0,1)$ and the FNN model designed for its estimation can produce all real numbers as the output, post-processing transformation needed to be applied to the response of the output neuron. Here, the sigmoid function, defined as:

$$\sigma(x) := \frac{1}{1+e^{-x}}, \tag{10}$$

was used. The function given in (10) projects all real numbers to the interval $(0,1)$.

To boost the FNN training process, the Adam optimization algorithm [60] was applied as it quickly converges to a minimum [61–64]. The mean squared error (MSE) was used to quantify the prediction error of the FNN model.

5. Simulation Study

The efficiency of the three methods (M1, M2, and M3) designed for H estimation is demonstrated in this section using computer simulations. Cholesky decomposition [65] was used for the generation of the FBM trajectories, since it allows to simulate outputs with extreme H parameter values (unlike Davies–Harte [66], which can fail to generate small samples for H parameters close to 1). Regarding the practical usefulness of the designed procedures, their efficiency can be expressed in the context of the length of the input trajectory required to estimate H with expected reliability. This study was performed for trajectories of various lengths, and the results are reported below.

The two statistical methods described in Section 3 were ready to use, whereas the designed feedforward neural network needed to be trained in advance, for which training and validation datasets were prepared.

The training dataset was formed with 1,572,64 FBM trajectories generated during computer simulations. This set consisted of vectors of different lengths (from 32 up to 1024) and referred to different H parameter values. For every trajectory length, $N \in \{32, 64, 128, 256, 512, 1024\}$, 262,144 trajectories were generated using computer simulations (262,144 × 6 = 1,572,864), each with the H parameter selected randomly from the uniform distribution $\mathcal{U}(0,1)$. Next, for each trajectory ξ_H, the ACVF (biased estimator (6), as introduced earlier) was calculated, resulting in a set of ACVFs of lengths N (N lags—$\tau \in \{0,1,\ldots,N-1\}$) with corresponding H parameters. Using the same procedure, the validation and test subsets were generated, each of a size of 196,608 (32,768 × 6 = 196,608).

The length of the ACVF vectors for the NN training was limited to 32 first lags (namely, $\tau \in \{0,1,\ldots,31\}$). This selection was preceded by the calculation of MAE prediction error, such as in Figure 2—more input samples do not decrease the error, whereas a smaller input size increases the MAE value.

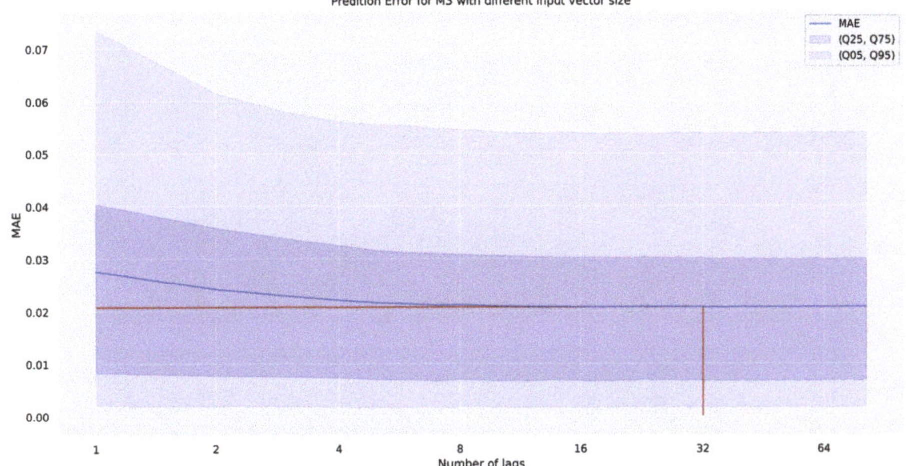

Figure 2. The MAE calculated for the M3 method when a different number of lags is used as the feedforward neural network (FNN) input (determining the size of the input layer of the FNN), depending also on the selected quantile; the number of lags (32) selected for use in the paper are marked with a red line.

To train the FNN (the architecture of the FNN as described in Section 4), input data were gathered into batches of 64 (the number of training examples used to calculate weight updates). The total number of NN parameters was 8385, trained for 13 epochs (the number of times that the training algorithm operated on the entire training dataset), for a total of 93 s $\times$ 13 epochs $\approx$ 20 min. The number of epochs was selected dynamically using the early stopping method [67]. Since the model did not improve the prediction error significantly after the third epoch, the training procedure could be stopped then (then, the training time would be 5 min). Calculations were performed on a PC with Intel Core i7 (3.7 GHz, 6 cores, 12 threads) and RAM of 64 GB.

The test dataset was used to compare the M1, M2, and M3 methods—for M1, the first lag was used ($\tau = 1$); for M2, the set of lags was $\tau \in \{0, 1, \ldots, \tau_{max}\}$ (if not stated differently, $\tau_{max} = 31$); M3 always used 32 lags ($\tau \in \{0, 1, \ldots, 31\}$). The metrics used were the absolute error and (for the aggregated results) the mean absolute error.

Figure 3 provides a comparison of the MAEs for the three methods (i.e., M1, M2, and M3) when dealing with trajectories of different lengths and with the diffusion characteristic (H parameter), grouped by the true parameter H into the bins [0–0.2), [0.2–0.4), [0.4–0.6), [0.6–0.8), and [0.8–1).

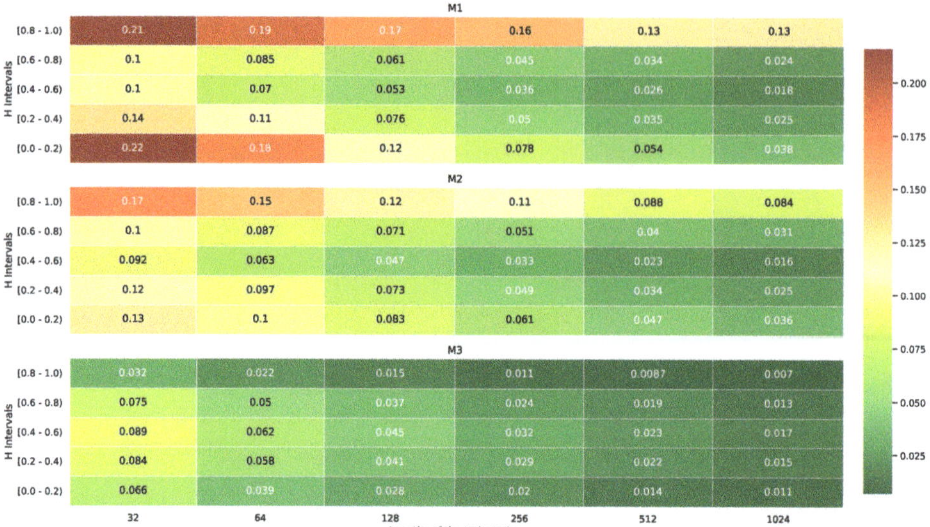

Figure 3. MAE heatmap for the M1, M2, and M3 methods depending on the length of the input trajectory and the value of the Hurst exponent applied during the computer simulations.

The conclusion is that the NN-based approach (M3) is more efficient than the other schemes studied in the paper, as it did not need long trajectories to estimate the Hurst exponent H with a minimized/minor error. For H close to 0 or 1, there was a significant difference in the MAEs between the considered methods—although the NN approach achieved a similar level of error to the other methods when fed samples of a size of 64 for the estimation, M1 and M2 struggled to deliver equivalent performance for longer inputs, i.e., up to 16-times longer trajectories, as was shown during the computer simulations. It is worth noting that the estimation error for the Hurst exponent in the diffusion case (i.e., $H \approx 0.5$) was similar for all of the considered approaches. In summary, since M1 estimated H reliably when using only the first lag, which was also true for the other methods, it was possible to distinguish between normal and anomalous diffusion using the information contained in the first lag or the first few lags.

To further compare these methods, the distribution of the (absolute) prediction error is shown in Figure 4. A logarithmic scale was applied to the y-axis in the boxplots to distinguish the performance of the following algorithms. The figure is divided into six parts, each reporting on the performance of the M1, M2, and M3 methods. In this way, it was possible to analyze how the distributions of the prediction error varied for each of these methods, the different lengths of the input trajectories, and also the different diffusion types (i.e., superdiffusion and subdiffusion); similarly to Figure 3, the input data were grouped evenly into five bins in reference to the true value of the H parameter.

Figure 4 depicts the spread of the estimation error and also reports on the number of outliers. The obtained results prove that the NN-based algorithm can more reliably estimate the H value for its following ranges (thus regimes of anomalous diffusion behavior) than in the two classical (M1 and M2) methods.

When it comes to the analysis of the multidimensional spread of the observed distributions, M3 performed similarly to the other methods. However, it is worth noting that M1 could not be used to obtain reliable results for some anomalous diffusion regimes, i.e., in several cases in Figure 4, the prediction errors were over 0.5 (which means that M1 could not distinguish between sub- and superdiffusion in these cases). There were also some cases in the presented study when all of the methods working with the smallest considered sample size (32) struggled to distinguish between subdiffusion and superdiffusion. Nonetheless, M3 applied to two-times longer trajectories (i.e., 64 samples inserted as the input) performed reliably and efficiently through the whole range of H values.

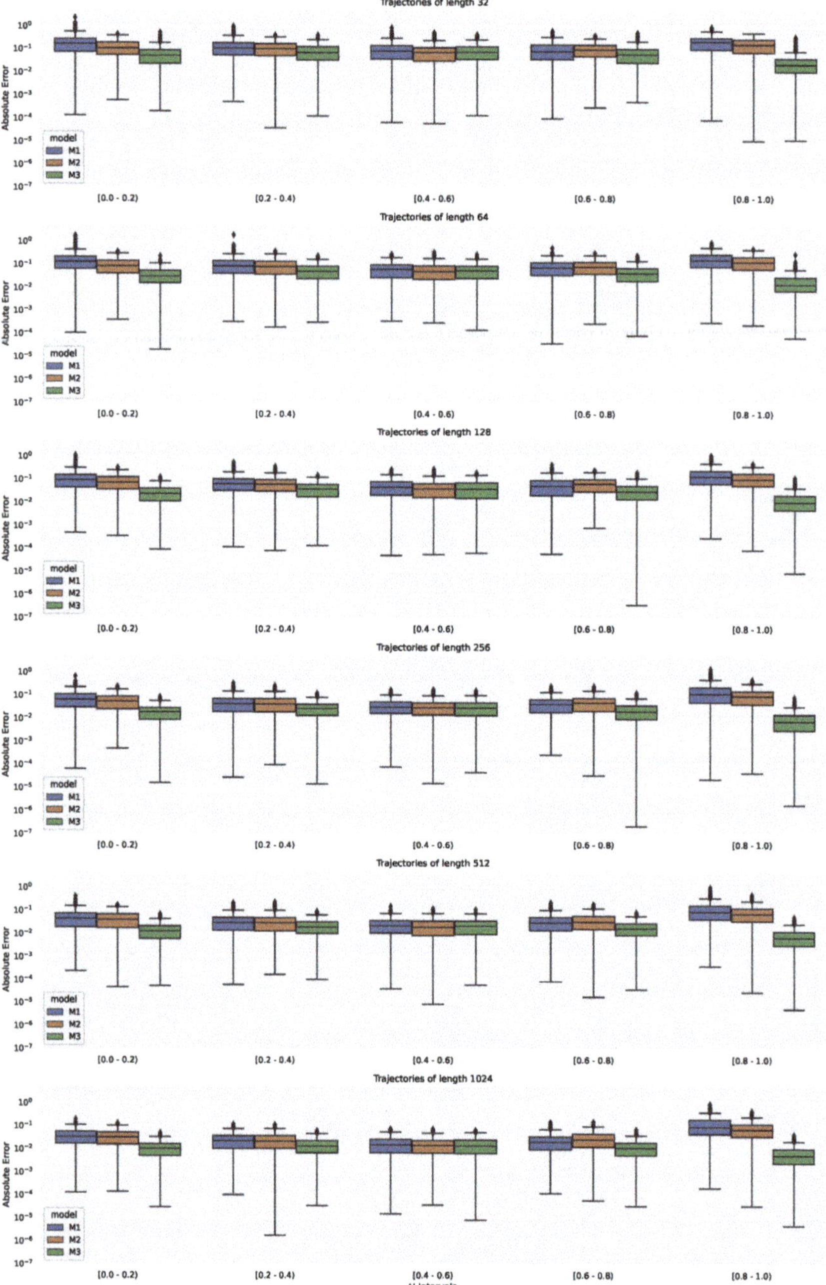

Figure 4. Absolute error calculated for the M1, M2, and M3 methods when fed with simulated input trajectories of different lengths and representing various modes of anomalous diffusion regimes (encoded with the H value).

It would be advantageous here to understand what makes M3 outperform the M2 algorithm. The most straightforward explanation relies on the fact that the input information contributes to outputs in various ways in M2 and M3. Namely, although M2 and M3 use the same input information, M2 explores each lag with the same importance, which is not the case for the FNN. The neural network-based method can learn inhomogeneous relationships contained in historical data, focusing on a specific subset of lags and leaving others with less influence on the observed output. This is exactly how the intelligent unpacking mechanism works in the M3 method.

This phenomenon can be clearly observed in Figure 5, where the MAE is compared for the methods working with trajectories of different lengths, for different H parameter values, and various numbers of lags used during the calculations. In the case of the M1 method, the results are presented only for $\tau = 1$, as there was no possibility to expand or shrink the set of lags here. M2 used lags up to τ_{max} for the Hurst exponent estimation, i.e., ($\tau \in \{0, 1, \ldots, \tau_{max}\}$) with $\tau_{max} \in \{2, 3, \ldots, 31\}$. The number of used lags could not be reduced for the M3 method since it required exactly 32 lags. To overcome this problem, a selected number of the first values of the calculated lags for the ACVF (this number is later referred to as non-zero lags) were extracted. The remaining lags were reset to zero during the calculations. This means that the relationships valid for the raw data were cut at the level of some lags, resembling the diffusion case $H = 0.5$ (i.e., there is no inter-dependency and the ACVF is equal to 0). All of these contributed to weakening the anomalous behavior (filling further lags with zero diminished/removed the long-range dependencies); thus, the reliability of the used methods for prolonged input data sequences could be improved, especially in the case of the M3 algorithm (see Figures 3 and 4).

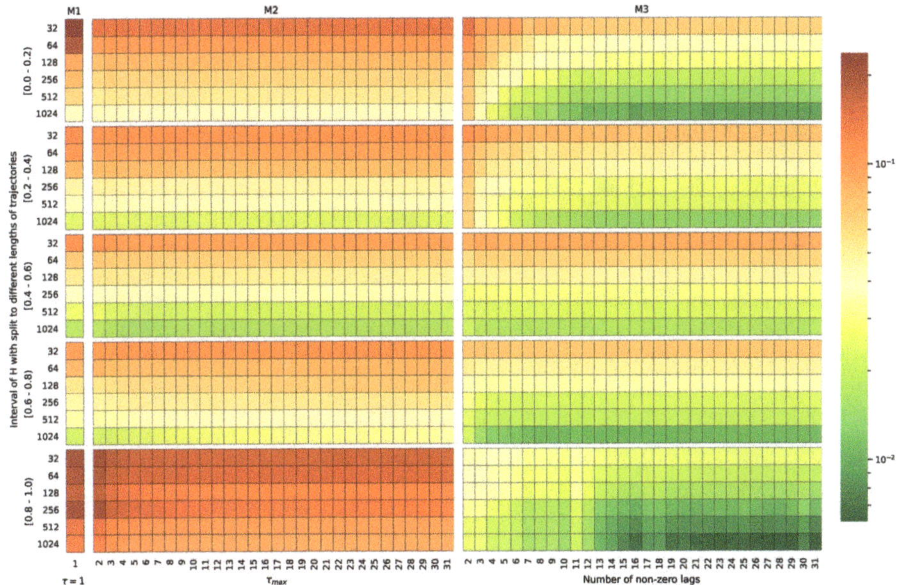

Figure 5. MAE heatmap calculated for the M1, M2, and M3 methods depending on the length of the input trajectory, τ_{max}, and for the following ranges of Hurst exponent values.

Figure 5 proves that using at least 20 lags was sufficient to minimize the error of H estimation in M3. However, the prediction error imperceptibly decreased for M2 with the addition of consecutive lags to the input. One of the explanations for this observation might be that the magnitude of the contribution to the observed outputs (here, the value of the Hurst exponent) was smaller for the higher-order lags than for the first few lags. This means that the feedforward neural network more efficiently unpacked the information about the anomalous diffusion process than the M1 and M2 statistical schemes.

6. Summary and Conclusions

Anomalous diffusion is a complex phenomenon observed in physical systems. This complexity is inherited in recorded data, which, in practice, can be additionally corrupted by measurement noise. Moreover, anomalous diffusion components can emerge from a bunch of other regimes manifested in the observed system (and thus, in recorded data). Unpacking and disentangling the information contained in such data is a challenge, especially because complex interrelations are typically encoded in a small number of data samples. The statistical modeling of anomalous dynamics is quite well established and is concerned with the approximation of nontrivial patterns by the anomalous diffusion exponent. The problem is that for complex systems and the limited a priori knowledge available, the statistical inference can bring about non-robust estimation, especially when noisy components occur in data.

In recent years, there has been increasing interest in the use of AI methods of data exploration in various areas of science and engineering, especially when dealing with complex systems and a limited amount of preliminary information. In this paper, we proposed the application of a simple feedforward neural network to the quantification of anomalous dynamics. Namely, we compared this AI-based approach with statistical modeling, which allowed us to conclude that a combination of these two data exploration methods can decrease the estimation error of the anomalous diffusion exponent.

Herein, we considered the FBM, as it is one of the classical Gaussian processes that can be used for the description of anomalous diffusion behavior. Moreover, the FBM for special cases reduces to the classical BM, and thus is also useful for the analysis of diffusion processes.

Our approach is based on the ACVF, which completely describes the zero-mean Gaussian process. Thus, we selected the sample version of this statistic as a base for the estimation methodology. Via a simulation study, we proved that the classical approaches that utilize the sample ACVF are effective; however, when the process under consideration becomes anomalous diffusion, then their efficiency decreases. Thus, there is a need to incorporate more advanced techniques. In this article, we addressed such problems and proposed a simple modification of the classical ACVF-based approaches through the inclusion of NN-based methodology. Our simulations showed that the introduced estimation method outperformed the other considered approaches, especially for short trajectory lengths. This message is crucial for practical applications, where the real trajectories may be relatively short. In this case, the method based on a combination of ACVF and NN techniques seems to be more effective in contrast to the classical algorithms. It should be highlighted that although the presented approach in this paper is based on the ACVF, this methodology can be applied to any other statistic that is crucial for the estimation of anomalous diffusion processes, e.g., mean squared displacement, p-variation, and ergodicity breaking parameter. Moreover, the potential applicability of the described approach is much wider than the anomalous diffusion field. It could bring new insight into studies of physical systems with various properties reflected and decoded in the corresponding statistical quantities, i.e., long-range dependence, ergodicity breaking, or self-similarity. Additionally, the FBM was considered here only as an exemplary anomalous diffusion process, and the presented approach can be used for any other models with anomalous diffusion behavior. The designed method can also also used not only for estimation purposes, but also for the classification of the anomalous diffusion model governing physical phenomena. It could be directly achieved by the NN learning various ACVF formulas that uniquely define different Gaussian processes. Finally, outside of the Gaussian world, other appropriate statistics that evidently and precisely detect proper non-Gaussian models can be effectively brought into the identification process. Thus, the introduced approach is universal in many areas and can be extended in various directions.

The combination of known statistical algorithms with the deep learning methodology is not a new idea in the area of anomalous diffusion phenomena analysis. Many authors have recognized that simple statistical methods in some cases seem to be inefficient for the proper identification or parametrization of the anomalous diffusion model, especially for recordings with short-length trajectories. In recent years, in the literature regarding anomalous diffusion processes, one can find the application of intelligent methods that enhance the classical approaches. In the physical sciences, deep learning methods have ound very interesting applications. We mention here the new approaches based on the artificial NN algorithms (i.e., [68–73]) or the general machine learning methods applied to fractional dynamics analysis (i.e., [74–77]). However, to the best of our knowledge, a combination of ACVF- and NN-based methods has not been presented in the context of anomalous diffusion analysis.

In a future study, we plan to analyze the influence of additive noise on the estimation results for a model disturbed by external force. The same problem has been considered previously (e.g., in [53]), where the effectiveness of the time-averaged mean square displacement-based approach for the H parameter estimation was analyzed for the FBM with additive noise. Another future study will be related to the combination of the other time-averaged statistics [78] and deep learning methodology for the estimation problem in the anomalous diffusion regime.

Author Contributions: Conceptualization, G.S., I.J., A.W.; methodology, D.S., M.B., I.J., A.W.; software, D.S., G.S., M.B.; writing-original draft preparation, D.S., G.S., M.B., I.J., A.W.; writing-review and editing, D.S., G.S., M.B., I.J., A.W. All authors have read and agreed to the published version of the manuscript.

Funding: This research received no external funding.

Conflicts of Interest: The authors declare no conflicts of interest.

References

1. Nezhadhaghighi, M.G.; Rajabpour, M.A.; Rouhani, S. First-passage-time processes and subordinated Schramm-Loewner evolution. *Phys. Rev. E* **2011**, *84*, 011134. [CrossRef] [PubMed]
2. Failla, R.; Grigolini, P.; Ignaccolo, M.; Schwettmann, A. Random growth of interfaces as a subordinated process. *Phys. Rev. E* **2004**, *70*, 010101(R). [CrossRef] [PubMed]
3. Gabaix, X.; Gopikrishnan, P.; Plerou, V.; Stanley, H. A Theory of Power Law Distributions in Financial Market Fluctuations. *Nature* **2003**, *423*, 267–270. [CrossRef] [PubMed]
4. Ivanov, P.C.; Yuen, A.; Podobnik, B.; Lee, Y. Common scaling patterns in intertrade times of U.S. stocks. *Phys. Rev. E* **2004**, *69*, 05610. [CrossRef] [PubMed]
5. Scher, H.; Margolin, G.; Metzler, R.; Klafter, J. The dynamical foundation of fractal stream chemistry: The origin of extremely long retention times. *Geophys. Res. Lett.* **2002**, *29*, 1061. [CrossRef]
6. Doukhan, P.; Oppenheim, G. (Eds.) *Theory and Applications of Long-Range Dependence*; Birkhäuser Boston, Inc.: Boston, MA, USA, 2003.
7. Golding, I.; Cox, E.C. Physical Nature of Bacterial Cytoplasm. *Phys. Rev. Lett.* **2006**, *96*, 098102. [CrossRef]
8. Mandelbrot, B.B.; Ness, J.W.V. Fractional Brownian Motions, Fractional Noises and Applications. *SIAM Rev.* **1968**, *10*, 422–437. [CrossRef]
9. Stanislavsky, A.; Burnecki, K.; Magdziarz, M.; Weron, A.; Weron, K. FARIMA modelling of solar flar activity from empirical time series of soft X-Ray Solar emission. *Astrophys. J.* **2009**, *693*, 1877–1882. [CrossRef]
10. Zeng, C.; Chen, Y.; Yang, Q. The fBm-driven Ornstein-Uhlenbeck process: Probability density function and anomalous diffusion. *Fract. Calc. Appl. Anal.* **2012**, *15*, 479–492. [CrossRef]
11. Chen, W.; Sun, H.; Zhang, X.; Korošak, D. Anomalous diffusion modeling by fractal and fractional derivatives. *Comput. Math. Appl.* **2010**, *59*, 1754–1758. [CrossRef]

12. Mura, A.; Pagnini, G. Characterizations and simulations of a class of stochastic processes to model anomalous diffusion. *J. Phys. A* **2008**, *41*, 285003. [CrossRef]
13. Piryatinska, A.; Saichev, A.I.; Woyczynski, W.A. Models of anomalous diffusion: The subdiffusive case. *Phys. A Stat. Mech. Its Appl.* **2005**, *349*, 375–420. [CrossRef]
14. Beran, J. *Statistics for Long-Memory Processes*; Chapman & Hall: New York, NY, USA, 1994.
15. Samorodnitsky, G.; Taqqu, M. *Stable Non-Gaussian Random Processes*; Chapman & Hall: New York, NY, USA, 1994.
16. Metzler, R.; Klafter, J. The random walk's guide to anomalous diffusion: A fractional dynamics approach. *Phys. Rep.* **2000**, *339*, 1–77. [CrossRef]
17. Metzler, R.; Klafter, J. The restaurant at the end of the random walk: Recent developments in the description of anomalous transport by fractional dynamics. *J. Phys. A* **2004**, *37*, R161–R208. [CrossRef]
18. Sato, K.I. *Lévy Processes and Infinitely Divisible Distributions*; Cambridge University Press: Cambridge, UK, 1999.
19. Wyłomańska, A. Arithmetic Brownian motion subordinated by tempered stable and inverse tempered stable processes. *Physica A* **2012**, *391*, 5685–5696. [CrossRef]
20. Wyłomańska, A. Tempered stable process with infinitely divisible inverse subordinators. *J. Stat. Mech.* **2013**, *10*, P10011. [CrossRef]
21. Wyłomańska, A.; Gajda, J. Stable continuous-time autoregressive process driven by stable subordinator. *Phys. A Stat. Mech. Its Appl.* **2016**, *444*, 1012–1026. [CrossRef]
22. Magdziarz, M.; Weron, A. Fractional Fokker-Planck dynamics: Stochastic representation and computer simulation. *Phys. Rev. E* **2007**, *75*, 056702. [CrossRef]
23. Magdziarz, M. Langevin Picture of Subdiffusion with Infinitely Divisible Waiting Times. *J. Stat. Phys.* **2009**, *135*, 763–772. [CrossRef]
24. Gajda, J.; Magdziarz, M. Fractional Fokker-Planck equation with tempered alpha-stable waiting times: Langevin picture and computer simulation. *Phys. Rev. E* **2010**, *82*, 011117. [CrossRef]
25. Gajda, J.; Wyłomańska, A. Fokker–Planck type equations associated with fractional Brownian motion controlled by infinitely divisible processes. *Physica A* **2014**, *405*, 104–113. [CrossRef]
26. Bodrova, A.S.; Chechkin, A.V.; Sokolov, I.M. Scaled Brownian motion with renewal resetting. *Phys. Rev. E* **2019**, *100*, 012120. [CrossRef] [PubMed]
27. Thalpa, S.; Wyłomańska, A.; Sikora, G.; Wagner, C.E.; Krapf, D.; Kantz, H.; Chechkin, A.V.; Metzler, R. Leveraging large-deviation statistics to decipher the stochastic properties of measured trajectories. *arXiv* **2020**, arXiv:2005.02099.
28. Fulinski, A. Anomalous Diffusion and Weak Nonergodicity. *Phys. Rev. E* **2011**, *83*, 061140. [CrossRef]
29. Beck, C.; Cohen, E.G.D. Superstatistics. *Physica A* **2003**, *322*, 267–275. [CrossRef]
30. Chechkin, A.V.; Seno, F.; Metzler, R.; Sokolov, I.M. Brownian yet Non-Gaussian Diffusion: From Superstatistics to Subordination of Diffusing Diffusivities. *Phys. Rev. X* **2017**, *7*, 021002. [CrossRef]
31. Sokolov, I.M. Models of anomalous diffusion in crowded environments. *Soft Matter* **2012**, *8*, 9043–9052. [CrossRef]
32. Klages, R.; Radons, G. (Eds.) *Anomalous Transport: Foundations and Applications*; Wiley: Hoboken, NJ, USA, 2008.
33. Hoefling, F.; Franosch, T. Anomalous transport in the crowded world of biological cellsE. *Rep. Prog. Phys.* **2013**, *76*, 046602. [CrossRef]
34. Metzler, R.; Jeon, J.; Cherstvy, A.; Barkai, E. Anomalous transport in the crowded world of biological cellsE. *Phys. Chem. Chem. Phys.* **2014**, *16*, 24128–24164. [CrossRef]
35. Sikora, G.; Burnecki, K.; Wyłomańska, A. Mean-squared displacement statistical test for fractional Brownian motion. *Phys. Rev. E* **2017**, *95*, 032110. [CrossRef]
36. Sikora, G.; Teuerle, M.; Wyłomańska, A.; Grebenkov, D. Statistical properties of the anomalous scaling exponent estimator based on time averaged mean square displacement. *Phys. Rev. E* **2017**, *96*, 022132. [CrossRef] [PubMed]
37. Taqqu, M.S.; Teverovsky, V.; Willinger, W. Estimators for Long-Range Dependence: An Empirical Study. *Fractals* **1995**, *3*, 785–798. [CrossRef]
38. Movahed, M.; Hermanis, E. Fractal analysis driver flow fluctuations. *Phys. A Stat. Mech. Its Appl.* **2008**, *387*, 915–932. [CrossRef]

39. Sikora, G. Statistical test for fractional Brownian motion based on detrending moving average algorithm. *Chaos Solitons Fractals* **2018**, *114*, 54–62. [CrossRef]
40. Carbone, A.; Kiyono, K. Detrending moving average algorithm: Frequency response and scaling performances. *Phys. Rev. E* **2016**, *93*, 063309. [CrossRef]
41. Arianos, S.; Carbone, A. Detrending moving average algorithm: A closed-form approximation of the scaling law. *Phys. A Stat. Mech. Its Appl.* **2017**, *382*, 9–15. [CrossRef]
42. Sikora, G.; Hoell, M.; Gajda, J.; Kantz, H.; Chechkin, A.; Wyłomańska, A. Probabilistic properties of detrended fluctuation analysis for Gaussian processes. *Phys. Rev. E* **2020**, *101*, 032114. [CrossRef]
43. Balcerek, M.; Burnecki, K. Testing of fractional Brownian motion in a noisy environment. *Chaos Solitons Fractals* **2020**, *140*, 110097. [CrossRef]
44. Bishop, C. *Neural Networks for Pattern Recognition*; Oxford University Press: Oxford, UK, 1996.
45. Jabłoński, I. Computer assessment of indirect insight during an airflow interrupter maneuver of breathing. *Comput. Meth. Progr. Biomed.* **2013**, *110*, 320–332. [CrossRef]
46. Kumar, R.; Srivastava, S.; Gupta, J.R.P.; Mohindru, A. Comparative study of neural networks for dynamic nonlinear systems identification. *Soft Comput.* **2019**, *21*, 103–114. [CrossRef]
47. Tutunji, T.A. Parametric system identification using neural networks. *Appl. Soft Comput.* **2016**, *47*, 251–261. [CrossRef]
48. Szymański, J.; Weiss, M. Elucidating the Origin of Anomalous Diffusion in Crowded Fluids. *Phys. Rev. Lett.* **2009**, *103*, 038102. [CrossRef] [PubMed]
49. Weiss, M. Single-particle tracking data reveal anticorrelated fractional Brownian motion in crowded fluids. *Phys. Rev. E* **2013**, *88*, 010101. [CrossRef] [PubMed]
50. Krapf, D.; Lukat, N.; Marinari, E.; Metzler, R.; Oshanin, G.; Selhuber-Unkel, C.; Squarcini, A.; Stadler, L.; Weiss, M.; Xu, X. Spectral Content of a Single Non-Brownian Trajectory. *Phys. Rev. X* **2019**, *9*, 011019. [CrossRef]
51. Brockwell, P.J.; Davis, R.A. *Introduction to Time Series and Forecasting*; Springer: New York, NY, USA, 1994.
52. Sikora, G.; Kepten, E.; Weron, A.; Balcerek, M.; Burnecki, K. An efficient algorithm for extracting the magnitude of the measurement error for fractional dynamics. *Phys. Chem. Chem. Phys.* **2017**, *19*, 26566–26581. [CrossRef]
53. Lanoiselee, Y.; Grebenkov, D.; Sikora, G.; Grzesiek, A.; Wyłomańska, A. Optimal parameters for anomalous diffusion exponent estimation from noisy data. *Phys. Rev. E* **2018**, *98*, 062139. [CrossRef]
54. Borodinov, N.; Neumayer, S.; Kalinin, S.V.; Ovchinnikova, O.S.; Vasudevan, R.K.; Jesse, S. Deep neural networks for understanding noisy data applied to physical property extraction in scanning probe microscopy. *Comput. Mater.* **2019**, *5*, 1–8. [CrossRef]
55. Gupta, S.; Gupta, A. Dealing with noise problem in machine learning data-sets: A systematic review. *Procedia Comput. Sci.* **2019**, *161*, 466–474. [CrossRef]
56. Gong, C.; Liu, T.; Yang, J.; Tao, D. Large-margin label-calibrated support vector machines for positive and unlabeled learning. *IEEE Trans. Neural Netw.* **2019**, *30*, 3471–3483. [CrossRef]
57. Hochreiter, S.; Schmidhuber, J. Long Short-Term Memory. *Neural Comput.* **1997**, *9*, 1735–1780. [CrossRef]
58. Pascanu, R.; Mikolov, T.; Bengio, Y. Understanding the exploding gradient problem. *CoRR* **2012**, *4*, 417.
59. Prajit, R.; Barret, Z.; Quoc, V. Searching for Activation Functions. *arXiv* **2017**, arXiv:1710.05941.
60. Diederik, P.; Jimmy, B. Adam: A Method for Stochastic Optimization. *arXiv* **2014**, arXiv:1412.6980.
61. Lample, G.; Ballesteros, M.; Subramanian, S.; Kawakami, K.; Dyer, C. Neural architectures for named entity recognition. *arXiv* **2016**, arXiv:1603.01360.
62. Sabour, S.; Frosst, N.; Hinton, G. Dynamic Routing Between Capsules. In *Advances in Neural Information Processing Systems 30*; Guyon, I., Luxburg, U.V., Bengio, S., Wallach, H., Fergus, R.,Vishwanathan, S., Garnett, R., Eds.; Curran Associates, Inc.: New York, NY, USA, 2017; pp. 3856–3866.
63. Dosovitskiy, A.; Fischer, P.; Ilg, E.; Hausser, P.; Hazirbas, C.; Golkov, V.; van der Smagt, P.; Cremers, D.; Brox, T. FlowNet: Learning Optical Flow With Convolutional Networks. In Proceedings of the IEEE International Conference on Computer Vision (ICCV), Santiago, Chile, 7–13 December 2015.

64. Mao, X.; Li, Q.; Xie, H.; Lau, R.; Wang, Z.; Smolley, S.P. Least Squares Generative Adversarial Networks. In Proceedings of the IEEE International Conference on Computer Vision (ICCV), Venice, Italy, 22–29 October 2017.
65. Watkins, D. *Fundamentals of Matrix Computations*, 2nd ed.; Pure and Applied Mathematics; Wiley-Interscience: New York, NY, USA, 2002.
66. Craigmile, P. Simulating a class of stationary Gaussian processes using the Davies—Harte algorithm, with application to long memory processes. *J. Time Ser. Anal.* **2003**, *24*, 505–511. [CrossRef]
67. Caruana, R.; Lawrence, S.; Giles, C. Overfitting in neural nets: Backpropagation, conjugate gradient, and early stopping. In Proceedings of the Conference: Advances in Neural Information Processing Systems 13, Denver, CO, USA, 8–13 December 2000; pp. 402–408.
68. Kowalek, P.; Loch-Olszewska, H.; Szwabiński, J. Classification of diffusion modes in single-particle tracking data: Feature-based versus deep-learning approach. *Phys. Rev. E* **2019**, *100*, 032410. [CrossRef]
69. Janczura, J.; Kowalek, P.; Loch-Olszewska, H.; Szwabiński, J.; Weron, A. Classification of particle trajectories in living cells: Machine learning versus statistical testing hypothesis for fractional anomalous diffusion. *Phys. Rev. E* **2020**, *102*, 032402. [CrossRef]
70. Bondarenkol, A.N.; Bugueva, T.V.; Dedok, V.A. Inverse Problems of Anomalous Diffusion Theory: An Artificial Neural Network Approach. *J. Appl. Ind. Math.* **2016**, *3*, 311–321. [CrossRef]
71. Bosman, H.; Iacca, G.; Tejada, A.; Wortche, H.J.; Liotta, A. Inverse Problems of Anomalous Diffusion Theory: An Artificial Neural Network Approach. *Inf. Fusion* **2017**, *33*, 41–56. [CrossRef]
72. Dosset, P.; Rassam, P.; Fernandez, L.; Espenel, C.; Rubinstein, E.; Margeat, E.; Milhiet, P.E. Automatic detection of diffusion modes within biological membranes using backpropagation neural network. *BMC Bioinform.* **2016**, *17*, 197. [CrossRef]
73. Bo, S.; Schmidt, F.; Eichhorn, R.; Volpe, G. Measurement of anomalous diffusion using recurrent neural networks. *Phys. Rev. E* **2019**, *100*, 010102. [CrossRef] [PubMed]
74. Munoz-Gil, G.; Garcia-March, M.A.; Manzo, C.; Martin-Guerrero, J.; Lewenstein, M. Single trajectory characterization via machine learning. *New J. Phys.* **2020**, *22*, 013010. [CrossRef]
75. Arts, M.; Smal, I.; Paul, M.W.; Wyman, C.; Meijering, E. Particle Mobility Analysis Using Deep Learning and the Moment Scaling Spectrum. *Sci. Rep.* **2019**, *9*, 17160. [CrossRef] [PubMed]
76. Wagner, T.; Kroll, A.; Haramagatti, C.R.; Lipinski, H.G.; Wiemann, M. Classification and Segmentation of Nanoparticle Diffusion Trajectories in Cellular Micro Environments. *PLoS ONE* **2016**, *12*, e0170165. [CrossRef]
77. Granik, N.; Weiss, L.E.; Nehme, E.; Levin, M.; Chein, M.; Perlson, E.; Roichman, Y.; Shechtman, Y. Single-Particle Diffusion Characterization by Deep Learning. *Biophys. J.* **2019**, *117*, 185–192. [CrossRef]
78. Maraj, K.; Szarek, D.; Sikora, G.; Balcerek, M.; Wyłomańsk, A.; Jabłoński, I. Measurement instrumentation and selected signal processing techniques for anomalous diffusion analysis. *Meas. Sens.* **2020**. [CrossRef]

Publisher's Note: MDPI stays neutral with regard to jurisdictional claims in published maps and institutional affiliations.

MDPI

Article

Extracting Work Optimally with Imprecise Measurements

Luis Dinis [1,2,*] and Juan Manuel Rodríguez Parrondo [1,2]

1 Grupo Interdisciplinar de Sistemas Complejos, Facultad de Ciencias Físicas, 28040 Madrid, Spain; parrondo@ucm.es
2 Departamento de Estructura de la Materia, Física Térmica y Electrónica, Universidad Complutense de Madrid, 28040 Madrid, Spain
* Correspondence: ldinis@ucm.es

Abstract: Measurement and feedback allows for an external agent to extract work from a system in contact with a single thermal bath. The maximum amount of work that can be extracted in a single measurement and the corresponding feedback loop is given by the information that is acquired via the measurement, a result that manifests the close relation between information theory and stochastic thermodynamics. In this paper, we show how to reversibly confine a Brownian particle in an optical tweezer potential and then extract the corresponding increase of the free energy as work. By repeatedly tracking the position of the particle and modifying the potential accordingly, we can extract work optimally, even with a high degree of inaccuracy in the measurements.

Keywords: confinement; information theory; Brownian particle; stochastic thermodynamics

Citation: Dinis, L.; Parrondo, J.M.R. Extracting Work Optimally with Imprecise Measurements. *Entropy* **2021**, 23, 8. https://dx.doi.org/10.3390/e23010008

Received: 30 October 2020
Accepted: 19 December 2020
Published: 23 December 2020

1. Introduction

Modern techniques have allowed for the manipulation of objects at the microscale. A paradigmatic example are colloidal particles trapped by optical tweezers. At this scale—the scale of Brownian motion—not only the motion of particles, but the energy fluxes, work, or heat, become stochastic. Nevertheless, the combination of manipulation and imaging or other detection techniques allow for some degree of control [1]. For instance, in driven systems, the external driving may be modified based on outcomes of measurements, as in feedback control, leading, for example, to (efficient) confinement in small region of space [2] or to the reduction of thermal fluctuations, i.e., cooling, a technique that is implemented in both classical or quantum systems [3,4]. Another application of feedback is an increase of the performance of certain motors operating at the microscale, such as Brownian ratchets or micro-motors [5–9].

Feedback exploits the information that is acquired through measurement as a thermodynamic resource. It is now known that the work needed to perform an isothermal feedback process, for a system in contact with an environment at constant temperature T, is bounded by the following extension of the second law of thermodynamics [9,10]:

$$W \geq \Delta F - kTI, \tag{1}$$

where ΔF is the free energy difference between the final and initial states of the process, k the Boltzmann's constant, and I is the amount of information that is gained in the measurement, quantified by the mutual information from information theory. Information is always positive (or zero) and, thus, in a cycle ($\Delta F = 0$) it is possible to extract work ($W < 0$) from a single thermal bath with measurement and feedback.

Equation (1) also shows that a given level of accuracy in the measurement, quantified by the mutual information, limits the amount of work that can be extracted in one feedback operation. Some especially tailored protocols saturate that bound (1) and they may be used to convert all of the information acquired into useful work. These are processes that are reversible under feedback [11–13]. In this article, we first review these protocols and show

how to use their special properties in order to extract energy with the same efficiency and even power when operating with higher measurement errors. In order to fix ideas, we use a well known model that we proceed to describe in the following section.

2. Model Description and Cycle Operation

As our system, we consider an overdamped Brownian particle that is in contact with a thermal bath, which acts as its environment. The particle feels a harmonic potential. This is a well proven theoretical model for an experimental system that was formed by a colloidal particle in water at constant temperature and trapped by optical tweezers. The potential $V_{\kappa,x_0} = \frac{1}{2}\kappa(x - x_0)^2$ has tunable parameters x_0, the position of the center of the trap, and κ, the stiffness. As the Brownian particle position fluctuates, the energy transfers and thermodynamic potentials become also fluctuating; in fact, they are stochastic variables. The framework to analyze energetics for these fluctuating systems in the mesoscale is stochastic thermodynamics [14–16] .We review the main concepts in the following. The Brownian particle may, due to a collision with the solvent, absorb some energy and climb the potential well. Or it may transfer energy back to the thermal bath via viscosity and go down in the potential. These energy transfers with the thermal bath constitute heat $\hat{Q}$ and, and since this energy can be stored as potential energy, this is the internal energy $\hat{E}$ of the particle. In our system then the internal energy is $\hat{E} = V_{\kappa,x_0} = \frac{1}{2}\kappa(x - x_0)^2$ [17–20]. We will use $\hat{a}$ to denote a stochastic variable and the regular letter a for the average over realizations, i.e., $E = \langle \hat{E} \rangle$. Another form of energy transfer is work $\hat{W}$: an external agent may modify the harmonic potential (changing the parameters) and increase or decrease the potential energy of the particle. If the internal energy depends on a parameter λ that is modified from λ_0 to λ_f then, formally, the definition of work is:

$$\hat{W} = \int_{\lambda_0}^{\lambda_f} \frac{\partial E}{\partial \lambda} d\lambda \tag{2}$$

This is best seen with an example. For instance, consider a fast increase of the stiffness of the potential from k_i to k_f. If the increase is very fast, so that the particle does not modify its position x during the time, the stiffness is changing, the energy of the particle increases by an amount $\Delta V = \frac{1}{2}(x - x_0)^2(k_f - k_i)$. This energy is supplied by the agent controlling the potential who has then performed a work $\hat{W} = \Delta V > 0$. Consequently, the particle is in a tighter parabola and the equilibrium dispersion of the position of the particle decreases, so that this is commonly referred to as a compression. If the stiffness is decreased, work is exerted on the agent by the system and, since the distribution of particle positions will eventually widen, this corresponds to an expansion.

With these definitions, energy is conserved and the first law is fulfilled either at the level of trajectories $\hat{\Delta E} = \hat{Q} + \hat{W}$ or as averages $E = Q + W$ [14–20].

In order to extract energy from the thermal bath, we propose the following cyclic operation in two stages:

1. Confinement of the Brownian particle by (repeated) measuring and feedback
2. Isothermal expansion

The system works as a motor if the work obtained in the isothermal expansion exceeds the work that is needed for confinement. During a compression, the free energy of the system increases (due to the entropy decrease). Using reversible feedback confinement [2], we can minimize the work that is needed for stage 1, which turns out to vanish, and extract all of the free energy increase of stage 1 as work during stage 2. Let us analyze each stage in more detail.

2.1. Optimal Confinement

The confinement of a system to a small region of the phase space (at constant temperature) implies a decrease of entropy of the system. For the entropy of a Brownian particle, we use the standard choice of Shannon's entropy, $S = -k \int \rho(x) \log(x) dx$, where $\rho(x)$

is the probability distribution of the particle position. With this choice, the second law of thermodynamics is fulfilled on average and the thermodynamic relation $F = E - TS$ is recovered for a system in contact with a thermal bath. Although, strictly speaking, this is a generalization of the free energy to non-equilibrium systems, in systems that are in contact to a thermal bath it plays a similar role as the standard thermodynamic free energy, and stochastic thermodynamics for our system closely resembles macroscopic thermodynamics [9].

Let us consider, for simplicity, that the internal energy change between the initial and final states of the confinement process vanishes (we will see later that this is the case in our particular system). A reduction of entropy then corresponds to an increase of free energy $\Delta F = \Delta E - T\Delta S$. This increase in free energy could then be extracted as work in an isothermal expansion. However, the whole process cannot operate as a motor, as this will defeat the second law (extracting work from a single thermal bath). Indeed, the second law states for the confinement

$$W_1 \geq \Delta F_1 \text{ (w/o. feedback)} \tag{3}$$

and then for the isothermal expansion back to the initial state ($\Delta F_2 = -\Delta F_1$)

$$W_2 \geq \Delta F_2 = -\Delta F_1 \text{ (w/o. feedback)} \tag{4}$$

so that $W_{\text{total}} = W_1 + W_2 \geq 0$ and the system dissipates energy into the thermal bath.

However, as explained above, when measuring and feeding back to the system, W is bounded by (1) instead. Thus, the work that is needed for the confinement may be reduced and the work output of the cyclic process ($\Delta F_{\text{cycle}} = 0$) may be negative:

$$W_{\text{total}} \geq -kTI \tag{5}$$

Notice that mutual information is always a positive quantity.

Following [2], we propose a reversible feedback confinement that can confine the particle with $W_1 = 0$ and, as will be shown later (see Equation (13)), without dissipating heat to the thermal bath, so that the increase in free energy that is produced by the confinement can later be completely recovered as work during a quasistatic expansion in stage 2.

For a system that is in contact with a thermal bath, a feedback process is reversible if the Hamiltonian is modified after the measurement, so that probability of the state of the system conditioned on the measurement outcome is the Gibbsian state of the new Hamiltonian. After a measurement, the probability to find a given state changes instantaneously, the new probability distribution takes into account the information obtained, and ut must be updated according to Bayesian inference. If the Hamiltonian also changes rapidly and the Gibbs state of the new Hamiltonian matches the posterior probability distribution, the system remains at equilibrium and no further evolution of the probability distribution ensues until a new measurement is taken.

In our model, we take the common assumption of Gaussian measurement errors. If the particle is located at a position x, then the measurement outcome m is Gaussian distributed around x and the dispersion σ_m quantifies the measurement error:

$$q(m|x) = \frac{1}{\sqrt{2\pi\sigma_m^2}} e^{-(m-x)^2/2\sigma_m^2} \tag{6}$$

After a measurement, the probability distribution of the position of the particle updates according to Bayes' theorem from the initial distribution ρ:

$$\rho'(x|m) = \frac{\rho(x)q(m|x)}{\pi(m)} \tag{7}$$

where $\pi(m) = \int dx q(m|x)\rho(x)$ is the marginal distribution of the measurement outcome.

For a Brownian particle in a time-independent potential, the equilibrium distribution is its corresponding Gibbs distribution:

$$\rho(x) \propto e^{-V(x)/kT} \tag{8}$$

In a harmonic potential, it is a Gaussian, centered in the trap position x_0 and with variance being given by $\sigma^2 = kT/\kappa$. It can be shown [7] that, after a measurement, the new distribution that is computed according to (7) remains Gaussian. If the initial distribution has mean $\bar{x}$ and standard deviation σ, after a measurement, then the distribution updates to a Gaussian with the mean and deviation given by [2]:

$$\bar{x}'(m) = \frac{\sigma_m^2}{\sigma^2 + \sigma_m^2}\bar{x} + \frac{\sigma^2}{\sigma^2 + \sigma_m^2} m \tag{9}$$

$$\frac{1}{\sigma'^2} = \frac{1}{\sigma^2} + \frac{1}{\sigma_m^2} \tag{10}$$

We can make the post-measurement distribution an equilibrium distribution by setting a new center of the trap position x_0' and stiffness κ', as

$$\kappa' = kT/\sigma'^2 \tag{11}$$

$$x_0'(m) = \bar{x}'(m) \tag{12}$$

Notice that $\kappa' > \kappa$; hence, the particle is more tightly bound or confined after this change. Additionally, Equation (10) implies $\sigma' < \sigma$, so that every measurement and feedback step further reduces the variance of the particle distribution.

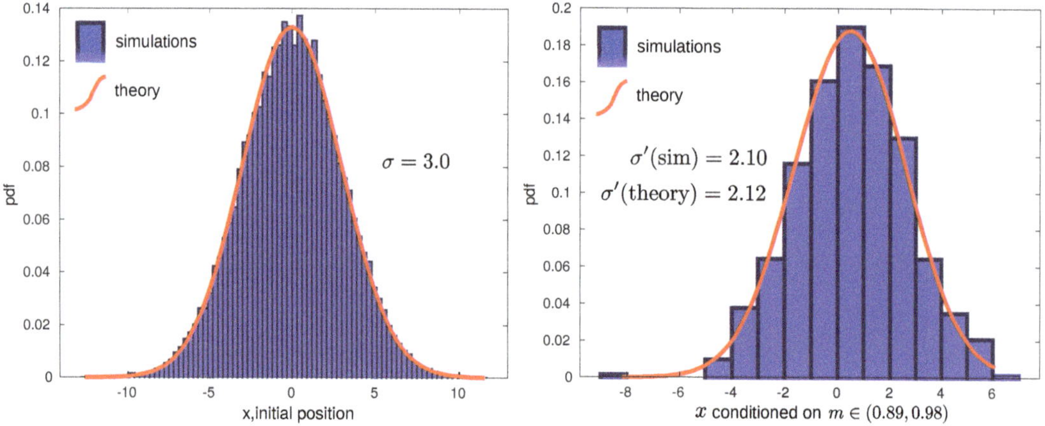

Figure 1. Reduction of variance after measuring. **Left**: Initial distribution. Histogram of 40,000 random Gaussian numbers centered in 0 with standard deviation $\sigma = 3.0$ (blue bars) and a theoretical Gaussian distribution with the same parameters (red continuous line). **Right**: Posterior distribution. Histogram of particle positions with measurement outcomes in a given interval (0.89, 0.98) (blue bars) and prediction according to Bayes' theorem (7) (red). Measurement outcomes were performed with measurement error $\sigma_m = 3.0$. Using $\sigma_m = 3.0$ and $\sigma = 3.0$ in (10) gives $\sigma' = 2.12$, which matches the sample standard deviation of 2.10.

In order to check this reduction of variance in simulation, we have computed the particle distribution after a measurement. For this, we first generate a large number of trajectories, starting from an initial equilibrium distribution for a harmonic potential centered in position $x_0 = 0$ and corresponding dispersion $\sigma = 3.0$, as depicted in Figure 1 (left). After some time interval, for each trajectory, we measure its position by generating a (Gaussian) random measurement outcome m around the actual position x with dispersion

$\sigma_m = 3.0$ (see the details in Section 5). We can then fix a small interval around a given measurement of the position $(m, m + \Delta m)$ of our choice, for instance (0.89, 0.98), and only check the realizations that gave a measurement in that interval. The distribution of the actual positions x of these particular realizations are distributed as in (7). In our case, a Gaussian with new reduced standard deviation $\sigma' = 2.12$ is given by (10). This can be seen in Figure 1 (right).

This process of measurement and feedback can be repeated and a new, more confined state could be achieved. Figure 2 (top) shows the confining effect of repeating this procedure.

In every measurement and feedback step, the trapped Brownian particle stays in equilibrium with the thermal bath at temperature T. Consequently, the average energy is not modified by the feedback process. The average internal energy E of a trapped particle in one dimension is given by the equipartition theorem, as $E = kT/2$. Because the process is isothermal, $\Delta E = 0$. On the other hand, always being in equilibrium, there is no relaxation of the particle distribution and the heat that is transferred from the heat bath vanishes on average $Q = 0$. Therefore, according to the first law, the average work done on the system also vanishes:

$$\Delta E_1 = Q_1 + W_1 = 0 \Rightarrow W_1 = -Q_1 = 0 \quad (13)$$

This has been checked in simulations, as shown in Figure 2 (bottom). Details about work computation during measurement and feedback can be found in Section 5.

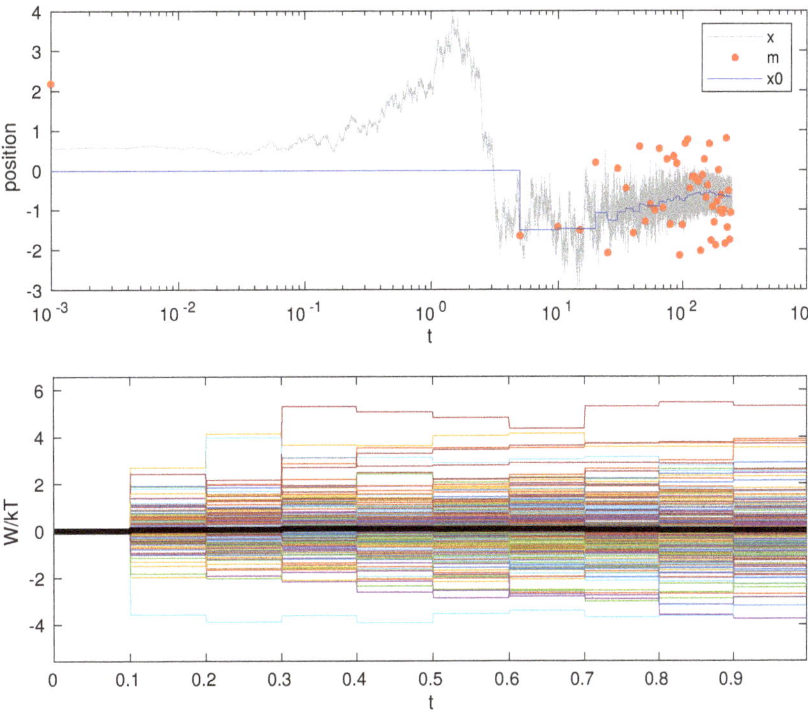

Figure 2. Confinement. **Top**: particle trajectory (gray line), measurement outcome (red dots) and trap center position (blue line). **Bottom**: Cumulative work for different realizations (color lines) and its average over 200 realizations (thick black line) for confinement in 10 measurement steps. See the simulation details in Section 5. Initial trap stiffness $\kappa = 0.1$ and position $x_0 = 0$. Initial condition is equilibrium with trap potential to avoid transient due to equilibration. Particle diffusion coefficient $D = 1$ and friction $\gamma = 1$.

In general, for other feedback protocols where the stiffness of the trap is suddenly changed, work is performed, on average [21], as in the simple example described after Equation (2). The feedback process used here is different (in addition to a sudden increase of stiffness, trap position is also modified in a precisely combined manner) and it is special in the sense that average work vanishes. As encoded in Equation (1), this can solely be achieved by using information regarding the position through measurement in the feedback (see Equation (9) for the new trap position). To see why this matters, consider our Brownian particle in a harmonic potential, where the observer happens to know that the particle is exactly at the bottom of the well. This would allow for this external agent to increase the stiffness of the potential well with an abrupt change, without performing work, since the energy of the particle is always zero at the bottom of the well, for any stiffness. The confining protocol is a refinement of this idea that works for any position of the particle, by displacing the bottom of the potential well towards the measured particle position and changing the stiffness in a suitable manner.

Furthermore, one can compute the mutual information that is obtained in the process of measurement and evaluate the increase in free energy ΔF_1 for the confinement stage. From the definition of mutual information:

$$I(m,x) = \int \pi(m) q(m|x) \log \frac{q(m|x)}{\pi(m)} \tag{14}$$

When considering that the measurement outcome distribution $q(m|x)$ and the marginal distribution $\pi(m)$ are Gaussian with variance σ_m^2 and $\sigma^2 + \sigma_m^2 = \sigma_m^2 \sigma^2 / \sigma'^2$, respectively, the information that is acquired in a measurement is

$$I(m,x) = -\frac{1}{2} \log \frac{\sigma'^2}{\sigma^2} \geq 0 \tag{15}$$

Mutual information intuitively measures the decrease in uncertainty of variable x if we know the value of m, or vice versa [22]. In our case, from (10), if the measurement error σ_m is very large then $\sigma' \approx \sigma$ and we extract almost no information from measuring ($I \approx 0$). Conversely, for infinite precise measurement $\sigma_m \to 0$, then $\sigma' \to 0$, and we obtain infinite information from a measurement, as an infinite precise description of a position would require an infinite number of bits to store it.

The entropy of a Gaussian of variance σ^2 is $S(\rho) = k \log \sigma \sqrt{2\pi e}$. In the measurement process, the distribution changes from a Gaussian of variance σ^2 to a Gaussian of variance σ'^2, and we have

$$\Delta S_{1\text{step}} = k \log \sigma' \sqrt{2\pi e} - k \log \sigma \sqrt{2\pi e} = k \frac{1}{2} \log \frac{\sigma'^2}{\sigma^2} = -k I(m,x) \tag{16}$$

Because $\Delta E = 0$, we finally obtain

$$\Delta F_{1\text{step}} = \Delta E - T \Delta S_{1\text{step}} = kT I(m,x). \tag{17}$$

This is valid for every measurement and feedback step while using the reversible feedback protocol. In a sequence of confinement steps with successive variances $\sigma_0, \sigma_1, \ldots, \sigma_n$, the total information is

$$I_{\text{total}} = -\frac{1}{2} \sum_{i=1}^{n} \log \frac{\sigma_i^2}{\sigma_{i-1}^2} = -\frac{1}{2} \log \frac{\sigma_n^2}{\sigma_0^2}. \tag{18}$$

σ_n^2 can be obtained from (10) by recursion, giving:

$$\frac{1}{\sigma_n^2} = \frac{1}{\sigma_0^2} + n \left(\frac{1}{\sigma_m^2} \right) \tag{19}$$

Finally, the free energy difference between the final and initial states in the confinement stage is

$$\Delta F_1 = kT I_{\text{total}} = \frac{kT}{2} \log \frac{\sigma_0^2}{\sigma_n^2} = kT I_{\text{total}}. \tag{20}$$

Every bit of information that is extracted in the measurement is turned into an increase of free energy during the confinement stage and it can be converted into useful work in the subsequent expansion.

2.2. Work Extraction by Isothermal Expansion

If an external agent changes the stiffness of the optical trap from κ_i to $\kappa_f < \kappa_i$, energy is recovered as work, as explained above. In a quasistatic process, the work done by the system is given by the free energy difference. Because stage 2 completes the cycle of operation of the motor ending in the initial state, we have $\Delta F_2 = -\Delta F_1$ and

$$W_{\text{total}} = W_1 + W_2 = 0 - \Delta F_1 = -kT I_{\text{total}}, \tag{21}$$

which corresponds to extracted work. In fact, it saturates expression (5) and it is the maximum possible work that can be extracted while using the information that was obtained in the measurements.

This result can also be recovered by the direct computation of the work of a process changing stiffness from κ_i to κ_f and while taking into account that, for a quasistatic process, one can use the equipartition theorem stating $\langle x^2 \rangle = kT/\kappa(t)$, with $\kappa(t)$ the instantaneous value of the stiffness. Subsequently, the average work during the expansion, according to (2), reads:

$$W_2 = \int_{\kappa_i}^{\kappa_f} d\kappa \frac{\langle x^2 \rangle}{2} = \frac{kT}{2} \int_{\kappa_i}^{\kappa_f} \frac{d\kappa}{\kappa} = \frac{kT}{2} \log \frac{\kappa_f}{\kappa_i} \tag{22}$$

The expansion starts at the end of the confinement process with a distribution of variance σ_n and ends at σ_0. Subsequently, while using the relation between stiffness and variance in the confinement stage (11), we have

$$W_2 = \frac{kT}{2} \log \frac{\kappa_f}{\kappa_i} = -\frac{kT}{2} \log \frac{\sigma_n^2}{\sigma_0^2} = -kT I_{\text{total}} \tag{23}$$

Notice that during both the confinement and expansion the system must be at equilibrium in order to transform every bit of information into useful work.

In practice, though, for a process changing the stiffness of the potential to be approximately quasistatic, it is enough that the time of the process is large compared to the inverse frequency of the trap given by $\nu = \kappa/\gamma$. This is the criterion that we have used for simulations. Additionally, it is worth noting that, even though the work in every realization of the expansion may differ in principle in a stochastic system, work is—in this particular example—a self-averaging quantity: for a quasistatic expansion, the total work obtained in any realization is very similar to its average value. The argument for self-averaging of the work is the following: from work definition (2), work in a single realization when expanding is $\hat{W} = \int \frac{x^2}{2} d\kappa$. If the expansion is very slow, in the time κ is modified a certain small amount, the particle position has time to fluctuate and sample the whole quasi-equilibrium distribution and x^2 approximately can be replaced by its average value (see the full computation in [14]).

Figure 3 depicts the complete diagram of the proposed cycle.

Finally, one could also define an efficiency η as the ratio between the extracted thermodynamic resource (work) and the thermodynamic resource consumed to make the engine run, in this case information. With this definition, this reversible feedback engine attains the maximum efficiency:

$$\eta = \frac{-W}{kTI} = 1 \tag{24}$$

as in a similar system [23], with just one measurement per cycle.

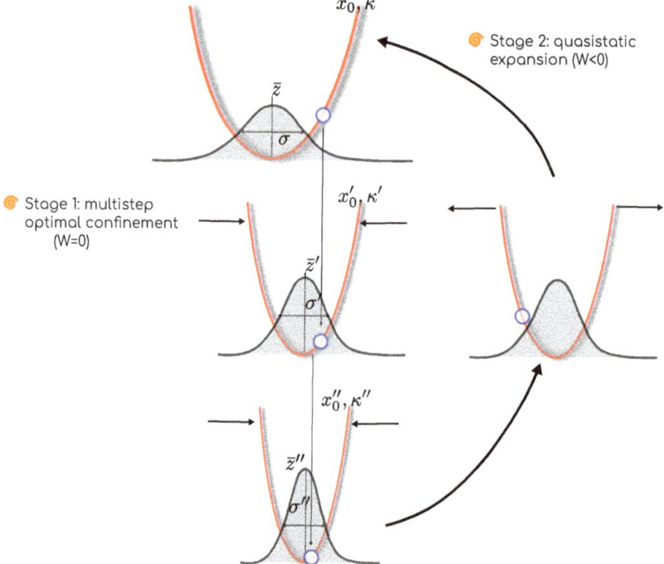

Figure 3. Cycle for extracting work from a thermal bath with inaccurate measurements.

3. Results

3.1. Work Is Optimal

We have performed computer simulations of the model system that is described above. Figure 4 depicts part of two consecutive cycles, each of them with a confinement stage that is composed of 10 measurement and feedback steps, followed by an isothermal expansion. The top panel depicts the particle position (gray line), trap center (blue line), and measurement outcomes (red dots), whereas the bottom panel shows the evolution of the stiffness along the cycle.

Figure 5 shows the cumulative work that was done on the system along the time of a single cycle. The thick solid line represents the average over 200 cycles. Every cycle consists of a confinement that is achieved by measuring the particle position 10 times and the subsequent isothermal expansion. Average work extracted ($W < 0$) by the end of the cycle approaches the expected result that is given by Equations (18), (19) and (21), marked with dashed black line. The shaded area represents the variance of the work, which is substantially large. As is apparent from the figure, most of the variance comes from the confinement step, with the quasistatic work being a self-averaging quantity. Finally, work that corresponds to two particular cycles is shown by thin blue lines.

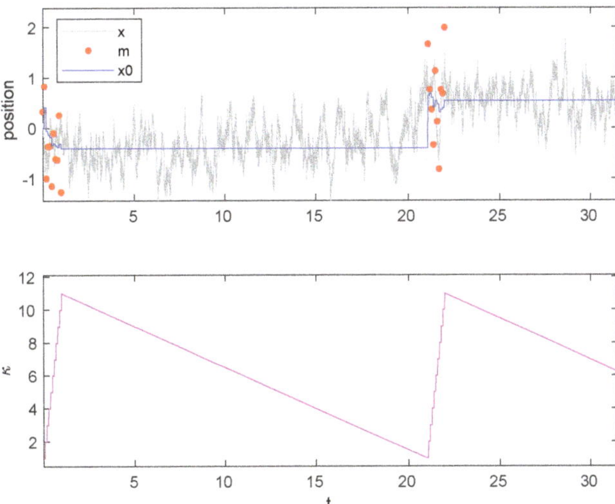

Figure 4. Trajectories. (**Top**) Particle trajectory (gray continuous line), trap center (blue continuous line), measurement outcomes (red dots). (**Bottom**) Stiffness evolution during the cycle. Every cycle starts with $\kappa = 1$, there are 10 measurement steps, followed by quasistatic expansion. $D = 1, \gamma = 1$.

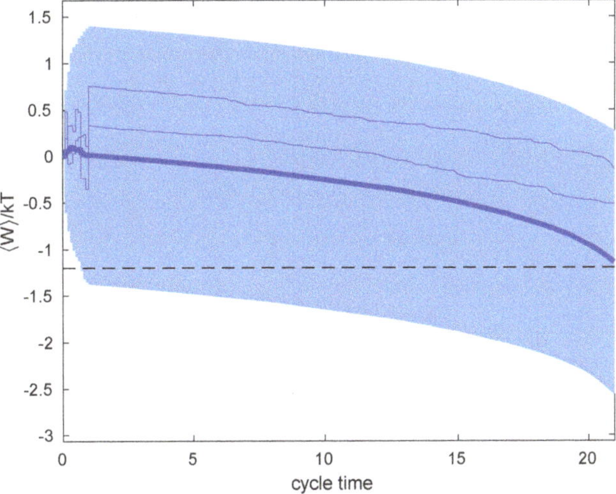

Figure 5. Average cumulative work along the confinement-expansion cycle (thick blue line) computed from 200 realizations. The shaded area corresponds to one standard deviation from the average. Thin blue lines represent cumulative in two representative cycles. Simulation parameters are: $\Delta t = 0.001$, time between measurements $\tau = 0.1$, number of measurements before expansion is 10, measurement error $\sigma_m^2 = 1$, initial stiffness of the trap $\kappa = 1$, diffusion coefficient $D = kT/\gamma = 1$, and drag coefficient $\gamma = 1$.

3.2. *Power and Efficiency with Higher Measurement Errors*

Consider two setups, A and B, with different measurement errors being given by variances σ_{mA}^2 and $\sigma_{mB}^2 = 2\sigma_{mA}^2$. Suppose that only one measurement step is performed in each system before the expansion. According to our discussion above, the measurement information that can be later transformed into work is smaller in system B than in A:

$$I_{B1} = \frac{1}{2}\log\frac{\sigma_{mB}^2 + \sigma_0^2}{\sigma_{mB}^2} = \frac{1}{2}\log\left(1 + \frac{\sigma_0^2}{2\sigma_{mA}^2}\right) < \frac{1}{2}\log\left(1 + \frac{\sigma_0^2}{\sigma_{mA}^2}\right) = I_{A1} \tag{25}$$

However, we can obtain as much information in system B with two measurements as in system A with one measurement. After two measurements, while using the reversible confinement protocol, the variance of the equilibrium distribution σ_{B2}^2 in system B is equal to the variance in system A after one measurement σ_{A1}^2:

$$\frac{1}{\sigma_{B2}^2} = \frac{1}{\sigma_0^2} + 2\left(\frac{1}{\sigma_{mB}^2}\right) = \frac{1}{\sigma_0^2} + \left(\frac{1}{\sigma_{mA}^2}\right) = \frac{1}{\sigma_{A1}^2} \tag{26}$$

Using (18), we obtain:

$$I_B = \frac{1}{2}\log\frac{\sigma_{B2}^2}{\sigma_0^2} = \frac{1}{2}\log\frac{\sigma_{A1}^2}{\sigma_0^2} = I_A \tag{27}$$

As explained above, this implies that the same work can be extracted in the subsequent quasistatic expansion. In fact, bothof the systems run with the same efficiency $\eta = 1$; hence, every bit of information is turned into work in the expansion. Furthermore, system B can also be run in principle at the same power as system A. During the confinement process, after the adjustment of the potential, the particle distribution is at equilibrium. No relaxation occurs, as explained previously. Therefore, a new measurement and feedback step could, in principle, be taken immediately after, in rapid succession. Thus, halving the time between measurements in system B as compared to system A ensures the same cycle time. As the work obtained is also the same, both of the systems operate with the same power. Figure 6 depicts this, where we show the simulation results for system A with one measurement and expansion and system B with two (faster) measurements and expansion. Approximately the same work is obtained in both systems. For reference, we have also marked the expected extracted work for a system with tge measurement error given by σ_{mB}, but using just one measurement.

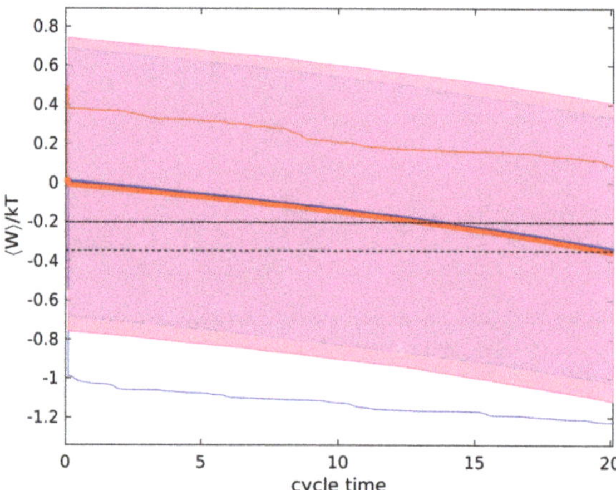

Figure 6. Average work extraction for two different measurement errors, using one measurement with variance $\sigma_{mA}^2 = 1$ (blue thick line), and using two measurements with variance $\sigma_{mB}^2 = 2$ (red thick line). Dashed line represents expected work extraction and fine dashed line corresponds to expected work extraction with just 1 measurement of variance σ_{mB}. Thin lines represent single realizations of the work in system A (blue) and B (red).

4. Discussion

Reversible feedback confinement is an optimal way of reducing the entropy of a system to be later used for work extraction. Nevertheless, it requires a high degree of control over the Hamiltonian, to adapt it to the new probabilistic state after the measurement. This might be a limitation for experimental realizations, although a low dissipation is expected, even if a similar or approximate protocol is implemented. Theoretically, the dissipation could be accounted for by using the Kullback–Leibler distance between the post-measurement particle distribution and the equilibrium distribution of the potential after feedback [24] , if they were different due to a less precise tuning of the potential.

In principle, for a measurement and feedback protocol, imprecision in the measurement, which will inevitably arise in an experimental setup, will limit the work extraction or power. Nevertheless, we have shown here that this limitation can be overcome by adding more measurement steps before the quasistatic expansion, as long as the reversible feedback confinement protocol is used. In principle, the application of this protocol is instantaneous. In practice, this means that the confinement may be applied in a very short time, limited maybe by the response time of the feedback mechanism or the measurement acquisition time. Thus, if the response times of measurement, feedback, and Hamiltonian modification are fast as compared to system's relaxation time, optimal work extraction is feasible, even with a high degree of inaccuracy in the measurement, while using repeated optimal feedback.

5. Materials and Methods

The confined Brownian particle evolves according to Langevin equation:

$$\gamma \dot{x} = -V'_{\kappa,x0}(x) + \xi(t), \tag{28}$$

with $\xi(t)$ Gaussian white noise $\langle \xi(t)\xi(t') \rangle = 2kT\gamma\delta(t-t')$, T bath temperature and k Boltzmann's constant. The potential $V_{\kappa,x0}(x)$ is defined above and it is controlled through measurement and feedback. Model simulations were performed in C language, solving the Langevin evolution equation with the Heun method for a stochastic differential equation [25]. We provide, in the following, some details on work computation, measurement, and feedback steps. For full details, the code is available here: http://seneca.fis.ucm.es/ldinis/code/extract_optimal_work.zip.

- Measurement. In order to perform a measurement in the simulation, a Gaussian number "r" with zero average and standard deviation 1 is generated. Subsequently, if particle position is x, the measurement outcome m is

$$m = x + \sigma_m r \tag{29}$$

  Notice that m is then distributed according to Equation (6)
- Feedback. Immediately after measurement, and using the measurement outcome m just computed, the potential parameters κ and x_0 are recomputed according to Equations (9) to (12). Notice that the old values need to be stored in an auxiliary variable for the work computation, as explained in the following step.
- Work computation during feedback process. According to its definition for a trajectory, work is the difference in the potential energy when the potential is changed. If $\kappa \to \kappa'$ and $x_0 \to x'_0$ as a result of measurement and feedback, then work is computed as

$$\Delta W = \frac{1}{2}\kappa'(x - x'_0)^2 - \frac{1}{2}\kappa(x - x_0)^2 \tag{30}$$

  This ΔW is added to a variable W that stores the cumulative work that was done along the whole simulation.
- After the feedback, evolution equation resumes with the new potential parameters.

- Work during expansion. Work is also performed as a result of the change in κ during an expansion. In the simulation, during the expansion stage, κ changes an amount $\Delta\kappa = \frac{\kappa_f - \kappa_i}{Nexp}$ in every time step, where $Nexp$ is the number of time steps of the expansion. Therefore, in a time step, a work

$$\Delta W = \frac{1}{2}\Delta\kappa(x - x_0)^2 \tag{31}$$

is performed. Again, this ΔW has to be added to the variable W, which stores the total or cumulative work of the whole process.

Author Contributions: Conceptualization, L.D. and J.M.R.P.; methodology, L.D.; software, L.D.; validation, L.D.; formal analysis, L.D.; investigation, L.D.; writing—original draft preparation, L.D. and J.M.R.P.; writing—review and editing, L.D. and J.M.R.P.; visualization L.D. and J.M.R.P.; supervision, L.D. and J.M.R.P.; project administration, L.D. and J.M.R.P.; funding acquisition, L.D. and J.M.R.P. All authors have read and agreed to the published version of the manuscript.

Funding: L.D. and J.M.R.P. acknowledge financial support from Ministerio de Ciencia, Innovación y Universidades grant number FIS2017-83709-R.

Data Availability Statement: Data for this study was generated using custom computer code. The code and instructions are available at http://seneca.fis.ucm.es/ldinis/code/extract_optimal_work.zip.

Conflicts of Interest: The authors declare no conflict of interest. The funders had no role in the design of the study; in the collection, analyses, or interpretation of data; in the writing of the manuscript, or in the decision to publish the results.

References

1. Touchette, H.; Lloyd, S. Information-Theoreitc Limits of Control. *Phys. Rev. Lett.* **2000**, *84*, 1156–1159. [CrossRef] [PubMed]
2. Granger, L.; Dinis, L.; Horowitz, J.M.; Parrondo, J.M.R. Reversible feedback confinement. *EPL (Europhys. Lett.)* **2016**, *115*, 50007. [CrossRef]
3. Cohen, A.E. Control of Nanoparticles with Arbitrary Two-Dimensional Force Fields. *Phys. Rev. Lett.* **2005**, *94*, 118102, [CrossRef] [PubMed]
4. Gieseler, J.; Deutsch, B.; Quidant, R.; Novotny, L. Subkelvin Parametric Feedback Cooling of a Laser-Trapped Nanoparticle. *Phys. Rev. Lett.* **2012**, *109*, 103603, [CrossRef] [PubMed]
5. Cao, F.J.; Dinis, L.; Parrondo, J.M.R. Feedback control in a collective flashing ratchet. *Phys. Rev. Lett.* **2004**, *93*, 040603. [CrossRef] [PubMed]
6. Cao, F.J.; Feito, M.; Touchette, H. Information and flux in feedback controlled Brownian ratchet. *Phys. A* **2009**, *338*, 112–119. [CrossRef]
7. Abreu, D.; Seifert, U. Extracting work from a single heat bath through feedback. *Europhys. Lett.* **2011**, *94*, 10001. [CrossRef]
8. Sagawa, T.; Ueda, M. Nonequilibrium thermodynamics of feedback control. *Phys. Rev. E* **2012**, *85*, 021104. [CrossRef]
9. Parrondo, J.M.R.; Horowitz, J.M.; Sagawa, T. Thermodynamics of information. *Nat. Phys.* **2015**, *11*, 131–139. [CrossRef]
10. Sagawa, T.; Ueda, M. Minimal Energy Cost for Thermodynamic Information Processing: Measurement and Information Erasure. *Phys. Rev. Lett.* **2009**, *102*, 250602. [CrossRef]
11. Horowitz, J.M.; Parrondo, J.M.R. Thermodynamic reversibility in feedback processes. *Europhys. Lett.* **2011**, *95*, 10005. [CrossRef]
12. Horowitz, J.M.; Parrondo, J.M.R. Designing optimal discrete-feedback thermodynamic engines. *New J. Phys.* **2011**, *13*, 123019. [CrossRef]
13. Horowitz, J.M.; Parrondo, J.M.R. Optimizing non-ergodic feedback engines. *Acta Phys. Pol. B* **2013**, *44*, 803–814. [CrossRef]
14. Sekimoto, K. *Stochastic Energetics*; Lecture Notes in Physics; Springer: Berlin/Heidelberg, Germany, 2010; Volume 799.
15. Seifert, U. Stochastic thermodynamics, fluctuation theorems, and molecular machines. *Rep. Prog. Phys.* **2012**, *75*, 126001. [CrossRef]
16. Van den Broeck C. Stochastic thermodynamics: A brief introduction. *Proc. Int. Sch. Phys. Enrico Fermi* **2013**, *184*, 155–193. [CrossRef]
17. Blickle, V.; Bechinger, C. Realization of a micrometre-sized stochastic heat engine. *Nat. Phys.* **2012**, *8*, 143–146. [CrossRef]
18. Blickle, V.; Speck, T.; Helden, L.; Seifert, U.; Bechinger, C. Thermodynamics of a Colloidal Particle in a Time-Dependent Nonharmonic Potential. *Phys. Rev. Lett.* **2006**, *96*, 070603.
19. Martínez, I.A.; Roldán, E.; Dinis, L.; Petrov, D.; Rica, R.A. Adiabatic Processes Realized with a Trapped Brownian Particle. *Phys. Rev. Lett.* **2015**, *114*. [CrossRef]

20. Martínez, I.A.; Roldán, E.; Dinis, L.; Petrov, D.; Parrondo, J.M.R.; Rica, R.A. Brownian Carnot engine. *Nat. Phys.* **2015**, *12*, 67–70. [CrossRef]
21. Schmiedl, T.; Seifert, U. Optimal Finite-Time Processes In Stochastic Thermodynamics. *Phys. Rev. Lett.* **2007**, *98*, 108301.
22. Cover, T.M.; Thomas, J.A. *Elements of Information Theory*, 2nd ed.; Wiley-Interscience: New York, NY, USA, 2006.
23. Bauer, M.; Abreu, D.; Seifert, U. Efficiency of a Brownian information machine. *J. Phys. A-Math. Theor.* **2012**, *45*. [CrossRef]
24. Kawai, R.; Parrondo, J.M.R.; Van den Broeck, C. Dissipation: The phase-space perspective. *Phys. Rev. Lett.* **2007**, *98*, 080602. [CrossRef] [PubMed]
25. Mannella, R. A Gentle Introduction to the Integration of Stochastic Differential Equations. In *Stochastic Processes in Physics, Chemistry, and Biology*; Freund, J.A., Pöschel, T., Eds.; Lecture Notes in Physics; Springer: Berlin/Heidelberg, Germany, 2000; pp. 353–364. [CrossRef]

Article

Diauxic Growth at the Mesoscopic Scale

Mirosław Lachowicz *,† and Mateusz Dębowski †

Institute of Applied Mathematics and Mechanics, Faculty of Mathematics, Informatics and Mechanics, University of Warsaw, ul. Banacha 2, 02-097 Warsaw, Poland; mateusz.debowski@mimuw.edu.pl

* Correspondence: M.Lachowicz@mimuw.edu.pl

† These authors contributed equally to this work.

Received: 27 October 2020; Accepted: 10 November 2020; Published: 12 November 2020

Abstract: In the present paper, we study a diauxic growth that can be generated by a class of model at the mesoscopic scale. Although the diauxic growth can be related to the macroscopic scale, similarly to the logistic scale, one may ask whether models on mesoscopic or microscopic scales may lead to such a behavior. The present paper is the first step towards the developing of the mesoscopic models that lead to a diauxic growth at the macroscopic scale. We propose various nonlinear mesoscopic models conservative or not that lead directly to some diauxic growths.

Keywords: diauxic growth; replicator equation; mesoscopic model; integro-differential equations

1. Introduction

In various processes in nature and social sciences, e.g., artificial neural networks, biology, medicine, and sociology, the logistic growth is observed in experiments. The logistic growth describes, at the macroscopic scale, the limited growth of a population. It is a typical way of modeling tumor growth—see e.g., [1–3] and references therein. It leads to the curve of S, or sigmoid, shape. In more mathematical terms a single inflection point is present. In some cases, however, a more complex behavior is observed. That was pointed out in 1949 by Monod—see [4], page 390—"*This phenomenon is characterized by a double growth cycle consisting of two exponential phases separated by a phase during which the growth rate passes through a minimum, even becoming negative in some cases*". Monod referred such a behavior to the growth of bacterial cultures and called it—diauxie. The similar effect was hypothesized in the analysis of a role for the CDC6 protein in the entry of cells into mitosis—see [5]. Based on the experimental data in [5], a new hypothesis that CDC6 slows down the activation of inactive complexes of CDK1 and cyclin B upon mitotic entry was formulated and the corresponding mathematical model was developed. Another example is the process of DNA melting in the case when the possible base pairs of AT (or TA) and of CG (or GC) appear in two separate groups composed only of AT and CG—see Figure 7.14, page 205, in [6].

In mathematical terms, we can refer to diauxic growth, if the corresponding increasing bounded function has more than one single inflection point. The first mathematical description of such a behavior is contained in [7].

One may note that the data of total cases of COVID–19, according to Johns Hopkins University, in September 2020, show the curves with more than one inflection points in cases of various European countries, like Spain, Italy, France, Germany, and UK. On the other hand, countries like Brazil, Chile, and South Africa display curves closed to the logistic growth (with only one inflection point).

The comparison between the logistic curve and the curve with diauxic growth is presented in Figure 1.

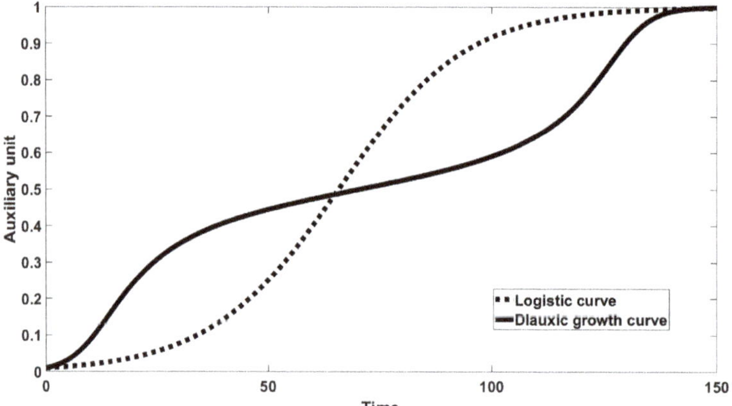

Figure 1. Comparison between logistic curve and diauxic growth curve.

In the present paper, we apply

Definition 1. *An increasing bounded and positive-valued real function is said to have a diauxic growth if its number of inflection points is bigger than one.*

Although the diauxic growth can be related, similarly to the logistic one, to the macroscopic scale one may ask whether the models on mesoscopic or microscopic scales (cf. [8]) can result in a diauxic growth. The present paper is the first step towards the developing of the mesoscopic models that lead to a diauxic growth at the macroscopic scale. We propose various nonlinear mesoscopic models, both conservative and not, which lead directly to some diauxic growths.

2. Replicator Equation

We consider the following replicator equations that occur in the multi-player games, see [7,9].

$$\frac{\mathrm{d}x}{\mathrm{d}t} = x(1-x)\mathcal{P}(x)\,, \tag{1}$$

where $\mathcal{P} = \mathcal{P}(x)$ is a polynomial. In [7] the following polynomials were considered

$$\mathcal{P}(x) = (x-a)^2 + \omega\,, \tag{2}$$

where $0 < a < 1, \omega > 0$ is a small number, and

$$\mathcal{P}(x) = (x-a)^2(x-b)^2 + \omega\,, \tag{3}$$

where $0 < a < b < 1$ and ω is a (small) number. The former refers to three players games whereas the latter to five players games. Both are related to two strategies $\uparrow$ and $\downarrow$ in an infinitely large population. The variables x and $1-x$ are the frequencies of strategies $\uparrow$ and $\downarrow$, respectively.

Consider the following payoff matrix in the case of a 3 players (for the sake of simplicity) game

$$\begin{array}{cccc} & \uparrow\uparrow & \uparrow\downarrow & \downarrow\downarrow \\ \uparrow & a_1 & a_2 & a_3 \\ \downarrow & b_1 & b_2 & b_3\,, \end{array}$$

where a_i, b_i, and $i = 1,2,3$, are the corresponding payoffs. The classical way of presentation is used. For example a_2 is payoff of the *"first player"* with strategy $\uparrow$ against the other players with strategies $\uparrow$ and $\downarrow$. Again, for the sake of simplicity, we assume that the payoffs are nonnegative.

Let now $\mu = \mu(t)$ and $\nu = \nu(t)$ be densities of players with strategies $\uparrow$ and $\downarrow$, respectively, cf. [10]. In terms of the averages payoffs of the two strategies their dynamics is defined by the system

$$\begin{aligned} \tfrac{\mathrm{d}}{\mathrm{d}t}\mu &= \mu\left(a_1\tfrac{\mu^2}{(\mu+\nu)^2} + 2a_2\tfrac{\mu\nu}{(\mu+\nu)^2} + a_3\tfrac{\nu^2}{(\mu+\nu)^2} - \kappa\right), \\ \tfrac{\mathrm{d}}{\mathrm{d}t}\nu &= \nu\left(b_1\tfrac{\mu^2}{(\mu+\nu)^2} + 2b_2\tfrac{\mu\nu}{(\mu+\nu)^2} + b_3\tfrac{\nu^2}{(\mu+\nu)^2} - \kappa\right), \end{aligned} \tag{4}$$

where, in addition to the net growth, we consider a linear death terms with rate $\kappa > 0$. We see that $x(t) = \frac{\mu(t)}{\mu(t)+\nu(t)}$ satisfies Equation (1) with

$$\mathcal{P}(x) = \hat{a}x^2 + 2\hat{b}x + \hat{x}\,, \tag{5}$$

where $\hat{a} = a_3 - 2a_2 + a_1 - b_3 + 2b_2 - b_1$, $\hat{b} = -a_3 + a_2 + b_3 - b_2$ and $\hat{x} = a_3 - b_3$. We refer to these statements throughout the paper.

3. Mesoscopic Model

We study the time–evolution of the probability density f. The function $f = f(t,u)$ is the distribution of an internal, microscopic state $u \in \mathbb{U}$ at time $t \geq 0$ of a (*statistical* or *test*) agent, $\mathbb{U}$ is a domain in $\mathbb{R}^d$, $d \in \mathbb{N} = \{1,2,\dots\}$. Such a description then has a mesoscopic nature. An arbitrary vector $u \in \mathbb{U}$ can be related to a biological state, activity, opinion (e.g., political opinion), a social state of a test agent, etc.—cf. [8,11–14] and references therein. The model has therefore a wide range of possible applications in various applied sciences, such as biology, medicine, social, or political sciences.

The time evolution is defined by the general nonlinear integro–differential Boltzmann-like equation, see [14] and references therein,

$$\frac{\partial}{\partial t}f(t,u) = Q[f](t,u)\,, \qquad t > 0\,, \quad u \in \mathbb{U}\,, \tag{6}$$

where

$$Q[f](t,u) = \int\limits_{\mathbb{R}^d} \Big(f(t,v)\, T[f(t,.)](v,u) - f(t,u)\, T[f(t,.)](u,v)\Big)\,\mathrm{d}v\,.$$

The nonlinear operator Q describes interactions between agents causing the change of state. The turning rate $T[f](u,v)$ measures the rate for an agent with state u to change it into v. A simpler equation, with two possible states only, was studied in [15]—see also [8].

The modeling process leads to a proper choice of the turning rate.

Case 1. *Let*

$$T[f(t,.)](u,v) = \beta(u,v)f^{\gamma}(t,v)\,, \qquad u,v \in \mathbb{U}\,,$$

where $\gamma > 1$ is here a given integer.

The rate of transition from state u to state v is proportional to the γ–th power of actual probability of state v. The higher is the probability, the larger is the chance of the change. The interaction kernel β corresponds to the tendency of agents to change a state. In particular, it may restrict the interactions only to states that are close each to other—see Ref. [16]. The (sensitivity) parameter γ describes the level of sensitivity of interactions: The greater is γ the more sensitive interactions are.

The models defined by Case 1 were proposed in [14], and then studied in various directions in [16–18]. Ref. [14] proposed results of global existence in the space homogeneous case for $0 < \gamma < 1$, whereas $\gamma > 1$ was considered in [16–18]. Assuming $\gamma = 1$ for symmetric β yields a trivial model. Thus it was excluded as it is stated in Case 1. The detailed information on the modeling leading to Case 1 can be found in [19] (see also references therein), where it was referred to the conformist society.

We consider the following equation

$$\frac{\partial}{\partial t} f(t,u) = Q[f](t,u) \qquad t > 0, \quad u \in \mathbb{U}, \tag{7}$$

with

$$Q[f](t,u) = f^{\gamma}(t,u) \int_{\mathbb{U}} \beta(v,u) f(t,v)\, \mathrm{d}v - f(t,u) \int_{\mathbb{U}} \beta(u,v) f^{\gamma}(t,v)\, \mathrm{d}v\,. \tag{8}$$

Independently we consider the following two, formally more general, kinetic equations

Case 2. *Let*

$$T[f(t,.)](u,v) = \underbrace{\int_{\mathbb{U}} \cdots \int_{\mathbb{U}}}_{\gamma\times} A\,(v,u,v_1,\dots,v_\gamma)\,\alpha\,(u,v_1,\dots,v_\gamma)\, \times$$
$$f(t,v_1)\dots f(t,v_\gamma)\, \mathrm{d}v_1 \dots \mathrm{d}v_\gamma\,, \qquad u,v \in \mathbb{U},$$

where γ is an integer.

Case 2 leads to

$$\frac{\partial}{\partial t} f(t,u) = \bar{Q}[f](t,u) \qquad t > 0, \quad u \in \mathbb{U}, \tag{9}$$

with

$$\begin{aligned} \bar{Q}[f](t,u) = & \underbrace{\int_{\mathbb{U}} \cdots \int_{\mathbb{U}}}_{(\gamma+1)\times} A\,(u,v,v_1,\dots,v_\gamma)\,\alpha\,(v,v_1,\dots,v_\gamma)\, \times \\ & f(t,v) f(t,v_1)\dots f(t,v_\gamma)\, \mathrm{d}v\, \mathrm{d}v_1 \dots \mathrm{d}v_\gamma \;\; - \\ & f(t,u) \underbrace{\int_{\mathbb{U}} \cdots \int_{\mathbb{U}}}_{\gamma\times} \alpha\,(u,v_1,\dots,v_\gamma)\, f(t,v_1)\dots f(t,v_\gamma)\, \mathrm{d}v_1 \dots \mathrm{d}v_\gamma\,. \end{aligned} \tag{10}$$

Case 3. *Let γ be an integer and*

$$
\begin{aligned}
&T[f(t,.)](u,v) = A_0(v,u)\,\alpha_0(v)\,f(t,v) + \\
&+ \sum_{j=1}^{\gamma} \underbrace{\int_{\mathbb{U}} \cdots \int_{\mathbb{U}}}_{j\times} A_j(v,u,v_1,\ldots,v_j)\,\alpha_j(u,v_1,\ldots,v_j)\,f(t,v_1)\ldots f(t,v_j)\,\mathrm{d}v_1 \ldots \mathrm{d}v_j\,. \\
&u,v \in \mathbb{U},
\end{aligned}
$$

Case 3 leads to

$$
\frac{\partial}{\partial t} f(t,u) = \tilde{Q}[f](t,u) \qquad t \geq 0, \quad u \in \mathbb{U}, \tag{11}
$$

with

$$
\begin{aligned}
\tilde{Q}[f](t,u) = &\int_{\mathbb{U}} A_0(u,v)\,\alpha_0(v)\,f(t,v)\,\mathrm{d}v - \alpha_0(u) f(t,u) + \\
+ \sum_{j=1}^{\gamma} \Bigg(&\underbrace{\int_{\mathbb{U}} \cdots \int_{\mathbb{U}}}_{(j+1)\times} A_j(u,v,v_1,\ldots,v_j)\,\alpha_j(v,v_1,\ldots,v_j) \times \\
&f(t,v) f(t,v_1)\ldots f(t,v_j)\,\mathrm{d}v\,\mathrm{d}v_1 \ldots \mathrm{d}v_j - \\
&f(t,u) \underbrace{\int_{\mathbb{U}} \cdots \int_{\mathbb{U}}}_{j\times} \alpha_j(u,v_1,\ldots,v_j)\, f(t,v_1)\ldots f(t,v_j)\,\mathrm{d}v_1 \ldots \mathrm{d}v_j \Bigg).
\end{aligned} \tag{12}
$$

The terms $A_j(u,v,v_1,\ldots,v_j)$ can be interpreted as the transition probabilities of changing from state v to u caused by interaction with agents with states v_1, v_2,...,v_j whereas $a_j(v,v_1,...,v_j)$ as rate of interaction between the agent with state v and agents with states v_1,...,v_j.

One may note that Equations (9) and (11), under suitable symmetry assumption, can be directly related with the dynamics of N interacting agents in the limit $N \to \infty$—see [8,13]. The former may be related to the interactions between γ agents, whereas the latter to interactions between j agents, with $j = 1,2,\ldots,\gamma$ and $j = 0$ corresponds to a stochastic change without any interaction—see [13]. One may note, however, that Equation (11) can be directly reduced to Equation (9) just taking $\alpha_j \equiv 0$ for each $j = 0,1,\ldots,\gamma-1$. On the other hand, thanks to the conservative properties, Equation (9) results in Equation (11) as well, under a suitable choice of A and α. For these reasons we concentrate on Equation (11) only.

The L_p-norm is denoted by $\|\,.\,\|_p$.

We may state the following local existence–uniqueness result for solutions to Equation (7).

Proposition 1. *Let $\gamma > 1$ and*

$$
\beta \in L_\infty(\mathbb{U} \times \mathbb{U})\,. \tag{13}
$$

If f_0 is a probability density such that $f_0 \in L_\infty(\mathbb{U})$, then there exists $T > 0$ such that the solution $f = f(t)$ to (7) exists and is unique in $L_\infty(\mathbb{U}) \cap L_1(\mathbb{U})$ on the interval $[0,T)$. The solution preserves positivity and L_1-norm (i.e., it is a probability density) on $[0,T)$. Moreover,

- *The solution, depending on initial data, is either global ($T = \infty$) or local ($T < \infty$).*
- *Under the additional assumption that β is a symmetric function—see [19]—the solution possesses all finite L_p-norms on $[0,T)$, $p > 1$, and the functions $t \mapsto \|f(t)\|_p$ are increasing for $t \in [0,T)$.*

The first part of proof follows from [14]—see [19]—based on the Lipschitz property of the corresponding operator. The rest follows by *a priori* estimates.

From [16–20] we see that the behavior of the solution to Equation (7) may be very complex and may lead to various interesting applications in biology, medicine, and social sciences.

In contrast to Equation (7) with $\gamma > 1$, Equation (7) with $\gamma = 1$ (for asymmetric β), as well as Equations (9) and (11) result in the global existence–uniqueness of solutions.

Proposition 2. *Let $\gamma = 1$ and Equation (13) be satisfied. If f_0 is a probability density then for any $T > 0$ the solution $f = f(t)$ to (7) exists and is unique in $L_1(\mathbb{U})$ on the interval $[0, T]$. The solution preserves positivity and L_1-norm (i.e., it is a probability density) on $[0, T]$.*

We consider the following conservative situation

Assumption 1. *Let γ be an integer and*

$$\begin{array}{l} A_j \geq 0\,, \quad \alpha_j \geq 0\,, \qquad \alpha_j \in L_\infty\left(\mathbb{U}^{j+1}\right)\,, \\ \int\limits_{\mathbb{U}} A_j\left(u, v, v_1, \ldots, v_j\right)\, du = 1 \quad \text{for all} \quad \left(v, v_1, \ldots, v_j\right) \in \mathbb{U}^{j+1} \\ \text{such that} \quad \alpha_j\left(v, v_1, \ldots, v_j\right) > 0\,, \qquad \forall\, j = 1, \ldots, \gamma\,. \end{array} \tag{14}$$

Proposition 3. *Let Assumption 1 be satisfied. If f_0 is a probability density then for any $T > 0$ the solution $f = f(t)$ to Equation (11) exists and is unique in $L_1(\mathbb{U})$ on the interval $[0, T]$. The solution preserves positivity and L_1-norm (i.e., it is a probability density) on $[0, T]$.*

Corollary 1. *The solutions in Propositions 2 and 3 are in $L_\infty(\mathbb{U})$ on every compact $[0, T]$ provided that $f_0 \in L_\infty(\mathbb{U})$.*

The proofs of Propositions 2 and 3 are standard and based on the Lipschitz property in $L_1(\mathbb{U})$—cf. [20]. Similarly Corollary 1 follows.

Moreover, we need the smoothness of the solutions. Let $W^{m,p}(\mathbb{U})$ and $C^m_B(\mathbb{U})$ be the Banach spaces—the classical Sobolev space (a subspace of $L_p(\mathbb{U})$) and the space of m–differentiable functions with the usual norms denoted by $\|\,.\,\|_p^{(m)}$ and $\|\,.\,\|_{[B]}^{(m)}$, respectively—see [21].

Let $X^{(m)} = W^{m,1}(\mathbb{U}) \cap C^m_B(\mathbb{U})$, $m = 0, 1, 2, \ldots$, and $\|\,.\,\|^{(m)}$ be defined

$$\|\,.\,\|^{(m)} = \|\,.\,\|_p^{(m)} + \|\,.\,\|_{[B]}^{(m)}\,, \qquad m = 0, 1, 2, \ldots\,.$$

In particular, for $m = 0$, we write $X = X^{(0)} = L_1(\mathbb{U}) \cap L_\infty(\mathbb{U})$ and $\|\,.\,\| = \|\,.\,\|^{(0)}$.

Proposition 4. *Let the assumption of Proposition 1 be satisfied and additionally $f_0 \in X^{(m)}$ and*

$$\int\limits_{\mathbb{U}} \beta(u, v) g(v)\, dv \in X^{(m)} \qquad \text{for each} \quad g \in X^{(m)}\,, \tag{15}$$

for some $m = 1, 2, 3, \ldots$. Then the solution $f = f(t)$ (given by Proposition 1) satisfies $f(t, .) \in X^{(m)}$ for all $t \in [0, T)$.

4. Macroscopic Behavior in the Conservative Case

In the present section we fix our attention on the behavior of the cumulative distribution function corresponding to the solution of a (mesoscopic) kinetic equation.

For simplicity we assume that $\mathbb{U} = [0, \infty[\equiv \mathbb{R}^1_+$ and $\gamma = 3$, however possible generalizations are straightforward.

We show that, for particular assumptions on the parameters $A_j, \alpha_j, j \leq 3$, of Equation (11), the solution $f = f(t,u)$ leads to the distribution

$$F(t,u) = \int_0^u f(t,\tilde{u})\, \mathrm{d}\tilde{u}\,, \tag{16}$$

that possesses a diauxic growth with respect to $t > 0$, for any sufficiently large $u > 0$.

Assumption 2. *We assume the interactions such that*

$$\alpha_j(u, v_1, \ldots, v_j) = j!\, \eta_j\, \chi\,(v_1 \leq u)\, \chi\,(v_2 \leq v_1) \cdots \chi\,(v_j \leq v_{j-1}) \quad \text{for all} \quad u, v_1, \ldots, v_j \in \mathbb{R}^1_+\,, \tag{17}$$

where $\chi\,(\text{true}) = 1$, $\chi\,(\text{false}) = 0$, η_j *are positive constants,*

$$\int_0^u A_j\,(\tilde{u}, v, w_1, \ldots, w_j)\; \mathrm{d}\tilde{u} = \chi\,(w_1 \leq u) \quad \text{for all} \quad u, v, w_1, \ldots, w_j \in \mathbb{R}^1_+\,, \tag{18}$$

$j = 1, 2, 3$ *(we keep in mind that* $\gamma = 3$*), with the standard convention, i.e., if* $j = 1$*, then* $w_1, \ldots, w_j$ *means* w_1*, if* $j = 2$*, then* $w_1, \ldots, w_j$ *means* w_1, w_2*, if* $j = 3$*, then* $w_1, \ldots, w_j$ *means* w_1, w_2, w_3*, and*

$$\alpha_0(u) = \eta_0 \quad \text{for any} \quad u \in \mathbb{R}^1_+\,, \tag{19}$$

$$\int_0^u A_0\,(\tilde{u}, v)\; \mathrm{d}\tilde{u} = \zeta(u) \quad \text{for all} \quad u \geq u_0 \quad \text{and} \quad v \in \mathbb{R}^1_+\,, \tag{20}$$

where $u_0 > 0$ *is a given constant and* ζ *is a increasing function such that* $\zeta(0) = 0$ *and* $\lim_{u \to \infty} \zeta(u) = 1$.

We may note, that Assumption 2 implies Assumption 1.

By Equation (17) and simple calculations, we obtain

$$\begin{gathered}\int_0^u \underbrace{\int_0^\infty \cdots \int_0^\infty}_{j\times} f(t,\tilde{u})\, \alpha_j\,(u, v_1, \ldots, v_j)\; f(t,v_1) \ldots f(t,v_j)\, \mathrm{d}v_1 \ldots \mathrm{d}v_j\, \mathrm{d}\tilde{u} \;= \\ \frac{\eta_j}{j+1} \left(\int_0^u f(t,\tilde{u})\, \mathrm{d}\tilde{u} \right)^{j+1}\,,\end{gathered} \tag{21}$$

for j equal 1, 2 and 3 , and any $f(t,\cdot) \in L_1\left(\mathbb{R}^1_+\right)$ and

$$\int_0^u f(t,\tilde{u})\, \alpha_0\,(\tilde{u})\; \mathrm{d}\tilde{u} = \eta_0 \int_0^u f(t,\tilde{u})\, \mathrm{d}\tilde{u}\,. \tag{22}$$

Moreover, for any $f(t,\cdot) \in L_1(\mathbb{R}^1_+)$ such that $\|f\|_1 = 1, j = 1,2,3$, by Equations (17) and (18), we have

$$
\begin{aligned}
&\int_0^u \underbrace{\int_0^\infty \dots \int_0^\infty}_{(j+1)\times} A_j(\tilde{u}, v, v_1, \dots, v_j)\, \alpha_j(v, v_1, \dots, v_j) \times \\
&f(t,v) f(t,v_1) \dots f(t,v_j)\, \mathrm{d}v\, \mathrm{d}v_1 \dots \mathrm{d}v_j\, \mathrm{d}\tilde{u} = \\
&j!\,\eta_j \int_0^\infty f(t,v) \int_0^\infty f(t,v_1) \chi(v_1 \le v)\, \chi(v_1 \le u) \int_0^{v_1} f(t,v_2) \dots \int_0^{v_{j-1}} f(t,v_j)\, \mathrm{d}v\, \mathrm{d}v_1 \dots \mathrm{d}v_j\, \mathrm{d}\tilde{u} = \\
&\mathbb{I}_1 + \mathbb{I}_2,
\end{aligned}
\tag{23}
$$

where

$$
\mathbb{I}_1 = \frac{\eta_j}{j+1} \left(\int_0^u f(t,\tilde{u})\, \mathrm{d}\tilde{u} \right)^{j+1},
$$

and

$$
\mathbb{I}_2 = \eta_j \left(\left(\int_0^u f(t,\tilde{u})\, \mathrm{d}\tilde{u} \right)^{j} - \left(\int_0^u f(t,\tilde{u})\, \mathrm{d}\tilde{u} \right)^{j+1} \right).
$$

Finally, for any $f \in L_1(\mathbb{R}^1_+)$ such that $\|f\|_1 = 1$, by Equations (19) and (20), for any $u > u_0$ we have

$$
\int_0^u \int_0^\infty A_0(\tilde{u}, v)\, \alpha_0(v)\, f(t,v)\, \mathrm{d}v\, \mathrm{d}\tilde{u} = \eta_0\, \zeta(u). \tag{24}
$$

By the above calculations, integrating Equation (11) with respect to u, we can see that any solution f of Equation (11), corresponding to an initial datum that is a probability density, is such that $x(t) = F(t,u)$ given by Equation (16), for any fixed $u > u_0$, satisfies the following equation

$$
\dot{x} = \mathcal{W}(x), \tag{25}
$$

where

$$
\mathcal{W}(x) = -\eta_3 x^4 + (\eta_3 - \eta_2)\, x^3 + (\eta_2 - \eta_1)\, x^2 + (\eta_1 - \eta_0)\, x + \eta_0 \zeta(u), \tag{26}
$$

where u is treated here as a (fixed) parameter.

Therefore, it is easy to see that the parameters of the model can be chosen in such a way that $t \to F(t,u)$ possesses a diauxic growth for any fixed sufficiently large u. We then obtain

Corollary 2. *Let Assumption 2 be satisfied and f_0 be a probability density on $\mathbb{U} = \mathbb{R}^1_+$. The solution $f = f(t,u)$ to Equation (11), given by Proposition 3, is such that the corresponding $F = F(t,u)$ given by Equation (16) has a diauxic growth with respect to t, for any sufficiently large $u \in \mathbb{R}^1_+$.*

5. Macroscopic Behavior in the Nonconservative Case

In order to adapt to a situation typical in game theory—cf. Section 2, we replace Assumption 1 by the following more general statement.

Assumption 3. *Let γ be an integer and*

$$\begin{array}{l} A_j \geq 0\,, \quad \alpha_j \geq 0\,, \qquad \alpha_j \in L_\infty(\mathbb{U}^{j+1})\,, \\ A_j\left(\,.\,,v,v_1,\ldots,v_j\right) \in L_1(\mathbb{U}) \quad \text{for all} \quad (v,v_1,\ldots,v_j) \in \mathbb{U}^{j+1} \\ \text{such that} \quad \alpha_j\left(v,v_1,\ldots,v_j\right) > 0\,, \qquad \forall\, j = 0,\ldots,\gamma\,. \end{array} \tag{27}$$

In this section, we deal with the macroscopic behavior derived by the mesoscopic structures defined in the previous section.

We decompose $\mathbb{U} = \mathbb{U}_* \cup \mathbb{U}^*$, where $\mathbb{U}_*$ and $\mathbb{U}^*$ are arbitrary (Lebesgue) measurable sets such that $\mathbb{U}_* \cap \mathbb{U}^* = \emptyset$, both with positive (Lebesgue) measures. For a given solution f of the mesoscopic equation we are interested in the behavior of

$$\int\limits_{\mathbb{U}_*} f(t,v)\,\mathrm{d}v \qquad \text{and} \qquad \int\limits_{\mathbb{U}^*} f(t,v)\,\mathrm{d}v \tag{28}$$

that can be related to $\mu(t)$ and $\nu(t)$, cf. Equation (4), as well as

$$\frac{\int\limits_{\mathbb{U}_*} f(t,v)\,\mathrm{d}v}{\int\limits_{\mathbb{U}} f(t,v)\,\mathrm{d}v}$$

that can be related to $x(t)$ in the macroscopic description, cf. Equation (1) with Equation (5).

Similarly to that of [22], we assume a direct dependence of the rate α_2 on the unknown function f in Equation (11). This is a Enskog-type of assumption known in kinetic theory—cf. [23] and references therein.

Assumption 4. *We assume*

1. $\alpha_0 = \alpha_1$ *and*
$$\alpha_2 = \alpha_2\left(f(t); v_1, v_2, v_3\right) = \frac{\kappa}{\left(\int\limits_{\mathbb{U}} f(t,u)\,\mathrm{d}u\right)^2},$$
2. $A_2 = A_2\left(u, v_1, v_2, v_3\right)$ *is such that*
   - *(a)* $\int\limits_{\mathbb{U}_*} A\left(u,v_1,v_2,v_3\right)\,\mathrm{d}u = \frac{a_1}{\kappa}$, *if* $v_1, v_2, v_3 \in \mathbb{U}_*$;
   - *(b)* $\int\limits_{\mathbb{U}_*} A\left(u,v_1,v_2,v_3\right)\,\mathrm{d}u = \frac{2a_2}{3\kappa}$, *if* $v_i \in \mathbb{U}^*$, *for some* $i = 1,2,3$, *and* $v_j \in \mathbb{U}_*$ *for each* $j = 1,2,3$ *such that* $j \neq i$;
   - *(c)* $\int\limits_{\mathbb{U}_*} A\left(u,v_1,v_2,v_3\right)\,\mathrm{d}u = \frac{a_3}{3\kappa}$, *if* $v_i \in \mathbb{U}_*$, *for some* $i = 1,2,3$, *and* $v_j \in \mathbb{U}^*$ *for each* $j = 1,2,3$ *such that* $j \neq i$;
   - *(d)* $\int\limits_{\mathbb{U}^*} A\left(u,v_1,v_2,v_3\right)\,\mathrm{d}u = \frac{b_1}{3\kappa}$, *if* $v_i \in \mathbb{U}^*$, *for some* $i = 1,2,3$, *and* $v_j \in \mathbb{U}_*$ *for each* $j = 1,2,3$ *such that* $j \neq i$;
   - *(e)* $\int\limits_{\mathbb{U}^*} A\left(u,v_1,v_2,v_3\right)\,\mathrm{d}u = \frac{2b_2}{3\kappa}$, *if* $v_i \in \mathbb{U}_*$, *for some* $i = 1,2,3$, *and* $v_j \in \mathbb{U}^*$ *for each* $j = 1,2,3$ *such that* $j \neq i$;
   - *(f)* $\int\limits_{\mathbb{U}^*} A\left(u,v,v_1,v_2\right)\,\mathrm{d}u = \frac{b_3}{\kappa}$, *if* $v, v_1, v_2 \in \mathbb{U}^*$.

Assume now that the payoffs $a_1, a_2, a_3, b_1, b_2, b_3$, see Section 2, are such that the corresponding Equation (1) with Equation (5) result in solutions that have a diauxic growth—cf. [7]. Then the kinetic Equation (11) leads to diauxic growth of (28) if Assumption 4 is satisfied. In fact

Theorem 1. *Let Assumption 4 be satisfied and $f_0 \in L_1(\mathbb{U})$ be nonnegative and such that*

$$\int_{\mathbb{U}_*} f_0(u)\,\mathrm{d}u > 0\,.$$

Then, for any $t > 0$, there exists a unique solution $f = f(t)$ of Equation (11) in $L_1(\mathbb{U})$. Moreover it is possible to choose the payoffs $a_1, a_2, a_3, b_1, b_2, b_3$ in such a way that (28) given by the solution $f = f(t)$ has a diauxic growth.

Proof. It is standard to see that the operator defined by the right hand-side of Equation (11) is locally Lipschitz continuous in $L_1(\mathbb{U})$. Then a local in time solution $f = f(t)$ exists in $L_1(\mathbb{U})$ and it is unique. It is also standard that the solution preserves nonnegativity of the initial datum. We observe that $\mu(t) := \int_{\mathbb{U}_*} f(t,u)\,\mathrm{d}u$ and $\nu(t) := \int_{\mathbb{U}^*} f(t,u)\,\mathrm{d}u$ satisfy Equation (4) on the interval of time of existence of the solution. Therefore $\frac{\mu(t)}{\mu(t)+\nu(t)}$ satisfies Equation (1) on the same time interval. By the form of Equation (4), we observe that any solution of Equation (4) must be bounded on any compact interval. This delivers an *a priori* estimate of the $L_1(\mathbb{U})$-norm of the solution, which concludes the proof. □

Remark 1. *For simplicity, we assumed at the beginning that all payoffs were nonnegative. It is easy to see that Assumption 4 can be easily modified to cover the case if any of payoffs is negative.*

6. Concluding Remarks

In the paper, we show that some mesoscopic models can produce a diauxic behavior on the macroscopic level. In such a case, the macroscopic picture is more complex that the usual one of a logistic-type, similar to the curve of cumulative normal distribution function (and thus related to the central limit theorem) with only one inflection point. The paper should be understood as the first step of description the relationships between the mesoscopic and macroscopic scales where new and interesting effects can appear. One may hypothesize that a complex but organized behavior on the level of micro-scale or meso-scale can lead to the diauxic macroscopic growth. This, however, still needs a new mathematical background.

Author Contributions: Conceptualization, M.L. and M.D.; methodology, M.L.; formal analysis, M.L. and M.D.; investigation, M.L. and M.D.; visualization, M.D.; supervision, M.L.; funding acquisition, M.L. All authors have read and agreed to the published version of the manuscript.

Funding: M. Lachowicz was supported by the National Science Centre, Poland, Grant 2017/25/B/ST1/00051.

Conflicts of Interest: The authors declare no conflict of interest.

References

1. Ledzewicz, U.; Munden, J.; Schättler, H. Scheduling of angiogenic inhibitors for Gompertzian and logistic tumor growth models. *Discret. Contin. Dyn. Syst. Ser. B* **2009**, *12*, 415–438. [CrossRef]
2. Foryś, U.; Poleszczuk, J.; Liu, T. Logistic tumor growth with delay and impulsive treatment. *Math. Popul. Stud.* **2014**, *21*, 146–158. [CrossRef]
3. Piotrowska, M.J.; Bodnar, M. Logistic equation with treatment function and discrete delays. *Math. Popul. Stud.* **2014**, *21*, 166–183. [CrossRef]
4. Monod, J. The growth of bacterial cultures. *Annu. Rev. Microbiol.* **1949**, *3*, 371–394. [CrossRef]

5. Dębowski, M.; Szymańska, Z.; Kubiak, J.Z.; Lachowicz, M. Mathematical model explaining the role of CDC6 in the diauxic growth of CDK1 activity during the M-Phase of the cell cycle. *Cells* **2019**, *8*, 1537. [CrossRef] [PubMed]
6. Thompson, C.J. *Mathematical Statistical Mechanics*; Princeton University Press: Princeton, NJ, USA, 1979.
7. Dębowski, M. Diauxic behaviour for biological processes at various timescales. *Math. Models Appl. Sci.* **2020**, 1–9. [CrossRef]
8. Banasiak, J.; Lachowicz, M. *Methods of Small Parameter in Mathematical Biology*; Birkhäuser: Boston, MA, USA, 2014.
9. Gokhale, C.S.; Traulsen, A. Evolutionary Multiplayer Games. *Dyn. Games Appl.* **2014**, *4*, 2270–2285. [CrossRef]
10. Durrett, R.; Levin, S. The importance of being discrete (and spatial). *Theor. Popul. Biol.* **1994**, *46*, 363–394. [CrossRef]
11. Ajmone Marsan, G.; Bellomo, N.; Gibelli, L. Stochastic evolutionary differential games toward a system theory of behavioral social dynamics. *Math. Models Methods Appl. Sci.* **2016**, *26*, 1051–1093. [CrossRef]
12. Bellomo, N.; De Nigris, S.; Knopoff, D.; Morini, M.; Terna, P. Swarms dynamics towards a systems approach to social sciences and behavioral economy. *Netw. Heterog. Media* **2020**, *15*, 353–368. [CrossRef]
13. Lachowicz, M. Individually-based Markov processes modeling nonlinear systems in mathematical biology. *Nonlinear Anal. Real World Appl.* **2011**, *12*, 2396–2407. [CrossRef]
14. Parisot, M.; Lachowicz, M. A kinetic model for the formation of swarms with nonlinear interactions. *Kinet. Relat. Models* **2016**, *9*, 131–164. [CrossRef]
15. Banasiak, J.; Lachowicz, M. On a macroscopic limit of a kinetic model of alignment. *Math. Models Methods Appl. Sci.* **2013**, *23*, 2647–2670. [CrossRef]
16. Lachowicz, M.; Leszczyński, H.; Topolski, K.A. Self-organization with small range interactions: Equilibria and creation of bipolarity. *Appl. Math. Comput.* **2019**, *343*, 156–166. [CrossRef]
17. Lachowicz, M.; Leszczyński, H.; Parisot, M. A simple kinetic equation of swarm formation: Blow-up and global existence. *Appl. Math. Lett.* **2016**, *57*, 104–107. [CrossRef]
18. Lachowicz, M.; Leszczyński, H.; Parisot, M. Blow–up and global existence for a kinetic equation of swarm formation. *Math. Models Methods Appl. Sci.* **2017**, *27*, 1153–1175. [CrossRef]
19. Lachowicz, M.; Leszczyński, H.; Puźniakowska-Gałuch, E. Diffusive and anti-diffusive behavior for kinetic models of opinion dynamics. *Symmetry* **2019**, *11*, 1024. [CrossRef]
20. Lachowicz, M.; Leszczyński, H. Asymmetric interactions in economy. *Mathematics* **2020**, *8*, 523. [CrossRef]
21. Adams, R.A. *Sobolev Spaces*; Academic Press: New York, NY, USA, 1975.
22. Lachowicz, M.; Quartarone, A.; Ryabukha, T. Stability of solutions of kinetic equations corresponding to the replicator dynamics. *Kinet. Relat. Models* **2014**, *7*, 109–119. [CrossRef]
23. Bellomo, N.; Lachowicz, M.; Polewczak, J.; Toscani, G. *Mathematical Topics in Nonlinear Kinetic Theory. The Enskog Equation*; World Scientific: Singapore, 1991.

MDPI
St. Alban-Anlage 66
4052 Basel
Switzerland
Tel. +41 61 683 77 34
Fax +41 61 302 89 18
www.mdpi.com

Entropy Editorial Office
E-mail: entropy@mdpi.com
www.mdpi.com/journal/entropy

www.ingramcontent.com/pod-product-compliance
Lightning Source LLC
LaVergne TN
LVHW070057170726
843515LV00007B/1559

* 9 7 8 3 0 3 6 5 3 4 8 5 5 *